THE EDGE OF OBJECTIVITY

THE EDGE OF OBJECTIVITY

AN ESSAY IN THE HISTORY OF SCIENTIFIC IDEAS

BY

CHARLES COULSTON GILLISPIE

PRINCETON, NEW JERSEY

PRINCETON UNIVERSITY PRESS

Printed in the United States of America
by Princeton University Press, Princeton, New Jersey

A passage from the Inaugural Lecture given by James Clerk Maxwell as Professor of Experimental Physics in the University of Cambridge, in October 1871, by way of

FOREWORD

❝WE ARE NOT HERE TO DEFEND LITERARY AND HISTORICAL STUDIES. *We admit that the proper study of mankind is man. But is the student of science to be withdrawn from the study of man, or cut off from every noble feeling, so long as he lives in intellectual fellowship with men who have devoted their lives to the discovery of truth, and the results of whose enquiries have impressed themselves on the ordinary speech and way of thinking of men who never heard their names? Or is the student of history and of man to omit from his consideration the history of the origin and diffusion of those ideas which have produced so great a difference between one age of the world and another?*

It is true that the history of science is very different from the science of history. We are not studying or attempting to study the working of those blind forces which, we are told, are operating on crowds of obscure people, shaking principalities and powers, and compelling reasonable men to bring events to pass in an order laid down by philosophers.

The men whose names are found in the history of science are not mere hypothetical constituents of a crowd, to be reasoned upon only in masses. We recognize them as men like ourselves, and their actions and thoughts, being more free from the influence of passion, and recorded more accurately than those of other men, are all the better materials for the study of the calmer parts of human nature.

But the history of science is not restricted to the enumer-

ation of successful investigations. It has to tell of unsuccessful enquiries, and to explain why some of the ablest men have failed to find the key of knowledge, and how the reputation of others has only given a firmer footing to the errors into which they fell.

The history of the development, whether normal or abnormal, of ideas is of all subjects that in which we, as thinking men, take the deepest interest. But when the action of the mind passes out of the intellectual stage, in which truth and error are the alternatives, into the more violently emotional states of anger and passion, malice and envy, fury and madness; the student of science, though he is obliged to recognise the powerful influence which these wild forces have exercised on mankind, is perhaps in some measure disqualified from pursuing the study of this part of human nature.

But then how few of us are capable of deriving profit from such studies. We cannot enter into full sympathy with these lower phases of our nature without losing some of that antipathy to them which is our surest safeguard against a reversion to a meaner type, and we gladly return to the company of those illustrious men who by aspiring to noble ends, whether intellectual or practical, have risen above the region of storms into a clearer atmosphere, where there is no misrepresentation of opinion, nor ambiguity of expression, but where one mind comes into closest contact with another at the point where both approach nearest to the truth."

CONTENTS

FOREWORD   vii

I. FULL CIRCLE   3

II. ART, LIFE, AND EXPERIMENT   54

III. THE NEW PHILOSOPHY   83

IV. NEWTON WITH HIS PRISM AND SILENT FACE   117

V. SCIENCE AND THE ENLIGHTENMENT   151

VI. THE RATIONALIZATION OF MATTER   202

VII. THE HISTORY OF NATURE   260

VIII. BIOLOGY COMES OF AGE   303

IX. EARLY ENERGETICS   352

X. FIELD PHYSICS   406

XI. EPILOGUE   493

BIBLIOGRAPHIC ESSAY   521

INDEX   545

THE EDGE OF OBJECTIVITY

CHAPTER I

FULL CIRCLE

IN THE YEAR 1604, Galileo Galilei formulated a law of falling bodies in a letter to his friend, Paolo Sarpi. "I have arrived at a proposition," he wrote, "which is most natural and evident, and assuming it, I can demonstrate the rest; namely, that spaces traversed in natural motion are in the squared proportion of the times, and consequently the spaces traversed in equal times are as the odd numbers beginning with unity. And the principle is this, that the naturally moving body increases its velocity in the proportion that it is distant from the origin of motion." This is a curious statement. For the first part is right, but one cannot explain how Galileo knew it, since it does not in fact follow from the principle, which is wrong. Under uniform acceleration, velocity varies directly as time, not distance, and any schoolboy learns the correct law by rote as either or both of two equations,

$$s = \tfrac{1}{2} \, gt^2 \quad \text{and} \quad s = \tfrac{1}{2} \, vt \, .$$

Even when he finally did get it right, Galileo could not so express it. Algebra had yet to be adapted to description of continuously developing quantities. He disposed only of the resources of ordinary language and of the geometry of Euclid and Archimedes. In 1632, after years of reflection and not a little frustration, he explained the law in *Dialogue on the Two Chief Systems of the World*, the great Copernican argument over which the Roman Catholic Church humiliated him; and there he repeated, "that the distances passed by the body departing from its rest are to each other in double proportion of the times in which those distances are measured."

To that, Sagredo, the receptive interlocutor, now responds, "This is truly admirable; and do you say there is a mathematical demonstration for it?" And Galileo gratifies the request he has invited by expressing the relation between velocity, distance, and time as a triangle. Falling

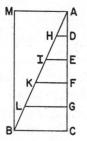

from rest at A, the body picks up speed through "infinite degrees of velocity." The time of fall is laid off on the vertical AC. Perpendiculars (DH, EI, etc.) represent the velocity after time AD, DE, etc., and the whole triangle is "the mass and sum of the whole velocity, with which in the time AC it passed such a certain space." Or, to put it otherwise, the area of the triangle ($\frac{1}{2}$ vt) measures the distance traversed. And to find the distances travelled by a body moving at uniform velocity (BC), the triangle may be doubled into a rectangle (ACBM).

But though perfectly correct, this must still seem painful and clumsy to the modern reader. Velocity appears as one variable and time as the other, whereas it has become customary to think of velocity rather as a ratio of distance to time. Moreover, it measures the linear distance s by an area. Nor does the geometry yet derive the law in the form which relates distance to acceleration ($s = \frac{1}{2}$ gt^2). In 1638, Galileo published his final and scientifically his finest work, *Discourses on Two New Sciences*. There at last he achieved an explicit statement of both common forms of the law. The discussions of the "Third Day" work towards a renewed demonstration of the velocity-time relationship, in more elegant geometrical form than in the *Dialogue*, and in less elegant language. Next, Galileo proved what until now he had only asserted: the distances are as the squares of the times. This was far more difficult. He had to formulate graphically what he called "uniformly difform motion," that is to say, a dynamical proposition involving acceleration in the essentially static forms of plane geometry.

He represented the "flow of time" by simple extension, the line AB, on which AD and AE measure any two intervals. To the right, the line HI stands for the path of descent at uniform acceleration, so that HL is the distance traversed in time AD, HM in AE, etc. These things being so, then "I say that the space MH to the space HL is in the duplicate ratio that time AE has to time AD." For, construct AC at any angle to AB. Then DO, EP, etc. will again represent maximum velocity at corresponding time. It followed from the previous (mean-speed) theorem that the spaces are equal which are traversed by one body at uniform acceleration from rest, and a second moving at a constant velocity which is one-half the maximum attained by the first.

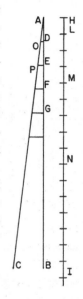

Thus, the distances of fall HL and HM would be equal to those traversed in times AD and AE at constant velocities one-half of DO and EP respectivly. But it had already been shown that the distances passed by two bodies in uniform motion are to each other as the product of the ratio of the velocities into the ratio of the times. Now, since EP is to OD as AE is to AD, then the ratio of velocities is in this case the same as the ratio of the times. "Therefore, the ratio of the spaces traversed is as the square of the ratio of the times. Q. E. D."

At this point, Salviati, who speaks for Galileo, stops the dialogue as if a light had dawned: "Please suspend your lecture for a moment while I speculate on a certain idea that has just now occurred to me." And he puts the two forms of the law together. AI represents time again, AF is at any angle, and C is the mid-point of AI. Then (to condense the argument a bit), if the body falls freely to C, BC will be the maximum velocity, and the distance will be measured by the rectangle of uniform velocity erected on the base EC equal to ½ CB.

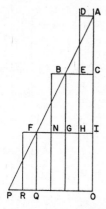

Moreover, if the body continued its descent at constant velocity BC, then in the interval CI it would cover twice the distance that it had described in time AC, starting from rest. But since the body is under uniform acceleration, its velocity during the time CI will increase by an amount FG equal to the parallel of the triangle BFG, which is equal to ABC. Then, adding to velocity GI (equal to BC) half of FG, which is the maximum velocity attained through acceleration, one gets the uniform velocity with which the same space would have been described in the time CI. And perhaps the drift is apparent without further paraphrasing. The rectangular areas which represent the space described increase in successive time intervals, "as the odd numbers beginning with unity, 1, 3, 5; . . . and in general, the spaces traversed are in the duplicate ratio of the times, i.e., as the squares of these times."

These figures represent the earliest integrations applied to developing physical quantities and may be taken, therefore, to symbolize the germ from which has grown a mathematical science, not alone of proportions, but of nature. For there was nothing novel about expressing uniform motion in the abstract as a ratio of change in geometrical quantities. Galileo's first triangle of motion was a mathematical commonplace, generally called the Merton Rule after the school of kinematic philosophy which flourished in that ancient Oxford College during the fourteenth century. Moreover, the mean speed theorem reduces to the law of acceleration, and needed rather to be stated helpfully than to be discovered. Everything, indeed, or nearly everything, that Galileo put together may be found in the writings of one or another of the late scholastics, in the aphorisms of Leonardo da Vinci, or in the works of some predecessor among the Renaissance mathematicians. But only Galileo, and he only after many a false start, developed the judgment and intuition and feel for the physical to select the elements of a physics from this *olla podrida* of mathematical techniques and philosophical assertions. His was the transforming touch of the mathematical physicist, the first of his kind, who would really change a situation instead of simply entering a discussion. That touch reveals itself thrice over in these passages. First, he derived the rule of uniform acceleration in a form applicable to freely-falling bodies. Then he included it in a general statement containing both the velocity-time and the acceleration-time-squared measures of path. Finally he applied it to the real case in nature and therein lay his genius.

For only Galileo would have given the discussion the turn it takes immediately after the last of these, his mathematical demonstrations. Simplicio, who upholds the Aristotelian case, bows before the force of geometry, "so that I am convinced that matters are as described, once having accepted the definition of uniformly accelerated motion. But as to whether this acceleration is that which one meets in nature in the case of falling bodies, I am still doubtful; and it seems to me, not only for my own sake, but also for all those who think as I do, that this would be the proper moment to introduce one of those experiments —and there are many of them, I understand—which illustrate in several ways the conclusions reached." And Galileo reports on the famous experiments on inclined planes which he had imagined, and some of which he may quite probably have actually performed.

❖

THESE WERE PORTENTOUS TRIANGLES. To the historian thinking broadly about the recent destiny and future prospects of western civilization, it may well appear that our own culture, in which whatever our temperament we are bound to live, is set off from those of Asia, Africa, and the world of antiquity by two fundamental factors. From one of these it emerged: its religious chrysalis was Christianity, investing history with the promise of fulfillment of a sort. The other it produced: the most dynamic, distinctive, and influential creation of the western mind is a progressive science of nature. Only there in the technical realm, indeed, does the favorite western idea of progress hold any demonstrable meaning. No one understands political power better than Machiavelli did. Picasso cannot conclusively be held a better or worse artist than Leonardo was. But every college freshman knows more physics than Galileo knew, whose claim is higher than any other's to the honor of having founded modern science, and more too than Newton did, whose mind was the most powerful ever to have addressed itself to nature.

In its early days, science was distinct from technology, springing rather from thought and philosophy than from craftsmanship. Nowadays, however, and indeed for the last century and more, science has merged ever more intimately with technology, so arming it with power, so enhancing its capacities, that no words, nor any fears or dreams, may exaggerate what depends upon the employment. Nor is the future of our own world of the West alone in play through this, its great invention. Perhaps the historian may be pardoned a single prophecy, if it comes at the beginning of a book before his tale has made him pompous. Anxious though our moments are, today is not the final test of wisdom among statesmen or virtue among peoples. The hard trial will begin when the instruments of power created by the West come fully into the hands of men not of the West, formed in cultures and religions which leave them quite devoid of the western sense of some ultimate responsibility to man in history. That secular legacy of Christianity still restrains our world in some slight measure, however self-righteous it may have become on the one side, and however vestigial on the other. Men of other traditions can and do appropriate our science and technology, but not our history or values. And what will the day hold when China wields the bomb? And Egypt? Will Aurora light a rosy-fingered dawn out of the East? Or will Nemesis?

Albert Einstein once remarked that there is no difficulty in understanding why China or India did not create science. The problem is rather why Europe did, for science is a most arduous and unlikely undertaking. The answer lies in Greece. Ultimately science derives from the legacy of Greek philosophy. The Egyptians, it is true, developed surveying techniques and conducted certain surgical operations with notable finesse. The Babylonians disposed of numerical devices of great ingenuity for predicting the patterns of the planets. But no Oriental civilization graduated beyond technique or thaumaturgy to curiosity about things in general. Of all the triumphs of the speculative genius of Greece, the most unexpected, the most truly novel, was precisely its rational conception of the cosmos as an orderly whole working by laws discoverable in thought. The Greek transition from myth to knowledge was the origin of science as of philosophy. Indeed, knowledge of nature formed part of philosophy until they parted company in the scientific revolution of the seventeenth century.

In our own world, science continues to be what it was in Greece, conceptual thought mediating between consciousness and nature. But it is also something more. It has become determinate instead of simply speculative.

For the scientific revolution reversed the direction in which information flows, and added body to the structure of communication. Greek science was subjective, rational, and purely intellectual. It started inside the mind whence concepts like purpose, soul, life, and organism were projected outward to explain phenomena in the familiar terms of self-knowledge. In those terms the success of an explanation depended only on its universality and capacity to satisfy the reason. Greek science scarcely knew experiment and never thought to move beyond curiosity to power. Modern science, on the other hand, is impersonal and objective. It takes its starting points outside the mind in nature and winnows observations of events which it gathers under concepts, to be expressed mathematically if possible and tested experimentally by their success in predicting new events and suggesting new concepts. Modern science has not abandoned rationality, but it is first of all metrical and experiential. Related to this is its association with technology as a continuation of that generalized thrust toward mastery of the world which began in the West with the Renaissance. Modern science, finally, seeks both to comprehend and control nature—though according to the positivist school of philosophers, whose persuasion dominates at the moment, comprehension is an illusory goal. For them prediction and control are everything.

A true revolution brings fundamental change through rebellion against constituted authority, but it is clear from the history of revolutions that to repudiate a debt is not to escape it. So it was that the creation of modern science in the Renaissance was at once a rebirth of Greek science and a bursting of its confines. To separate the new from the old in the Renaissance is always difficult, for humanists steeped in classical learning found antique words for new ideas. It is, however, no falsification of a complex situation—it is rather a first approximation toward resolving it—to say that science stirred into new life under the inspiration of Plato working against the cramping of learning within a fossilized Aristotelianism.

By Galileo's time, the science and authority of Aristotle had led the western mind a long way to a dead end. Aristotle's was the most capacious of philosophies. In principle it explained everything, dealing rather in reasons than structures, and preferring categories over abstractions. For example, Galileo could describe mathematically how a stone would fall under ideal conditions. He could not say why it fell. Aristotle's physics, on the contrary, could not measure its motion. But this was not to be expected in a real world of friction and complexity where ideal conditions never occur. Aristotle could do more important things. He could explain why a stone fell, why sparks flew upward, and why the stars ran round in their courses.

Beneath its physical manifestation as translation, Aristotelian motion is metaphysical, an instance of change, an evidence of imperfection. Change is the act of things realizing their potentialities in a world striving ever to fulfill its creator's will toward order, which is toward the good, so far as its corruption permits. In an orderly cosmos there is by definition a place for everything. Heavy things belong at the bottom. To say that the stone falls because it is of the class of things which are heavy constitutes, therefore, a full explanation. So, too, fire rises because it is light. The locus of that element is in the ethereal region, with air below it, water below that, and crude earth massed round the center. But what of an arrow? Here a distinction of motions must be introduced. Its motion is not natural but forced, not orderly but disorderly and violent. It must have a cause. Logic requires effects not to outlast their causes. Therefore, every motion against nature presupposes a moving agent, and demands explanation.

What, then, moves the arrow after it has parted from the bow-string? In a philosophy which is nothing if not universal, to have no answer would be to have no science, and after some hesitation, Aristotle meets the dynamical difficulty with the air. It is the surrounding medium closing in behind the projectile which urges it along its way. There can, therefore, be no motion in a void. There cannot even be natural motion, for in this case the medium serves to retard the body. In a vacuum a stone would fall instantaneously. Since that is absurd, nature knows no void, and the world must be a plenum, finite and by later standards rather small. But the inadmissibility of the void goes deeper than abhorrence of the vacuum. It goes all the way to the foundations. There is no such thing as place in a void, and the goal of this philosophy was to define the right and necessary place for every species of being according to the purpose that it served. Nor can there be existence in the nothing. In a void no stone could tell where to go, no flame find the way to leap. The very notion of direction or order would become meaningless. To admit the void is to accept the reign of chaos, wherein whirl is king, in lieu of our own world full of meaning.

So it is throughout Aristotle's physics. It was a serious physics, a consistent and highly elaborated ideation of natural phenomena. It started from experience apprehended by common sense, and moved through definition, classification, and deduction to logical demonstration. Its instrument was the syllogism rather than the experiment or the equation. Its goal was to achieve a rational explanation of the world by showing how the myriad subordinate means are adapted to the larger end of order. Its operations were suited to these interests. Direct and minute observation, classification of forms by species, analysis of how the part serves the whole—these are useful acts up to a point in natural history, as the description of life and its environment was called until the nineteenth century. Not till then was biology ready to transcend the Aristotelian sense of purpose in nature and follow physics into objectivity. Aristotelian physics, too, had immense humane advantages denied to that which has supervened since Galileo. It easily fell in with a sense of Providence in nature. As the physical system sheltering the world view of Islam, Judaism, and Christianity, it became the scientific orthodoxy of all three religions which shaped the West in its emergence from the dark centuries after Rome. For Aristotelian physics made sense of the world and strengthened the hands of men of God and all those striving to redeem civilization, culture, and truth from barbarism.

There was only one trouble. It was wrong. For however congenial Aristotelian physics was to the self-knowledge of the minds that elaborated it, nature is not like that, not an enlargement of common sense arrangements, not an extension of consciousness and human purposes. She is more elusive, more coquettish perhaps and infinitely more subtle, hiding her ways from the merely dogged or the worthy, and only occasionally yielding to the truly curious those glimpses of great order and altogether inhuman beauty which are the reward of him who strikes the right note, and all the reward he seeks—that and fame. But who, asked James Clerk Maxwell two millennia later, "who will lead me into that still more hidden and dimmer region where Thought weds Fact, where the mental operation of the mathematician and the physical action of the molecules are seen in their true relation?" For the order is mathematical and the notes harmonious, Platonic rather than Aristotelian.

Not that Plato and Aristotle differed on all fundamentals. They were teacher and pupil. To both, nature appeared as the creation of artful mind, and order as the expression of rationality. Both took a humanistic rather than a naturalistic view of science. By explanation, both meant identification of what lofty purpose would reveal divine intelligence. But Aristotle addressed himself to physical and biological contrivance, and Plato to ideal being. Unlike Aristotle, Plato did not make a science. He inspired one—in Archimedes perhaps, and much later in Galileo certainly. His influence was less and more, the spell he cast over posterity at once sterilizing and stimulating: sterilizing in that he takes truth out of the world of things, stimulating in that he identifies ideal simplicity with mathematical reality. He speaks poetically to an aesthetic vision of nature, but never to common sense, which in puzzling he offends. What is truth, then, and what the good? They are the eternal and the perfect, being not becoming. The real is the ideal, and change the mirror of corruption. For Aristotle had simply transposed Plato's metaphysics into physical terms so as to make a distinction between cosmology and physics, the one concerned with the heavenly regions beyond the moon, the other with our sublunary sphere where different laws obtain, where everything is mortal and contingent. Thus was the uniform cosmos of the earliest Greek philosophers dichotomized, and the chance missed of laying down a single science of heaven and earth. In science, Aristotelian kinematics was the most influential consequence of this distrust of change as *prima facie* evidence of imperfection. Only one motion is perfect in Plato, in Aristotle, and after them down to Kepler and the seventeenth century, namely circular motion, that by which the heavens go, for only in circles can motion occur as changelessness.

"God is always geometrizing," Plato is supposed to have said. This, along with admiration for circles, Plato drew from the school of Pythagoras, in whose semi-legendary figure science retreats into a prehistoric mélange of myth, mysticism, and mathematics. In the search conducted by the pre-Socratic philosophers for the principle of unity in nature, the Pythagoreans hit upon the assertion that nature is made of number and that numbers have shape. In their eyes, the world is actually made of lines, triangles, squares, cubes, and circles, even as the nineteenth-century physicist might think it made of ninety-two varieties of material atoms. Numbers contain the form of things, at once real and ideal. In numbers lie the clean, eternal structures beneath the welter of appearance. Discovery of the irrationality of the square root of two, and of incommensurable quantities in general, is always described as having shocked this faith. But the shock helped, for it led to appreciation of geometric ratios like the Merton rule of motion, which would give Galileo his mathematical description of falling bodies. Indeed, the Pythagoreans may themselves have been the ones to make the first statement of physics as we know it. They studied stringed instruments and found the relationship between the length of a vibrating string and the pitch it emits, and thus they expressed the experience of harmony as a geometrical quantity. Nevertheless, like the influence of Plato after them, and in fact through Plato, their legacy moved down into the underworld of science, as well as out into its sane and wholesome reaches. As the misbegotten twin of a mathematical physics, they spawned the secret mania of numerology. Nostradamus was their heir as well as Galileo. Rosicrucianism is their progeny as well as relativity.

The two greatest of the Greek philosophical traditions laid ancient science under one final limitation. It was impossible in principle for either Plato or Aristotle to have made a mathematical physics because they agreed that mathematics and physics do not fit, and differed only over which was at fault. For Plato, mathematical relationships are eternal, ideal, and therefore real and true. But there is no such certainty, indeed no certainty at all, in the

world of things, and physics is at best a "likely story."
For Aristotle, on the contrary, it is physics which deals with
the real. Mathematics is true, to be sure, but only in the
abstract. The world itself is made of qualities and forms
and fine distinctions which may not be expressed in the
precise, quantitative, absolutely unreal terms of mathematics. Ontologically speaking, ancient science fell between these stools. Archimedes might have retrieved it.
His was a scientific intellect of the very highest order. In
discernment and power, he was the peer of an Einstein.
Everyone will recognize that the law of the lever, and
more generally the principles of simple machines, represent just that marriage of geometry to physical objects
which both Plato and Aristotle held to be impossible.
Archimedes arrived at the concept of statical moment by
abstracting from physical weight and combining its quantity with geometrical length in Euclidean ratios. Reciprocally, the Archimedean determination of the center of
gravity of geometric figures introduced physical intuition
into a mathematical problem. But Archimedes came late,
the lamp was burning low, and his best pupil was Galileo,
seventeen hundred years later, who set statics into motion
to found our science of mechanics.

SO FAR AS PHYSICS WAS CONCERNED, the scientific revolution occurred on two levels, cosmology and dynamics. It
would be complete only when Newton's law of gravity
united knowledge of heaven and earth, separate since Aristotle, into a single theoretical science of matter in motion.
On both levels, the revolutionary inspiration came from
Platonic criticism playing a ray of mathematical realism
upon the vast mass of verbal distinctions into which Aristotelian natural philosophy had proliferated in the late
Middle Ages, and from which it endlessly drew theological

tidbits out of nature. Ultimately, dynamics proved the deeper level of scientific thought. But cosmology was the more dramatic.

The Renaissance inherited a complex situation in cosmology. The Aristotelian model of the universe was the well-known Chinese nest of crystalline spheres concentric about a logy, corrupt, and stationary earth. Each carried like luminous studs the moon, the sun, a planet, or the pattern of the fixed stars. Spin was communicated to the whole complex by an outermost shell. Beyond lay bliss. Borne in on every hand, by theology, poetry, literature, and philosophy, this gave the educated man his gross picture of the cosmos. It made sense of the diurnal motion of the heavens, but was no use to astronomers who had to follow and predict the visible motions of the planets along the zodiac. They advance at varying velocities, slowing at times to a pause and retreating a little before going forward once again. These inequalities are the projection onto each planet of the orbital motion of the earth. Appearances were further complicated by slight variations in latitude and by the uncertainty of all data. It is a myth that the ancients were accurate observers. Least important as a source of error was the actual ellipticity of orbits, for the amount by which they do depart from the circular was within the margin of error that this astronomy had to tolerate. The gravest irregularities, the retrograde motions of the planets, might have been saved by a heliocentric model of the solar system. The Pythagoreans are said to have believed in a central sun, and Aristarchos of Samos certainly proposed such a theory in the third century before Christ. But it was stillborn, and instead, astronomers saved the immobile earth for common sense by using in practice the purely geometric astronomy of Claudius Ptolemy.

Ptolemy compounded the apparent movements of each

heavenly body from combinations of circular figures. He employed three devices, epicycle, eccentric, and equant. The epicycle places the planet on a small circle, the center of which describes a large circle called the deferent about the ultimate center of motion. The eccentric makes that center of motion in a particular case some point apart from the center of the earth. The equant is a point

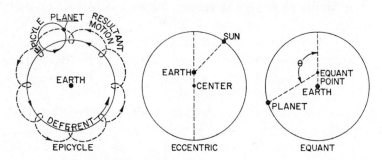

EPICYCLE   ECCENTRIC   EQUANT

other than the center so taken that there is a uniform rate of increase in the angle formed by the diameter through the equant and a line joining it to a point on the circumference. Combinations of epicycles on epicycles, epicycles on eccentrics or equants, and whatnot, to the number of seventy-odd distinct constructions, gave an account of the anomalies. The geometrical virtuosity was admirable. It saved the phenomena, and the proto-positivist physics of the likely story asked nothing more. Moreover, it worked well enough to sustain the calculation of the calendar from Roman antiquity to the sixteenth century, when at last the accumulation of error exceeded the tolerable. Indeed, navigation is still practised as if the earth were at the center of the great display of celestial bearings.

Any revolution has deep roots in culture but begins with some definite act, often meant to purify corrupt practices and restore what some conservative radical imagines as a pristine state of things. In 1543 Nicolas Koppernigk, Copernicus as he latinized himself, published *De revolutionibus orbium coelestium libri sex*. It is an extremely difficult book. Nor have many or perhaps any modern writers, and certainly not the present one, combined in their own understandings the sympathy, the scholasticism, the latinity, the astronomy, the trigonometry, and the gothicism which would permit them to penetrate the true spirit of Copernicus's life-work. But it does seem tolerably clear that astronomically he was just such a puritanical reactionary, at whose hand the old forms lost not only accretions but their rationale, and began giving way to new.

No scientist, in any case, has ever addressed himself to problems of greater magnitude relative to the state of knowledge. His theory rearranged the solar system and set the immobile earth to spinning daily on its axis while revolving annually about the sun, which replaced it at the center. That is a great and gross difference, and no denigrations must be allowed to obscure it, whether they refer to the conservatism of his mathematics, the obsessive circularity of his kinematics, or the monkish timidity of his life. Stock objections blocked assent for about a century, and to rehearse them will suggest how strongly his ideas had to swim upstream against the tide of common sense. Nor, in the absence of the principle of inertia and the composition of motions, is it any wonder that he tended to falter and keep his notions to himself. For on a moving earth we ought to feel ourselves rushing through the air. A stone dropped from a tower should land to the west of it. Cannonballs fired west should travel farther under the same charge than those fired east. The earth should whirl itself to pieces. We should all fly off like pebbles from a sling. "Those experiences," wrote Galileo just ninety years later, "which overtly contradict the annual

motion of the Earth, have so much more of the appearance of convincingness, that I cannot find any bounds
for my admiration how reason was able in Aristarchos
and Copernicus to commit such a rape upon their senses
as, in despite thereof, to make herself mistress of their
belief." Even deeper, though less seriously intuitive than
our feeling of immobility, was the prejudice against the
unworthiness of the earth to be moving in circles through
the heavens, that being a motion suited only to aethereal
and immaterial bodies.

A Polish scholar, Copernicus was educated in Cracow,
Bologna, and Padua. His father had died when he was a
boy in Thorn, and he became the charge and protégé of
an uncle, Lucas Watzelrode, an ecclesiastical statesman
and a considerable figure in the Baltic world where the
jurisdictions of Poland, the Church, and the military order
of the Teutonic Knights mingled in late medieval confusion. Watzenrode became sovereign bishop of Ermland,
an ecclesiastical principality with a narrow outlet to the
Baltic at the cathedral town of Frauenburg. There Watzenrode arranged a canonry or prebend in the chapter
for his nephew. For years Copernicus held his prebend
as an absentee, loath like many a Pole and German before
and since to return from Italy, where he went in 1496
immediately after his election. At Bologna and at Padua
Copernicus was educated as a humanist in classics, medicine, geometry, and astronomy, and his intellectual world,
perhaps even his spiritual world, was that of Greek and
Latin antiquity. Later, he did into Latin a book of epistles
on moral, pastoral, and mildly amorous subjects by one
Theophylactus, an obscure Byzantine writer of the seventh
century. But only in astronomy did Copernicus display the
taste to select from antique learning that which would
truly enter into the Renaissance movement to renew the
world. And if like a good humanist he invoked in Aristarchos the license of antiquity, his originality is no more

ambiguous than that of the whole culture of the Renaissance in its Janus posture. For Copernicus made heliocentricity part of science and not just speculation. With him it enters into the structure of the progressive knowledge of nature. His work added a determinate element. The whole history of science is his vindication, and criticism, therefore, must try to say what that element was.

There is evidence that he experienced the simplicity and elegance of the Aristarchan idea as early as 1505 or

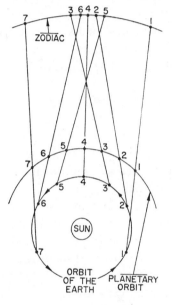

It may be illustrated in principle by an idealized diagram showing how the motion of the earth would "save" seven successive observations of one of the outer planets.

1506. The preface to his *Revolutions* says that he had kept the idea by him, not just for nine years as the Pythagoreans had enjoined, but for four times nine. It may have come to him like some conversion, just as he was about to depart for home. He had then spent his ten years in the sun, and in the open intellectual climate of Italy. If the great idea did occur in such fashion, it makes an affecting picture. He never left Poland again. There he spent the rest of his life, on that Baltic shore, living at his uncle's residence in crabbed little Heilsburg until Watzenrode's death in 1512, thereafter inhabiting a tower room in the crenellated wall of crabbed little Frauenburg. He would take an occasional hand in the Graustarkian governance and diplomacy of Ermland. He would peer through the mists and pore over Ptolemy and the tables of astronomy in a mood of gathering distaste. For Copernicus studied the figures, not the stars. And he worked at his system, which he intended to be more than an alternative set of geometrical hypotheses. He meant it as a real system of astronomy, to supplant the Ptolemaic devices used by astronomers, navigators, and calendars and to be, moreover, a true representation of the physical relations of sun, earth, moon, and planets.

Inevitably, the facts and figures were recalcitrant. Many were quite wrong. The execution of the work could not bear out the grand simplicity of its leading idea. That he stuck to that idea in the teeth of all the difficulties, daunted but not defeated by discrepancies, might be taken, perhaps, as a testimonial to the rationalizing role of theory in science, and to the virtue of a faith in some ultimate reason in things. It is no wonder that on completion of the work in 1532 or thereabouts, he hesitated to publish, and alluded now and again to legendary Pythagorean injunctions to secrecy. But he had not refrained from all communication. Quite early, probably in 1512, he circulated a sketch of his theory in manuscript. *The Commentariolus* it is called. News of his ideas spread in the gossipy world of Renaissance scholarship, widely enough for Luther to pass his famous remark about some fool who "went against Holy Writ." Then in 1539 there arrived in Frauenburg one Georg Joachim Rheticus (as he called himself after his birthplace, though his real name was von Lauchen), a young professor of mathematics from Wittenberg, powerfully attracted to the rumors of heliocentricity, and all eagerness to learn what substance they might hold. Copernicus admitted him to scientific intimacy. Together they reviewed the data. Rheticus published the first authorized account, the *Narratio Prima*, at Dantzig in 1540.

Its success, or perhaps the urgings of Rheticus and other friends, persuaded Copernicus to commit his full treatise to the press. It is divided into six books. Book I contains a general description of the system of the world: the sun and not the earth is the center; what appears as the daily rotation of all the heavens is the effect of the earth's spinning on its axis; the annual motion of the sun is the appearance of the actual revolution of the earth annually; the advances and retrogressions of the planets are projections of the same cause combined with their own motions; the distance from the sun to the earth or planets is very small compared to the remoteness of the fixed stars. Book II contains a star catalogue compiled from ancient and more recent astronomy, from which Copernicus computed the elements of the year. The remaining four books contain detailed mathematical theories, which is to say geometric devices and trigonometrical methods, for predicting the motions of the planets referred to the earth and to the sun, together with the real motions of the earth and moon. Printing was completed in Nuremberg in 1543, under the supervision of Andreas Osiander, a friend of Rheticus and a Lutheran theologian. He took the precaution, or the liberty, of adding a disarming preface to the effect that the author claimed for his system only mathematical convenience, not truth. This was incorrect. Copernicus was a mathematical realist in the Pythagorean tradition according to which figure and number contain the structure of things. But he never read the preface which said that his book was not true, nor the book itself, for he had been stricken by cerebral haemorrhage and lay on his deathbed when the first printed copy reached Frauenburg.

The Copernican *Revolutions* is a great work of science. Like all great works, it has had to survive a certain tradition of belittling scholarship, which in this case seems to imply that there was no real reason to prefer the heliocentric to the geocentric model, the two being geometrically interchangeable. Nor is the merit of Copernicus very evident when it is made to rest on simplicity in computation and economy in celestial motions. What with his reliance upon Ptolemaic data, his humanist's deference for antique learning, and his own belief in circularity as the heavenly pattern, Copernicus saved nothing significant in motions out of the long years during which he struggled through thickets of computation. He had the misfortune, indeed, to complicate his problem unnecessarily by adding to the rotation and revolution of the earth a third motion, a top-like wobble, whereby the equinoxes precess (which did require some account) and the north pole points to Polaris on both sides of the orbit (which did not). Nor will the superiority of Copernicus appear in a game of counting epicycles between Ptolemy's astronomy and his, the low score winning.

Copernicus, indeed, cannot satisfy such critics, for his criterion of elegance was different. It was not by eliminating epicycles that he thought to simplify and rationalize the procedures of astronomy: rather, it was by discerning the structure in things which befits the foundation of order. That foundation was the circle, the perfect figure. And it is the principle of circularity rather than of economy which conveys the inwardness of his vision of the world. Facts are pesky. Copernicus would subdue them by the epicycle. He would arrange them by the eccentric. And he did have to place the center of the orbit of the earth at a geometric point outside the sun in open space. But what had mortally insulted him in Ptolemaic astronomy was the equant. This saves the appearance of variable velocity by differentiating between uniform and circular motion. Copernicus thought it a cheat. In order to eliminate the equant, he added eccentrics and epicycles, and thus he lost to circularity motions that he had gained by sending the earth around the sun.

The superiority of the Copernican system, therefore, was conceptual rather than actual. Its vindication lay in the future when the data would be perfected, not in the past whence came its inspiration. Kepler might conceivably have proved that the sun goes around the earth in an ellipse. In a sense it does. But he could not have proved that Mars goes around the earth in an ellipse, because it does not. Moreover, there were important empirical respects in which the Copernican system actually was the simpler. With all the complications in detail, it exhibited one grand regularity which the Ptolemaic did not. The periods of revolution of the planets followed the same order as their distance from the center—the greater the radius, the longer the year. Given the complications of both systems, the effective radius was no simple line, of course, and the comparison might be made somewhat differently. In the Copernican system the angles subtended at the outer planets by the orbit of the earth exactly equalled the Ptolemaic angle subtended at the earth by the respective epicycles. Thus, in the Copernican system the relative contribution of the epicycle to the theory of each planet diminishes with the distance from the sun, while that of the deferent increases. As for the inner planets, it was a suspicious feature of the Ptolemaic construction that their deferents both equal that of the sun. Further, the Copernican order explained much more naturally why the sun and moon never appear to reverse their direction along the zodiac. And finally, it was far more reasonable—as Copernicus insisted—to think of the earth spinning daily, than to imagine what velocity such a rotation would impose upon the fixed stars.

These solid reasons will seem more persuasive in retrospect than will the Copernican aesthetic. Copernicus had steeped himself in the Pythagorean cult, Christianized in neo-Platonism. The sun itself had been the object of Pythagorean worship. In the tradition of Christian mysticism, illumination became the light of truth permeating the soul. And for Copernicus no arrangement was thinkable but that the sun, the lamp of the world, should occupy its center:

> In the middle of all sits the Sun on his throne. In this loveliest of temples, could we place the luminary in any more appropriate place so that he may light the whole simultaneously. Rightly is he called the Lamp, the Mind, the Ruler of the Universe: Hermes Trismegistus entitles him the God Visible. Sophocles' Electra names him the All-seeing. Thus does the Sun sit as upon a royal dais ruling his children the planets which circle about him.

But there is something meretricious about all this by the sixteenth century, or so it seems, and it is fortunate for the reputation of Copernicus that it does not have to rest upon his attainments in literature or philosophy.

Thus far our concern is with geometric conservatism tempered by a touch of sun-worshipping superstition. What, then, was determinate? What was there in the celestial gyroscopics of Copernicus to speak with such authority to the physical intuition of a Kepler or a Galileo? And ultimately it was just that, the physical element in his imagination, all wrong though it was, which seems to mark Copernicus out from the antique, whence he took data and techniques, and to lead forward into science. For his theory associated real physical events with a mathematical formalism. He needed something to make the earth turn, the earth and the celestial orbs, which he thought to be actual structures. He never looked beyond geometry, and based his kinematics as well as his computations upon circles. "Rotation is natural to a sphere," he boldly wrote, "and by that very act is its shape expressed." Put a globe in space, it will spin. This may not help much in the interpretation of nature, but it is the key to the interpretation of Copernicus. There was, moreover, one very important respect in which Copernican astronomy did enter into the development of physics, even if only critically. It was subversive of Aristotelian order. In a Copernican universe no physics could survive which depended on a central earth as the locus of the heavy. Not that many persons saw the point, or could yet relate celestial motions to terrestrial physics. But for cosmology too, the implications went beyond the choice of geometrical systems. Where does the world begin or end, if the earth is not the center? How deep are the stars set into space, if they are not pinpricks on the dome of the cosmos? Copernicus never said. One objection to his theory was that the stars show no annual parallax to the unaided eye. He met it by moving them out in hypothesis to where a parallax would be unobservable. And though the imagination could scarcely fail to travel on to an infinity of space and worlds (and created beings?), he wisely left that alarming prospect for more adventurous philosophers to discuss, one of whom, Giordano Bruno, burned for his temerity in the first year of the new century.

❖

COPERNICUS FOUND his worthiest readers in Johannes Kepler and in Galileo. "I have confessed to the truth of the Copernican view," wrote Kepler, "and contemplate its harmonies with incredible ravishment." In Kepler science wore for a time the guise of music. His quality has been compared to Mozart's. There is that same sweet

sense of proportion, that same ear for the felicity of exact
quantity. The laws which bear his name were only the
three most enduring chords in the mathematical rhapsody
over nature that occupied all his life. By the first, the
planets revolve in ellipses about the sun at one focus. The
second defines the quantity which does grow uniformly
while velocity varies: the radial vector between sun and
planet sweeps out equal areas in equal times. Kepler had
these to publish in *Astronomia Nova* in 1609. The third
he found only in time for his last book, *De harmonice
mundi*—Harmony of the World—in 1619: the squares of
the periods of the planets are proportional to the cubes
of their mean distances from the sun.

Until Kepler, circularity had been the basis of cosmic
order, things eternally rounding back on themselves. What,
then, must have been the combination of imagination,
devotion to fact, and faith in a deeper order, which nerved
him to break the habit that had governed astronomy since
its creation, to stretch the solar system into its true shape,
and to put it on a mathematical footing more abstract
than the perfection of the circle. His was one of the great
elastic feats of the human mind, comparable in its novelty
only to the enunciation of relativity, which also altered
the fundamental shape science finds in nature. No scientist
has ever taken his reader so utterly into his confidence as
Kepler did. Were it not for his rare innocence, it would
be embarrassing. He tells everything—his personal feelings, the humiliations of a horrible childhood, the indignities of his health, the defects of his wife, the follies of his
own aggressions. If anyone has ever been objective about
himself, Kepler was that man. Too myopic to observe the
stars, he had to calculate upon them. Nor does he conceal
anything of his science, his inspirations and his guesses,
his false starts and disappointments, his stupidities and
mistakes, his final and beautiful triumphs. He poured him-

self onto the page in a torrent of scientific consciousness, in a wealth of imagery and symbolic possibility flashing along on a great dark tide of trigonometric drudgery. He was a quite extraordinary person, and if (as has been said) he lifted the spell of circles, he casts his own spell over all who study him. "There is hardly a page in Kepler's writings—some twenty solid volumes in folio—that is not alive and kicking," observes Arthur Koestler, who has experienced that enchantment, and whose interesting book, *The Sleepwalkers* (of which the same is true), is its expression.

Kepler combined extravagance in emotion with meticulous devotion to fact, and unlike Copernicus, he had the facts. Or rather he obtained them, after the failure of a first book. He had studied theology and amused himself with cosmology at Tübingen, a Lutheran University. His own leanings were Calvinist, toward a God of predictable quantity, perhaps. He took a post as Provincial Mathematicus and teacher of astronomy at the Catholic University of Gratz in Styria. There he began teaching in 1594, at the age of twenty-three. Two years later he published an idea about the organization of the universe which came to him at the blackboard. The notion was that the five perfectly symmetrical solids of geometry concentric with the sun form the mathematical skeleton of the solar system. The orbit of Mercury is inscribed in an octahedron, which is circumscribed by the orbit of Venus, which is inscribed in an icosahedron, which is circumscribed by the orbit of Earth, and so on through the dodecahedron, tetrahedron, and cube, in and around which fit the remaining orbits. Kepler's *World Mystery* is a vision of the organization of a Copernican cosmos in depth, and a further syncretism of Christianity and Pythagorean religiosity. It is also wrong. It does not fit. But though he soon recognized its failure in detail, Kepler never doubted that thus to seek the harmonic and geometric proportions of reality was to know God, "whom in the contemplation of the universe, I can grasp, as it were with my very hands." And in 1600 he met the man who had the figures which he thought would fit his model to the facts. This was Tycho Brahe, just then taking up residence as Imperial Mathematicus to the half-mad Rudolf II in the Hapsburg capital at Prague.

Tycho was one of those admirable and indispensable laborers in the vineyard whose mission it is to observe or experiment upon nature with all the accuracy that ingenuity can command, but to whom the higher quality of theoretical insight is denied. In 1572 a magnificent nova startled all Europe and demonstrated change in heaven. But Tycho would not be persuaded of the Copernican system. At most he would concede a compromise, by which moon, sun, and sphere of fixed stars go around the earth, while the five planets go around the sun. This saved both the immobility of the earth and the most serious astronomical evidence for Copernicanism—the relationship of the period and radius of deferent to distance from the sun. Tycho was a Danish aristocrat, a rogue nobleman who had abandoned the violent preoccupations of his class for astronomy, though without attaining serenity. His King granted him the island of Hveen. There he constructed a fine observatory, Uraniborg, over which he ruled in feudal splendor for twenty years. Along the walls he had great quadrants constructed, one on a radius of fourteen feet. They brought the art of celestial observation to the highest accuracy of which the human eye was capable unaided by the telescope. The best of Tycho's data were reliable down to one minute of arc, a tenfold improvement over what had gone before. Moreover, the usual practice of astronomy discontented Tycho. Astronomers would take observations at the nodes and syzygies, and interpolate the orbits between these points. He or his assistants, on the contrary, followed the planets nightly for over two decades, and pricked their observations on a great brass-sheathed globe which became their record.

He took his records with him when he quarreled with the King of Denmark and transferred his allegiance to the Emperor in Prague. There Kepler joined him, in retreat before a wave of Catholic suspicion of Protestants in Styria. Their relations of patron and supplicant were compounded in difficulty by those of waning monarch and heir apparent. For Tycho knew that there was in Kepler a theoretical power of the highest order, and sought to bind him to use the treasury of observations to establish the Tychonic theory. But he died in 1601, leaving Kepler to appropriate the data from his legal heirs and to succeed him as Imperial Mathematicus, a post which often went unpaid what with the habitual arrears in which the Hapsburg chancellery conducted its affairs.

Even before Tycho's death, Kepler had been at work upon the theory of Mars, the most intractable of the planets because of its greater eccentricity, and also (fortunately) that for which the information was the fullest, since it is the nearest beyond the earth and is not lost in the sun like Venus and Mercury as a morning or an evening star. Kepler assumed two conditions, by virtue of which he started his thinking at one remove beyond the Copernican and toward the Newtonian. The first was geometrical—that the planes of the earth's orbit and of Mars' intersect in the center of the sun. The second was a related assumption, but physical and therefore more significant, since otherwise this could have been only a reversion to motion by the equant. When the *Astronomia Nova* finally appeared, its subtitle would be *Physica Coelestis*—Physics of the Heavens. From the beginning, Kepler's embryonic feel for the mechanical led him to invest the sun with the power that causes motion in the planets. His object was celestial dynamics and not just a kinematics, his goal a force law and not just a law of motion. To the power in the sun he opposed in equilibrium another in each planet, the equal and unending contest between the two determining its orbit. (Kepler was a combative thinker, and saw his long pursuit of the true law as a personal contest between himself and Mars; or perhaps one should say a sportsmanlike thinker, since the game was conducted like chess, at a high pitch of tension but without acrimony.) It was reasonable to suppose that the power in the sun diminished with distance, and that an inverse proportion would govern the relation of the planet's velocity to its distance from the source of power.

And now emerges one of the tantalizing coincidences with which Kepler's intellectual career abounds. Among his physical discoveries was the optical law that the intensity of illumination on a surface diminishes as the square of the distance from the source of light. (This, like the law of gravity itself, derives rather from solid geometry than from the nature of light, and may be taken as an indication of how fruitful was the supposition that the world is really made on geometrical forms.) And since Kepler's sun radiates power as well as light, science seems to tremble here on the brink of the Newtonian cosmology. But then one realizes that it could not take the plunge. Kepler's *anima motrix* in the sun weakens at a linear rate with distance. Moreover, it is not a radial force. It is a tangential drag, suggested to Kepler's mind by the analogy with magnetism. Kepler had been profoundly impressed by *De magnete* which William Gilbert published in 1600. Gilbert treated of loadstones and compass needles and beyond them of the suggestion that the earth itself is a great magnet aligning all little magnets by affinities. So thought Kepler on his grander scale of the cosmos, where affinities run between bodies related by their intrinsic natures. Such are the sun and the planets, and such the moving spirit which emanates from the sun and sweeps the planets along their appointed rounds. And in the absence of the principle of inertia, therefore, the *vis motrix* will appear less as an anticipation of gravitational force, requiring chiefly to be redirected at right angles, than as an expression of the ancient Aristotelian instinct that motion presupposes a mover.

Nevertheless, this was a physics, and it gave body to Tycho's numbers. For years Kepler wrestled with those figures, striving to surprise in them the secret of Mars. He had already tacitly decided that the geometric and physical description of the orbit must be one, and that he had to do with an actual curve through space and not (like Ptolemy or Copernicus) with epicyclic combinations by which he might calculate. The figures for Mars in opposition to the sun gave him his point of departure. Using them, he must find the direction relative to the fixed stars of the line of apsides, the eccentric location thereon of the sun, and the radius. No rigorous solution was possible. It was a geometer's nightmare, to be dealt with by main force of approximation and trial — and error, too. Kepler spares his reader none of the hopes or disappointments. "If thou"— he tells whoever is still with him — "art wearied with this tiresome method of computation, have pity on me, who had to go through it seventy times at least, with an immense expenditure of time; nor will it astound thee that the fifth year has almost passed since I encountered Mars. . . ." Then when at last he thought he had the values right, he verified the orbit — and found disagreements amounting to eight minutes of arc between certain observed positions and those predicted by his theory.

That was a very small amount. Before Tycho it would have been undetectable. And nothing so testifies to Kepler's conscience as his sacrificing six years of work to eight minutes of arc. For this tiny failure was not to be taken as an instance of the tiresome intractability of nature, to be fudged over in order to finish the job, but as an opportunity to learn something new. "If God has sent us an observer like Tycho," wrote Kepler, "it is in order that we should make use of him." Those eight minutes turned out, indeed, to be the fault which broke the circle. And their relation to the discovery of Kepler's laws illustrates that often it is not the big problems which refine the scientific understanding, but the minuscule discrepancies: not the logical structure of Aristotelian natural philosophy, but the deep, small problem of what makes the arrow fly; not some gross breach of Newtonian physics into which Einstein stepped, but the almost undetectable absence of the aether drag. And Kepler now began to think that perhaps the orbit might be some figure other than a circle.

Pursuing that possibility, he found his so-called second law (of equal areas in equal times) before the first law (of elliptical orbits). He had, indeed, been using it as a

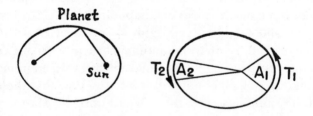

calculating device before enouncing it as a governing principle, and the purported derivation is a comedy of errors with a happy ending rewarding virtue in the truth. He hesitated for a time, thinking that his difficulty might lie in this method of computing. He had in fact reason to doubt it, for the assumption that the velocity varies inversely as distance from the eccentric sun holds true only for the ends of the apsidal diameter, where the sun-planet vector and the tangent are perpendicular. Moreover, when he did decide to commit his fortunes to the equal area theorem, he committed a yet more fearful solecism in demonstrating it. The time, he said, to traverse an infinitesimal arc will be proportional to the distance from the sun. The sum of these distances, therefore, measures the area swept out in the orbit, and any particular segment of the area will be measured by the time the sun-planet vector requires to traverse it—hence, equal areas in equal times. And this reasoning is absolutely false, of course. It is if he were to have said, $dr \sim dA$, and therefore $\Sigma r = A$. Kepler must certainly have known better. But he also

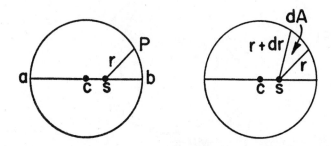

knew that the result was right, and that the two errors cancel; and he pressed on, having left Mars to itself for a bit while investigating more closely the motion of the earth. For it was in relation to the latter, considered from a fixed point in the orbit of Mars, that Kepler stated the law of equal areas.

The data encouraged him in his abandonment of tradition, for they seemed to give an oviform curve falling symmetrically inside the circle along two sides. But what curve? At first he supposed it to be the profile of an egg, with the sun nearer the pointed end. He had yet to think of a figure around two geometrical foci and one center of force. Nor will it do to follow step by step as he tacked and turned, using the right answer before he knew he had it. Investigating the geometrical properties of his ovoid was a lamentable task, to which conic sections gave no access, and he was driven to use the ellipse as a manageable approximation to the "true" curve. Thus, he came upon another of his coincidences. The maximum width of the crescent which would have filled the ovoid to a semicircle on one side was .00429 times the radius. Quite independently of this measurement, Kepler determined that the maximum angle formed at Mars between the sun and the center of the orbit is 5° 18'. He was living with all these numbers, and it struck him that the secant of that angle is

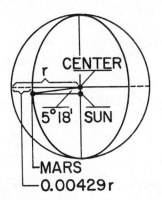

1.00429. This could not be chance. What he did not know—tantalizing is too feeble a word this time—is that this relation is one of the conditions defining an ellipse, and that he had made an arithmetical mistake in the testing. Thus, he was driven off onto a theory of librations—a kind of breathing of the orbit in and out—whereby the resultant motion is compounded of the orbital revolution and a radial oscillation by the diameter, only to find, after going back to the egg, then returning in desperation to the ellipse, which he now constructed geometrically with the sun in one focus, that this was the identical figure at which he had already arrived trigonometrically months and months before. For it must be remembered that all this was done without analytical geometry, and without regular tables of logarithms.

It is too much, one feels on reading Kepler; no one should have had to go through this. And it was worse than it seems. For Kepler's laws could not be to him what they are to science, expressions reducible to a unified system of dynamics. Lacking analytical geometry, they did not follow one out of the other. For Kepler himself they were simply snatches of melody in search of a symphony. They held but little of the fulfillment which had deluded him for the moment in his youthful vision of the five perfect solids. He had still to perceive the geometric structure of the cosmos in depth. His ellipses were a sorry substitute for circles, after all—a "cartful of dung," he called them, in a fit of disgust at their irrationality, before he had found anything to relate the two-dimensional shape of the orbits to the structure of the solar system in depth. Meanwhile, the news came of Galileo's telescope and the moons of Jupiter. Rudolf was deposed, for whom Kepler had cast horoscopes—but only when they would do no harm. He believed that his own life had been affected by the configurations of the planets at his conception and at his birth, but hesitated to lend himself to a statesman's retreat from reality into astrology. He obtained a post as provincial mathematician in Linz, the dullest of Austrian cities. And there he set himself to composition of *De harmonice mundi*, the last of his great books, of which the writing was interrupted by the necessity to defend his horrible old mother, who had been indicted as a witch, perhaps with some reason.

The harmonies of which he treated are ideal geometric proportions according to which God had created all the manifestations of being. Just so the octave, the fifth, and the fourth which we perceive as harmonies are given by invariable ratios between the lengths of instrumental strings. In geometry itself there are concords (the regular polygons and solids) and discords (irregular shapes which cannot be constructed by rule, like the seven-sided figure). Plane geometry deals with symbols of the two-dimensional material world. The sphere stands in its three-dimensional perfection for the Holy Trinity. The plane section of the sphere contains the dual aspect of man—body and spirit.

Thus, in cosmology Kepler was bound to search for a reason in things deeper (in every sense) than his ellipses. Where is the harmony in the world? He tried the parts in every way, like a blind man with a jigsaw puzzle. Do the planetary periods follow a harmonic series, investing the third dimension of time with meaning? But he could find no regularity in that progression. Perhaps the sizes or volumes of the planets hold the secret? There was no evidence. Perhaps some ratio hides in the variation of solar distance as between the successive planets? Nothing there. Perhaps in the relation between greatest and least speeds? Or between mean speeds? Now he felt a little closer. He moved in imagination to the sun, and examined the least and greatest values of the angular velocities of each of the planets. And now (he saw) his faith was founded. For Saturn the ratio is 4:5, the major third; for Jupiter it is the minor third; for Mars, the fifth: thus in the music of the spheres, played to the mind's ear, Saturn and Jupiter take the bass part, Mars the tenor, Earth and Venus the alto, and Mercury the soprano.

Such was the route by which Kepler came upon his third law, which at last established the connection he had ever been seeking between a planet's motion and its distance, between the operation and the structure of the solar system. It was a very unexpected correlation, that the squares of any two periods are to each other as the cubes of the mean distances. Nor could it have then been found but by repeated trials. Indeed, Kepler's reader will himself experience a certain difficulty in finding it among the fantasies and ecstasies of this last book. But it is not, perhaps, quite true to say that Kepler himself saw no more in it than in his Pythagorean flights of musical geometry. The preface to the chapter which reveals this relationship contains his famous and often quoted *Te Deum*:

> Eighteen months ago the first dawn rose for me, three months ago the bright day, and a few days ago the full sun of a most wonderful vision; now nothing can keep me back. I let myself go in divine rage. I defy the mortals with scorn in this, an open confession. I have stolen the golden vessels of the Egyptians to make out of them a holy tabernacle for my God, far from the frontiers of Egypt. If you pardon me, I will be delighted; if you are angry with me, I shall bear it. Behold the die is cast, I am writing a book for my contemporaries or—it does not matter—for posterity. It may be that my book will await its readers for a hundred years. Has not God Himself waited six thousand years for someone to contemplate His work with understanding?

✧

NEITHER COPERNICUS NOR KEPLER ever broke with the medieval or the baroque in their mathematicism. Their cosmos was built according to a geometrical structure, as a cathedral may be which is embellished with gargoyles and figures of saints. But the Platonic forms were hung about with these Gothic excrescences of the North, nor was it always easy to tell whether the rich animistic symbolism pertained to the decoration or the fabric.

In Galileo there is no such doubt. His Platonism was

in the Archimedean rather than the Pythagorean tradition, and his thought and work brought to bear on nature the clean, clear light of Italian classicism. The Latin genius speaks out in Galileo. His is the passionate objectivity of Machiavelli, which says that wishes do not signify—this *is* how the world works. He stripped from the skeleton of the cosmos the obscuring layers of sentience and pious moral and edifying lesson, and left as object of the search the hard, straight bones of Euclidean dimension, Platonism bleached bare, sterilized of its mystical nonsense in the Tuscan sun. A loftier humanism replaces anthropomorphic sentimentality in Galileo's natural philosophy. In the *Dialogue* Sagredo retorts on the common prejudice which attributes it "to natural bodies as a great honour and perfection that they are impassible, immutable, inalterable, etc: as, conversely, I hear it esteemed a great imperfection to be alterable, generable, mutable, etc:"

"The more I delve into the consideration of the vanity of popular discourses, the more empty and simple I find them. What greater folly can be imagined than to call gems, silver, and gold noble and earth and dirt base? For do not these persons consider that, if there were as great a scarcity of earth as there is of jewels and precious metals, there would be no king who would not gladly give a heap of diamonds and rubies and many ingots of gold to purchase only so much earth as would suffice to plant a jessamine in a little pot or to set a tangerine in it, that he might see it sprout, grow up, and bring forth goodly leaves, fragrant flowers, and delicate fruit? It is scarcity and plenty that make things esteemed and despised by the vulgar, who will say that here is a most beautiful diamond, for it resembles a clear water, and yet would not part with it for ten tons of water. These men who so extol incorruptibility, inalterability, etc., speak thus, I believe, out of the great desire they have to live long and for fear of death, not considering that, if men had been immortal, they would not have had to come into the world. These people deserve to meet with a Medusa's head that would transform them into statues of diamond and jade, that so they might become more perfect than they are."

And it epitomizes Galileo's lifelong position that it should have been he who distinguished between primary and secondary qualities in bodies. The primary he defined as properties essential to metrical description—length, width, weight, figure. The secondary—color, taste, odor, texture— are those which are modes of perception in us, rather than real essences permeating matter. The difference is that between object and subject.

Galileo Galilei was born in Pisa in 1564, the son of a Florentine musician. He studied mathematics and natural philosophy and discovered, so it was said, the isochronism of the pendulum when he was nineteen by considering the swing of the great lamp in the nave of the cathedral. In 1592 he moved to Padua. His lectern may still be seen preserved in that university, where Vesalius and Galileo lectured, and Copernicus and Harvey studied, and which has a juster claim than any other to recognition as the nursery of modern science.

There was little novelty in addressing oneself to the failure of the Aristotelian theory of motion. Galileo's kinematics derived from the Merton tradition and his dynamics from a related fourteenth century school, that of Jean Buridan and the University of Paris. They were dissatisfied with the absurdity of an explanation which would sustain the flight of heavy cannon balls by virtue of the air pressure in their wake. This made the air at once resistant and propellant. And instead they proposed as the cause of motion an impetus impressed upon the missile. The conception wears something of the aspect of momentum. Impetus varied proportionately to velocity and the amount of prime matter. Nevertheless, though a great step forward, the theory proved incapable of objectifying motion. Impetus remained an indwelling quality consumed in the process that it caused. Nor did it work a conceptual separation between motion and the missile.

What was original, therefore was Galileo's ultimate conception of motion, not his criticism of Aristotle. Indeed, so original was it that it may be taken as one of those exceedingly rare events, a true mutation in ideas, a break with the past. It altered man's consciousness of a real world outside himself in nature. This new world is to be grasped rather by measurement than by sympathy. In it an Archimedean science is possible, not just of statics contemplating things at rest, but of dynamics studying to know things in motion. One may say with some confidence what was revolutionary in Galileo's law of falling bodies: It was that he treated time as an abstract parameter of a purely physical event. This enabled him to do what no Greek had done, to quantify motion in number. Galileo spent twenty years wrestling with the problem before he got free of man's natural biological instinct for time as that in which he lives and grows old. Time eluded science until Galileo. The dimensionality of space is evident at a much lower level of abstraction. Again and again, Galileo himself tried to find a general expression for the velocity of a body in relation to the distance it had fallen. And always he failed.

Ultimately he expressed velocity in relation to the time of fall, and he succeeded. But how difficult this idea was! The magnitude of the conception is hidden from us by its very success. Thanks to Galileo, we are all bound to be physicists to some extent, in virtue of what seems like simple consciousness. It is ever the lot of scientific innovation thus to disappear into the drabness of common sense, and we need an effort of imagination to move back into the scholastic definition of motion. The necessity may itself be taken as a measure of Galileo's achievement. History is easier than discovery, after all. To return to Aristotle is to move into a world of human association and (in scholasticism) of religious comforts. This demands only patience and tact, whereas to look forward out of the Renaissance into Galileo's world was to stand alone peering into a nature deprived of sympathy and all humane association. That required both courage and power of abstract thought, which, one of the greatest of gifts, goes against the grain in all but the rarest temperaments. For sentiment rebels against the condition that nature sets the natural philosopher. This is that science communicate in the language of mathematics, the measure of quantity, in which no terms exist for good or bad, kind or cruel, and that she abandon our language of will and purpose and hope—abandon or denature or impoverish it, turning force, for example, from personal power into mass-times-acceleration.

It must prove very interesting to the historian of science that a leading man of letters of our own time should write upon his subject. Arthur Koestler's *The Sleepwalkers* has already been mentioned. In no other book do Copernicus and Kepler come to life as there. Moreover, the author has given himself more trouble to master the texts than many professional scholars have done who have written of the scientific revolution. Nor is it meant as denigration of a distinguished piece of work if one historian of science alludes to the interpretation as an instance of the offense that science does give to the literary intelligence. It is an instance over which the scientist and the educator would do well to pause. They might ponder it amidst all the wishful slogans about some ultimate unity in science and the humanities. For Kepler is the hero of the story precisely in the degree that his was a "sleepwalking" genius—unconscious, irrational, and gestalt-like. His was a divine madness, his an expression of Pythagorean myth emergent from the collective subconscious of man the eternal artist, man one with nature: "One of the points that I have laboured in this book" writes Koestler, "is the unitary source of the mystical and scientific modes of experience; and the disastrous results of their separation." Inside minds like Kepler's, "We find no abrupt break with the past, but a gradual transformation of the symbols of their cosmic experience—from *anima motrix* into *vis motrix*, moving spirit into moving force, mythological imagery into mathematical hieroglyphica—a transformation which never was, and one hopes, never will be entirely completed."

For Koestler, Galileo is the villain because he sought to force that transformation, and to make that break: "He was utterly devoid of any mystical, contemplative leanings, in which the bitter passions could from time to time be resolved; he was unable to transcend himself and find refuge, as Kepler did in his darkest hours, in the cosmic mystery. He did not stand astride the watershed; Galileo is wholly and frighteningly modern." He would reduce all nature to his primary qualities of size, figure, number, and motion, and classify everything else as subjective and secondary. Thus began the "Fatal Estrangement" between science and ethics, which has left us in a world in which we have no place. There we drift toward a state of nihilism where ignorant technicians clash by night. For scientists have absolved themselves of moral responsibility and eschewed understanding in favor of measurement. It is in Galileo that we are for the first time asked to abstract our judgment of achievement from our judgment of character. Thus Koestler—nor does the point depend upon his hostile estimate of Galileo's character.

Now then, one may react differently to Galileo, rather in admiration at his daring than in revulsion from the cruel edge of objectivity. One may feel that it is precisely the interest and the merit of Galileo that he does put it up to us to accommodate ourselves to reality, instead of indulging the primitive illusion that there is a refuge in nature or in myth or in collectivity. Nevertheless, the historian will feel confirmed by this response of a more sensitive temperament in his own instinct that the intellectual history of modern science does begin with Galileo, and with the transformation he wrought in the conception of motion. Things in motion, after all, are what science studies.

In any case, we are all brought up in that idea of motion as a relative and persistent state of bodies which is implicit in Galileo and explicit in Newton. Taking it for granted in our early education, we never learn to see how contrary it is to experience, wherein motion always does wear out, nor how very abstract the idea is. It *is* abstract. Galilean motion has no existence apart from bodies moving in relation to each other, but has no effect upon those bodies. It is no longer a process. It is a pure relation to be described by a mathematical expression. The world of experience never displays this ideal relationship. But though the actual event always falls short, nevertheless the ideal or the mathematical is the way to grasp it. We are to think about the perfect fit and not about the wrinkles in the garment. No longer is local motion a special, temporal case of the metaphysics of change. No longer is it one kind of development by which things realize the ends their natures prescribe, as the boy grows into the man and the man strives to lead the good life. No longer is it the manifold expression of a cosmic tension toward order and the good. There is no meaning henceforth in motion, but only quantity: the abstract quantity, velocity, which has one component in time and one in distance. Nor are these dimensions what we reach through and live through. They are lines in Euclid.

Had Galileo restricted himself to mechanics, he would

never have given that offense which raised him up to become one of the great tragic figures in the history of human creativity. But he was no mere technician, no bloodless mathematician. He was a humanist, a man of charm and taste and literary grace. He had the full gift of civilization. Wine, he used to say in the nicest of definitions, was light held together by moisture—"Il Vino era Luce impastata con humore." He was far more than a physicist. He was a natural philosopher who saw beyond the crucial little problem of falling bodies, about which it is difficult to feel strongly, to the great vision of a universe and a science in which the real is described by the ideal, the concrete by the abstract, matter by mind. "The book of nature is written in mathematical characters," he declared in *Il Saggiatore*. And there lies the heart of the issue between natural philosophy as understood by Galileo, and scholastic philosophy as understood by educated Christendom in general, and by Robert Cardinal Bellarmine in particular, whose mind was behind the acts of the Roman Curia. Bellarmine was a Jesuit, a soldierly statesman of his Church. He produced the catechism it still enjoins. In its opinion he was a saint. He was, at all events, a tenacious statesman, bringing continuity to papal policy in successive reigns and accepting the responsibility for seeing the Church and civilization itself safely through the philosophical confusions and civil disorders created by the Reformation and its aftermath of conflict.

Copernicanism became involved in those confusions through no fault of Bellarmine's. For him, the attitude to be adopted towards this new cosmology was a matter of policy, not truth. He already had the truth, which it was his mission to defend, and it did not lie in the mathematical description of nature. It lay in the whole structure of Catholic thought, to which geometry was quite irrelevant. Nor could the professors and the guardians of scholastic thinking allow it any relevance. Good Aristotelians, their real world was contingent, compounded of qualities and meanings and fine distinctions and physical objects not to be understood mathematically. In the heavens everything is perfect, and physics does not treat of perfection. Rome had felt no hostility towards the heliocentric hypothesis during Copernicus's lifetime, nor in the decades since his death. On the contrary, in 1536 Cardinal Schoenberg, an intimate of the papal court, wrote Copernicus urging him to publish, for the honor of their compatriots and the benefit of the learned world.

Nor did Galileo anticipate theological opposition when he turned on his own account to Copernicus. It was the philosophers, the professors, and the pedants of Padua whom he was bound to provoke and among whom he must walk warily. From his earliest maturity, Galileo's instinct for theoretical elegance told him that Copernicus was right. But he could scarcely think to establish a physical picture which could not be proved by a mathematical philosophy of nature which few believed. Then in 1610 there came to hand the telescope. This was something which anyone could understand *grosso modo*, and in which everyone would be interested. The big facts leaped to view. The moon was no celestial jewel, but a pitted mass of dead rocks. Out beyond, there circled around Jupiter its satellites, the "Medici planets" as Galileo cannily wished to call his finds, a miniature Copernican system for all to see.

It took Galileo time to learn that some would decline to look. "What shall we make of this," he wrote to Kepler—il signor Gleppero as he appears in one document—when certain scholastic philosophers did refuse to peer through the telescope, "shall we laugh or shall we cry?" For worldly affairs are only to be mastered in their ordinariness and complexity, and Galileo had a quality of scornful innocence about them, a kind of self-righteousness, which is characteristic of minds capable of great sophistication when addressed to nature, where elegance and simplicity are goals of thought. His difficulties with the Church did not arise from any innate hostility of religion to science or any opposition between truth and knowledge. They were the consequence of that failure which in some degree is bound to afflict communication between the man of science and the man of the world. How could Galileo imagine that men would not wish to be told how the world is made, just because to learn might be disturbing? How could the ecclesiastics in Rome, responsible for faith, morals, and the order of civilization, understand Galileo's passion for overturning the settled structure of natural philosophy against both common sense and orthodoxy? They were utterly ignorant of mathematical reasoning, the cardinals and monks, nor could they possibly appreciate the force it carries.

The drama between science and the church, therefore, unfolds with that inevitability which is tragic because it arises from the characters of men rather than the necessities in things. In 1610 Galileo returned from Padua to Florence at the invitation of the Grand Duke. There his views were soon denounced from the pulpit by two monks, self-appointed guardians of morality of the type which makes no distinction between originality and infidelity. Echoes came back from Rome. Galileo took alarm, not for himself, but lest the authorities, through inability to judge what science held for the future, should require Catholics to choose between faith and reason.

To Rome he went, therefore, to educate the authorities. Like many who need education, they did not know it and resented the implication. "He is passionately involved in this opinion," wrote back the Florentine ambassador in anxiety to the Grand Duke, "as if it were his own business, and he does not see and sense what it might involve; here the monks are all powerful, and they do not like him; so that the climate of Rome is getting very dangerous for him, and especially in this century; for the present Pope, who abhors the liberal arts and his kind of mind, cannot stand all these novelties and subtleties." And the only reward of Galileo's effort was an injunction in 1616 from the Congregation of the Index declaring the Copernican doctrine erroneous. It was not to be "held or defended," though it might still be used confidentially as a mathematical device.

Then in 1623 the old Pope died, to be succeeded by Maffeo Barberini as Urban VIII, and Galileo took heart. For the new Pope had the reputation of a humanist. And now Galileo proposed to write a book, "full of geometry, astronomy, and philosophy," not ostensibly to prove the Copernican hypothesis, but in the Platonic form of *Dialogue on the Two Chief Systems of the World*. The discussion was to remain hypothetical. It is doubtful that he really meant to deceive the Pope. But his wit and knowledge were too much for his prudence. Simplicio, the Aristotelian, is overwhelmed and made to accept defeat after defeat protesting, with a kind of humiliated affability, that he remains convinced of his point of view, though he cannot make it prevail. The deportment of Salviati, the Platonist-Copernican, wears that air of divine malice which is what gives offense in a superior understanding where it is likely to appear as arrogance. There was simplicity in Galileo, as there is in science, but no humility. The attempts of Sagredo to hold the balance are as patently unconvincing as the disclaimer of the preface where Galileo says "I have personated the Copernican in this discourse, proceeding upon an hypothesis purely mathematical."

Indeed, the reader is left to feel that only a fool could reject the Copernican universe or fail to perceive the elegance of the mathematical structure of being. This was certainly not what the Pope had authorized, since he did fail to perceive it, and it is not surprising that Galileo should have been summoned to Rome in 1633 and forced by all the fearful authority of the Inquisition to abjure in the name of Catholic truth his belief that the earth moves. There are those who would have had him choose the part of martyr. They forget that he was a Catholic, loyal to his church when its servants erred. And another element obscures the inevitability of a conflict between science and religion and removes this most embarrassing episode in the intellectual history of Rome from the realm of truth to that of politics. Galileo's condemnation turned on his failure to obey an injunction not to discuss the Copernican system. Such a document was introduced into evidence, to which he was refused access. But he could not have disobeyed it. He had never seen it. For it seems to have been a forgery slipped into the file by certain priests among his enemies.

Galileo's abjuration degraded all concerned, reducing scientist and churchman to the least common denominator of politician. But what was intellectually a larger retribution awaited Galileo's *Dialogue*. For the triumph over Aristotelianism of that great book is literary rather than scientific. In his anxiety to establish a mathematical science uniting heaven and earth in a single physics, Galileo pressed eagerly forward over difficulties. Had he instead stopped to resolve them, he would not have had a unified cosmos with which to replace Aristotle's. But he would have followed the logic of his own law of falling bodies, and classical physics might bear his name instead of Newton's.

Motion is again the problem, the silver cord of all who could not make the break with Greece. Though Galileo took motion as natural and made it persist, thus opening the way to conservation and force laws, he failed, nevertheless, to formulate the principle of inertia, the first of Newton's laws and the cornerstone of classical dynamics. This was because when forced to choose between cosmic order and the absolute mathematicization of nature, he too chose order. For the function of science in Greece was to explain the universe in a single rationale, and not simply to generalize some limited set of phenomena. A universe which can be explained, a cosmos which we can fit, must be finite, and even Galileo never quite confronted the prospect of infinity. Consequently natural motion for Galileo, inertial motion, is that motion which neither rises nor falls, which is equi-distant from the center of the earth. It is, therefore, circular motion. No longer the center of the cosmos, the earth remains the center of motion.

Galileo had what he needed to break the circle. His own law of falling bodies contained the parabolic trajectory of projectiles. But in the third day of the *Dialogue* he failed to draw the deduction and made a body falling from rest follow a semi-circle to the center of a rotating and hypothetically permeable earth. He had what he needed. Kepler sent him the *Astronomia Nova* containing the law of elliptical orbits. And the physical force holding a planet in an elliptical orbit is mathematically identical with the force that makes earthbound projectiles follow a parabola. Both curves are conic sections. But he paid no attention. He wished to make Copernicus, not chaos, prevail. With rectilinear inertia but without gravity, things will fly off in straight lines to infinity. Had Galileo abandoned circularity, had he thought on Kepler's work which he admired but found no time for, he might have made inertia rectilinear. He might have united heaven and earth by the force of gravity. But he did not: for physical proof of the rotation of the earth, he adduces the tides, sliding around in the basins of the sea like broth in a wobbling cup.

Indeed, he knew he did not have motion right in the *Dialogue*. And a final achievement awaited this extraordinary man. Neither his humiliation by the Holy Office, nor his inner sense of having overreached himself in his magnum opus, led him to abandon science or accept senility. Confined by order of the Pope to his house above Florence in Arcetri, with failing eyesight and in faltering health, he wrote the truly scientific *Discourses on Two New Sciences*. One of these is what we call strength of materials. That he should present statics in such guise is an indication of how he had solidified Archimedean abstractions in physics proper. The other is "local motion." And now he does treat it geometrically as if it persists along the tangent, without betraying undue guilt at the implication that its departure from the circumference is undetectable and insignificant at ordinary distances. (This is the same approximation which Archimedes makes in considering the radially directed force of a weight at the extremity of a lever as perpendicular to the beam.) No great cosmic or philosophical questions intrude into this unimpassioned treatise. The same names conduct the discussion. But it is about as controversial and stirring as some freshman lecture on mechanics, of which, indeed, it is the ultimate source.

The crowning irony of Galileo's career is that the failure of the great *Dialogue* should be so much more interesting than the success of the unobjectionable *Discourses*. And even here he foreshadows the dilemma with which history confronts us when it considers science as a creation of the human spirit. For Koestler is right in a sense. The success of science does stand in an inverse relation to its appeal. And one must differ from his judgment, not in denying the inhumane character of science once created, but in rejecting the implication that the dilemma may be escaped by lamenting science, or softening it, or changing what it says about nature. That way lies the romantic sentimentality of Slavic and Teutonic culture. That way lies surrender of the measure of independence which science by its determinate toughness has won for scholarship and thought. That way lies, finally, the unforgivable vulgarity which resents what surpasses the common understanding, and which ends in the darkness at noon foretold by Galileo in an autograph note written in the margin of his own copy of the *Dialogue*:

> In the matter of introducing novelties. And who can doubt that it will lead to the worst disorders when minds created free by God are compelled to submit slavishly to an outside will? When we are told to deny our senses and subject them to the whim of others? When people devoid of whatsoever competence are made judges over experts and are granted authority to treat them as they please? These are the novelties which are apt to bring about the ruin of commonwealths and the subversion of the state.

CHAPTER II

ART, LIFE, AND EXPERIMENT

PHYSICS has been the cutting edge of science since Galileo, and its mathematicization in dynamics was, therefore, the crucial act in the scientific revolution. But modern science is a complex enterprise, compact of elements both intellectual and social. The strands which the genius of the seventeenth century twisted together, it drew from diverse areas of the Renaissance. Naturalism, for one, had expressed itself in art long before it became an essential element of the scientific outlook. Empiricism, secondly, may be more explicitly exemplified in the life sciences than in physics, and there it was early coupled with a half-heroic, half-petulant, and wholly self-conscious revolt against authority. Thirdly, the technological achievements of Renaissance navigator, engineer, and artisan were realities behind the Baconian philosophy, which makes science an inductive systematization of observations or experiments performed upon nature rather than of concepts, and which holds up as the reward for understanding nature the power to control its forces.

For Leonardo da Vinci, as for many a Renaissance humanist, there was a whole world in man. In his eyes science and art were both illumination—the reality of the great world suffusing the consciousness of the little in the act of perception. This was not to be, of course. The two modes of grasping nature are different, the one particular and concrete, the other general and abstract. Nevertheless, in Leonardo's infinite capacity to see and draw the structure in things, he presaged all unwittingly that acceptance of the innocence of nature which takes it as the morally passive object of inquiry.

To the mind sensitive to cultural pathos, there is nothing more moving in the history of our technical tradition than the pencil of Leonardo forever pinning nature to the page. What are we to make of him? Of those notebooks teeming with designs of devices yet unborn, submarines and armored cars, machine guns and planned cities? Of the aphorisms that seem to foretell organizing concepts like inertia and evolution in all branches of science, from astronomy through mechanics and geology to biology? Of the clear assumption that to illustrate the structure of a bat's wing is to understand the dynamics of flight preparatory to flying—that to see is to know is to act? The historian must answer that he can make nothing of them, and that no one else could. Leonardo's science remains the marvelous jottings of a unique spirit. Nor were the contents of his notebooks known until the end of the eighteenth century, when they were carried back to Paris by the cultural raiders sent to Italy by the French Republic in the wake of Napoleon's armies.

It is true that Leonardo's studies of draperies and anatomies, of horses stamping and rearing, of weapons and fortifications, of architecture and the strength of materials, express a mind which perceived geometric form in nature. But his geometry was palpable, not abstract. His was the mind of an engineer addressing itself to mechanisms which are identical with their specifications, which fit themselves utterly. His conception of mathematics was what the subject would be if those plaster and string models which illustrate curves in space and figures of solid geometry, were themselves solid geometry.

It is tantalizing, but it did not lead to science in general. Leonardo's humanism and naturalism did, however, lead into one particular science, the anatomy of the human body. In an early stage, the descriptive sciences had to presuppose that things are what they seem to observation. Accordingly, the artistic technique of recording three-dimensional observations on a plane surface was an essential instrument. That technique was an application of the geometry of perspective. An attempt to imagine a scientific anatomy illustrated by a medieval artist will suggest its importance. And on the other hand, the study of anatomy was stimulated by the artist's need to master the subject in order to achieve the natural representation demanded by the taste of the High Renaissance. Leonardo warns the student of the difficulties:

> Though possessed of an interest in the subject you may perhaps be deterred by natural repugnance, or, if this does not restrain you, then perhaps by the fear of passing the night hours in the company of these corpses, quartered and flayed and horrible to behold; and if this does not deter you then perhaps you may lack the skill in drawing essential for such representation; and even if you possess this skill it may not be combined with a knowledge of perspective, while, if it is so combined, you may not be versed in the methods of geometrical demonstration or the method of estimating the forces and strength of muscles, or perhaps you may be found wanting in patience so that you will not be diligent.

✧

So far as is known, Andreas Vesalius, who distinguished himself above all other Renaissance anatomists, had never read Leonardo's precepts nor ever seen Leonardo's practice in anatomical drawing. But his work looks as if he had been acting on that inspiration, which is only to say that Leonardo epitomized but did not cause the crossing of art and science in naturalism. The year 1543 saw one of those publishing coincidences which serve the history of ideas as chronological pegs. It was the date both of Copernicus's *On the Revolutions of the Celestial Orbs* and Vesalius's *On the Fabric of the Human Body*. But how different is the anatomical treatise, not only in subject matter but in manner and appearance. The reader's eye is not repelled by the crabbed computations of astronomy, but is invited by the bold clear typeface of Italian printing. The evidence is presented, not in forbidding trigonometrical tabulations, but in stunning woodcuts of the human body, which are so clearly the work of an old master that they have been attributed (though most tendentiously) to Titian. And if the sheets on which the great muscular figures posture gracefully are placed side by side, it is apparent that they are displaying the physical structure of man against a continuous Renaissance landscape. This has even been identified. It lies in the countryside of Petrarch, near Abano Terme, not far southwest of Padua, where Vesalius worked and taught. There he had access to Venice, and to the workshop of Titian, if not to Titian himself. Like the style of the Venetian school, the culture of the Renaissance was already a little full blown by 1543. But under the encroaching shadow of the baroque, the work of Vesalius established a permanent residence for naturalism in science, just as at the very last moment of the Renaissance, and as its final triumph, the work of Galileo was to embody Platonism in physics.

The sciences of life, therefore, find their place in the scheme of a scientific revolution. The impression is difficult to avoid, however, that it was a subordinate place. Despite the very evident appeal of Vesalius's subject, or perhaps because of it, his achievements were of a lower intellectual order than those of Copernicus or Galileo. His were not the ideas which changed man's conception of the world, or even of himself. Nor did those of any biological scientist before Darwin. Generally, the deepening of theory in the physical sciences preceded the widening of fact, whereas the sciences of life developed in the reverse order. When the transformation of biology did come in the nineteenth century—not till then!—it took the form, bound to be something less than revolutionary, of an assimilation of biology to the objective posture of physics.

The disadvantageous comparison of the science of living nature to physics must not be pushed too far, for the material, if not more difficult, was at any rate more incoherent. Nor were generalizations lacking. The movement of thought from Vesalius's anatomy to Harvey's demonstration of the circulation of the blood is as interesting for the evolutionary structure of theory as any episode in the history of physics. The limitation of Harvey's achievement was in its scope, not its merit. In the theory of gravity Newton could unite Kepler's planetary laws with Galileo's mechanics in a mathematical science of matter in motion that encompassed all of physics. But the circulation of the blood united only anatomy and physiology. This was as near as biology could come in generality to physics, and it left innumerable fragments of information and superstition strewn across the vast wastelands of medicine and natural history, unorganized by any objective concepts.

It is, indeed, indicative of the inchoate nature of these subjects that the word "biology" had to await the nineteenth century to be coined. In the sixteenth and seventeenth centuries the subjects it was to embrace scarcely had an independent existence. Anatomy and physiology were rather aspects of medicine than science, and medicine was oriented more toward art and therapy than knowledge. Although human anatomy was studied more by analogy to animals than from cadavers, this practice was the source rather of error than of comparative anatomy, which does not antedate the eighteenth century. Natural history, for its part, was pursued rather in the spirit of the bird-watcher or the moralist than the investigator. Etymologically, the term means simple description of nature. Zoology was the source of fables, botany of medicinal herbs, and mineralogy of ores. Nor was the mineral as distinct from the animal and vegetable kingdoms as might be supposed, for minerals were thought of as bred in the womb of the earth to be ranged by species in categories of form.

In all fields the attitude to Aristotle and antiquity was ambivalent. There was criticism in detail, and a kind of ritual resentment of authority. In part this was a wholesome striving for originality, an assertion of the imperative of seeing for oneself. But mingled with this was the less worthy element of jealousy that those unsure of themselves feel, less for the mistakes of authority, than for the superiority which earns it. There was, as a consequence, no such clean break with antiquity as is represented by the law of falling bodies, but only a girding against it. For part of the difficulty in biology was that Aristotle's methods really did suit its problems for a very long time. Taxonomy, the classification of organisms, had to be the first step in ordering the millions of forms of life. Considerations of purpose, the teleological analysis of function, dominated biology right down to Darwin. The attempt to answer the question why? carried the biologist much further into his science than it did the physicist; or perhaps one should say that it became an obstacle much later. For all these reasons, therefore, biology was the less radical of the two great branches of science, and so it is, perhaps, that throughout history biologists have been more likely to be men of humane temper than have their mathematical colleagues, whose minds dwell on the abstract and the exact rather than on life and the flesh.

Vesalius lived a somewhat puzzling life. What the spirit of his career actually was is less clear than in the case of anyone of comparable stature in the history of science.

He was born in Brussels in 1514 into a family which had originated in the Rhenish town of Wesel (hence the surname) and which had a long medical tradition. He studied first at Louvain and then at Paris, where he hated his teachers. Indeed, he always expressed that violent scorn for his professors which is likely to seem (at least in the eyes of their alarmed successors) one of the less attractive Renaissance conventions. He went back to Louvain to submit his doctoral dissertation and on to Padua to complete his studies. There the degree of M.D. was awarded him in 1537, and on the very next day he was named professor of surgery by the Venetian Senate. He was then 23 years old. He taught for five or six years only, and published his course in 1543. Then, his great book in print and his reputation assured, he abandoned anatomy and teaching to accept appointment as court physician to the Emperor Charles V and to spend the rest of his life tending the ailments of that powerful and unhealthy monarch, who felt more secure in ignoring medical advice when Vesalius was by him to deal with the consequences. Whether Vesalius is to be counted a scholarly inquirer, therefore, or a careerist, is a question as difficult to avoid as to answer.

He was, at any rate, a great success as a teacher. In those six years he worked out and put into practice the tactics of anatomical demonstration. Since his time the subject has been corrected in many details and subordinated to a scientific biology. But in its substance it has not changed essentially. Vesalius's book was not a work of ideas. Perhaps, therefore, there was no point in his continuing to teach once it was printed, for it put the anatomical theater between covers. To the squeamish, indeed, that might even seem the best place for it. The tourist may still visit the old anatomical theater in the University of Padua, built only fifty years after Vesalius taught there. It is much as it was then. But the term "theater" is misleadingly spacious. For the room is a tiny, airless pit, oval in form and scarcely thirty feet across. Around the sides run shallow galleries in which one can barely stand. What must the atmosphere have been when these were packed with scores of sweating students, some of whom would surely faint or vomit, all jostling and craning to see down on the slab in their midst where the professor was dissecting the putrefying cadaver of some thief or beggar who would have been notably unsavory even when alive.

The success of Vesalius's course and of the book which embodied it was compounded of three elements: the authority of its information, the method of exposition, and the systematic approach. None of these was wholly novel, and Vesalius's essential contribution was the comprehensive skill with which he wove them into a corpus of anatomical practice rather than originality in any single detail or method. Vesalius himself made a great point of learning anatomy from bodies rather than books. And it is true that Greek humanism in antiquity and Christian teaching in the Middle Ages had created a powerful repugnance for opening the human body even in death. Nevertheless, Vesalius was far from having been the first anatomist to look inside his subject. Queen Elizabeth allowed the medical school at Cambridge three criminals a year. At the University of Bologna there was a standing rule in the fourteenth century that the medical students might procure cadavers for dissection, provided they did not belong to people who lived within thirty miles of Bologna. Indeed, the problem of the inadequate supply of bodies, like that of their rapid decomposition, was a handicap but not an absolute obstacle to research.

More obstructive was the spirit in which dissections were performed, for it was less one of inquiry than of demonstration—intended to show the artist or the student how the body is made rather than to subject it to functional analysis. Far from being nonexistent, dissection was almost routine by Vesalius's time, and he criticized his predecessors' lectures not for omitting demonstrations, but for confiding them to an assistant. Traditionally, the professor of medicine would stand up to his lectern droning out some text from Galen on the heart while the ignorant menial down below would grapple in the body laid out at his feet and hold up the liver by way of illustration. The frontispiece to *De Humani Corporis Fabrica* points the moral by representing Vesalius descended from his chair discoursing and demonstrating directly from the corpse, while the nobility and men of note look on. Two laboratory assistants have been demoted and are seen beneath the table scuffling squalidly, while a monkey at the left is both a symbol of medicine and a criticism of the source of too much anatomical misinformation. Vesalius's innovation, then, is the method of actual demonstration which has become the technique of much teaching of natural knowledge. It is no accident, but a conscious philosophy of communication. Vesalius evidently worked very closely with the artist in order to assure integration of the text and plates and to perpetuate the graphic realism of his method in print. His book is no casual commentary but a work of reference and a manual of technique. For he had that fine hand of the surgeon in which the probe or scalpel becomes like an outgrowth of sense.

De Fabrica, finally, is a highly systematic work in which the system is no arbitrary scheme but the Vesalian analysis of the body itself. The structure of the object gives structure to the study. Skin and muscle are flayed away layer after layer to reveal each level of organization: the muscular system itself, the vascular system, the nervous system, the respiratory organs, the abdominal tract, the architecture of the skeleton, and the articulation of the joints.

Particular organs, the heart and lungs, the brain, the genitalia, are dismounted and dismembered and depicted, mistakenly on occasion (the uterus looks like some Freudian nightmare), but generally with an accuracy and precision never before seen. The plates combine the scientist's eye for detail with the artist's eye for effect. The importance of the work, therefore, derived from its influence as a whole. Not only is it a model anatomical treatise, it was the first treatise in the history of any science in which all the relevant facts were put down in order and from nature. There were no new theories, but here were all the facts. Here was a technique, laborious and detailed, but available for mastery. One knew what one had to do to become an anatomist. Here was a body of information which irresistibly invited comparison with the traditional account of the subject. And so, even though Vesalius was a Galenist in physiology, the effect of his work was to ruin the composure of the old science, to teach the lesson that the whole corpus of anatomy must be reformed by meticulous observation, and eventually by that independent manner of thought of which Vesalius, its founder, was not himself quite capable.

Anatomy necessarily moves from description to function, and despite Vesalius's contempt for second-hand observation, Galen remained the lawgiver of physiology. No one could conceive of the body alive and working in the objective mood of science. There is no physician of classical antiquity of whom we know more, and none of any time whose influence so touched every detail. The legacy of Hippocrates, his master, was that of a school, not a man, and of a naturalistic philosophy of medicine, not a body of medical art. Galen's place in the history of medical science was altogether comparable, indeed, to Aristotle's in physics. If possible, he went beyond Aristotle in his commitment to teleology as the path to understanding.

His was a teleology which compounded Aristotle's rational interest in the purpose of objects with Plato's mystique about the divinity of perfect plans. It is often said that posterity looked at the body with Galen's eyes. It was not that simple—Vesalius certainly looked with his own—but it is easier to look than to see differently, and there was no alternative to the illusion of understanding created by explanations like that which makes the humble heel as perfectly adapted for its role in the human body as is the sun for giving life to the body of the world.

The function of the heart is crucial for any physiological theory which would discern the mechanism of the body. But Galen followed Plato on the sovereignty of the liver. His admiration for the rich capacities of that organ echoes certain strange and difficult passages in the *Timaeus*. In Galen the liver mediates between the three major involuntary processes of life: digestion, respiration, and the beating heart. From the stomach it draws through the intestinal veins the "white chyle," the product of the digestive "coction" or cooking of food. By a second coction the liver makes blood of chyle, and of this blood a large part is sucked up the *vena cava* by the expanding stroke or diastole of the heart into the right auricle. Then the heart contracts. Some blood is pushed out the pulmonary artery to irrigate the lungs, and the rest is extruded right through the central wall of the heart, the septum, to the left side. There it mixes with vital spirits of the air piped directly from the lungs down the pulmonary vein, which in Galen is a pneumatic tube, not a blood vessel. Hence, there are two kinds of blood. That which had been transmuted by vital spirits—further quickened, according to some accounts, by animal spirits issuing from the brain—was a life-giving fluid, bright red and foaming with spirit, which surged up and down the body through the arterial vessels in a cyclic ebb and flow like the rhythm of the lungs or the tides. At the same time, the venous blood from the liver carried dark nutriment more sluggishly about the body. In neither case was there a circulation.

Here was a complex and consistent explanation of the body. Like Aristotle's physics it made sense of its subject, it was wrong, and it was not refutable by the techniques of which anatomy disposed. But like the question of what moves the projectile, there were certain little difficulties, certain handles to dissent waiting to be grasped, which would turn out to have such an unexpected leverage as would overset the entire structure. There was, first, the statement that the main action of the heart is suction, whereas anyone who had ever opened the chest cavity of a laboratory animal could see that diastole is a relaxation of muscle and systole, a tensing. There was, secondly, the direct passage of air from lungs to heart, whereas dissection frequently showed these vessels, not only to be full of blood, but to be structurally identical with blood vessels and very unlike bronchial tubes. Galenists might and did say that such phenomena were accidents of dissection, derangements produced by the shock of incision or of death. Nor were they specious in these views, but only faithful to that assumption of order in their subject which is one of the preconditions of science. For science had yet to develop that assurance which takes each occasion to modify or rearrange the elements of order as a widening or deepening of confidence rather than a weakening.

Their faith was tested most shrewdly by the routing of blood right through the central wall of the heart. That it passed the test for centuries is the most notable testimony to the ordering power of the Galenic system. For the septum is a thick, tough piece of muscle. Vesalius investigated the pits himself and could find no passages: "None of these pits penetrate from the right ventricle to the left; . . . therefore, indeed, I was compelled to marvel at the activity of the Creator of things, in that blood should sweat from right ventricle to the left through passages escaping the sight." And what is interesting in retrospect is how investigators refrained from finding the answer, which is that if blood cannot pass directly from one side of the heart to the other, it must go round through the lungs. When one reads the memoirs, it almost seems to shriek at them from their own discoveries. But they went on fitting those findings into Galen's pattern right up to publication of William Harvey's theory of the circulation in 1628. (The chronological span from Vesalius to Harvey, it will be noticed, is exactly the same as that from Copernicus to Galileo.)

The conservatism created by the illusion of understanding may be illustrated by the failure to exploit two major discoveries of that interval, by either of which a bold mind might have anticipated Harvey. The lesser circulation of the blood, that is to say its transit from right ventricle to left auricle by way of the lungs, was described in print in a very curious book of 1553. Its author was a Spaniard, Miguel Serveto, one of the more implausible figures in the history of science. He had studied law at Toulouse, where he had also developed an interest in theology, particularly in natural theology, and his book dealt with that subject under the title *Christianismi Restitutio*. It argues in favor of penetrating beyond the vanities and corruptions perpetrated by the theologians of all centuries to the simple word of God which may be read in nature. Even Moses does not escape Serveto's censure. This was heresy, of course, and Serveto was burnt at the stake by order of the responsible theologians—not in this case Catholic theologians, but John Calvin and the elders of Geneva, which righteous city was no host to blasphemy.

Serveto's description of the passage of the blood from the heart through the lungs and back to the heart occurs in a page and a half of *Restoration of Christianity*. This passage was his chief scientific writing. It is possible that he knew an Arabic description of the lesser circulation, contained in the writings of Ibn al-Nafis. Serveto had studied anatomy at Paris. A discussion of the Holy Spirit in the Trinity led him to write of the three spirits of the blood, which conveys soul throughout the body. There is, he said, no communication through the septum. Instead, "the subtle blood, by a great artifice, passes along a duct through the lungs; prepared by the lungs, it is made bright and transfused from the pulmonary artery to the pulmonary vein. Then in that vein it is mixed with air during inspiration, and purged of impurity on expiration." What with the notoriety attendant on his burning, Serveto's book was often read, and even accepted, though not widely. But it never served, apparently, to suggest to anyone influential the possibility that all the blood circulates. All it did was to substitute a pulmonary detour for sweating through the septum as a means of getting blood from one side of the heart to the other and burnished by vital spirits in the transit.

A second discovery might have been even more suggestive. One of Vesalius's successors in the chair of anatomy at Padua was Fabrici d'Acquapendente, a supporter years later of Galileo. In 1603 he published an anatomical work describing certain valves he had discovered in the veins. Vesalius had never noticed them, but they are not, in fact, difficult to observe even with quite simple techniques. Their significance now seems absolutely obvious, for they permit venous blood to flow in one direction only, toward the heart. But Fabrici had learned his anatomy well. Their purpose, he said, is that of microcosmic floodgates. They check the flow away from the heart and restrain the outward ebb so as to prevent all the blood from collecting in the feet or hands. By his time, therefore, anatomists had all the information needed to get the dynamics of the body right. They had even tried ligaturing veins, and had observed the accumulation of blood on the side away from the heart, a phenomenon explained by pointing out that the blood took alarm and pressed off in the wrong direction. The subject, therefore, was in a state where it needed, not new facts, but a new approach.

✧

THE ANSWER was born of the first faithful marriage on equal terms of inquiry and empiricism. William Harvey's *On the Motion of the Heart* is a classic of inductive science. Galileo, to make a contrast, conducted most of his experiments in his head and on paper. They are beautiful examples of models in physical thought, but not of experimental science. When he did perform one, it was usually to demonstrate properties of the behavior of bodies which he had already deduced in good Archimedean fashion from geometric consideration of motion or equilibrium. In Harvey, too, scientific creation was the elegant expression of a temper combining imagination with impatience, and what provoked impatience in both minds was error, not the discipline of systematic thought. But Harvey was the first to use observation and experiment to find out something fundamental instead of to demonstrate it. In his work experience and its artificial reproduction in the laboratory were first instrument of inquiry, and then arbiter of theory.

Harvey's family were minor gentry in Kent. He was educated first in Canterbury, then at Cambridge, and in 1599 went on to Padua as a medical student. Both Galileo and Fabricius were teaching then, and we know he studied under Fabricius. There was always something both grave and imperative about his writing and his bearing—qualities intensified, perhaps, by Paduan manners. He was back in London in 1602, where (it is said) he always carried the little dagger affected by Paduan students in their dress. His practice was a success. He was appointed to Saint Bartholomew's Hospital in 1609 and served as physician extraordinary to James I and Charles I. In 1615 the Royal College of Physicians of London appointed him Lumleiian lecturer, on which foundation he was to discourse annually of anatomy and surgery. He combined research with practice, and published his book in 1628. He had then, he wrote, been developing his views for years in his lectures. The subject was involved in much confusion: "Those persons do wrong who while wishing, as all anatomists do, to describe, demonstrate, and study the parts of animals, content themselves with looking inside one animal only, namely man—and that one dead." And it would appear that Harvey had been dissatisfied with the physiological doctrine of Padua and had been studying the heart ever since, through ever wider and more meticulous inquiry involving frequent examinations of the insides of many different living animals and the collation of many observations.

His book is a truly beautiful argument. Absolutely appropriate facts are arranged with perfect art. It would be almost as difficult to withhold assent from his demonstrations as from a geometric theorem. "All true and fruitful Natural Philosophy," wrote Francis Bacon, "has a double scale or ladder, ascendant and descendant, ascending from experiments to axioms and descending from axioms to the invention of new experiments." Harvey was a scientist rather than a methodologist and meant his book to convey the correct physiology of the heart. But in retrospect it may be more instructive to read it as an object lesson in inductive reasoning. Each chapter makes a simple point illustrated by multiple instances. First it is shown from vivisection of cold-blooded animals (chosen for their slow heart-beat) that the contraction of the heart muscle is a pumping stroke. Then it is demonstrated how the pulsation, first of the auricles, afterward of the arteries, synchronizes with the rising thrust of the ventricles which lifts the heart against the chest wall. Now we are ready to agree to the local action of the heart. This in turn prepares us to draw more out of the induction than the instances that went into it. It leads to consideration of the pulmonary circulation which must be a function of the structure of the heart, for animals without lungs exhibit hearts with only one ventricle. Moreover, in the mammalian embryo the two ventricles act as one since the lungs are not yet in use. Occasionally Harvey betrays the authentic intolerance of science for stupid error. The pulmonary circulation is still denied by anatomists who make blood seep through the septum: "But damme there are no pores and it is not possible to show such."

Having shown the heart transferring the blood from the veins to the arteries by filtering it through the lungs, Harvey is ready to state his thesis in its most general form. He postulates the completion of the circuit from arteries to veins in the major circulation throughout the body.

The remaining matters, however (namely, the amount and source of the blood which so courses through from the veins into the arteries), though well worthy of consideration are so novel and hitherto unmentioned that, in speaking of them, I not only fear that I may suffer from the ill-will of a few, but dread lest all men turn against me. To such an extent is it virtually second nature for all to follow accepted usage and teaching, which, since its first implanting, has become deep-rooted; to such extent are men swayed by a pardonable respect for the ancient authors. However, the die has now been cast, and my hope lies in the love of truth and the clear-sightedness of the trained mind.

In substance, Harvey's arguments are quantitative and anatomical. The decisive one turns on the amount of blood. From the capacity and rate of the heart and its beat, he calculates that in one hour the heart pumps a weight of blood greater than that of a man—far more blood than could possibly be created out of any amount of food one could eat, be the liver never so active. It is impossible to say where it came from or where it went except on his theory.

Harvey now starts his descent of the ladder of induction, which verifies by the prediction of consequences. In form he presents his demonstrations as suppositions imposed by the hypothesis. In being confirmed by experiment, these then serve to validate the hypothesis, promoting it (as some might say) to the rank of theory. So the arteries are made as blood vessels, not air tubes, and observations of the valves in the veins and appropriate ligatures in both sets of vessels establish the direction of flow. Therefore,

> Since calculations and visual demonstrations have confirmed all my suppositions, to wit, that the blood is passed through the lungs and the heart by the pulsation of the ventricles, is forcibly ejected to all parts of the body, therein steals into the veins and porosities of the flesh, flows back everywhere through those very veins from the circumference to the centre, from small veins into larger ones, and thence comes at last into the vena cava and to the auricle of the heart; all this, too, in such amount with so large a flux and reflux—from the heart out to the periphery, and back from the periphery to the heart—that it cannot be supplied from the ingesta, and is also in much greater bulk than would suffice for nutrition.
>
> I am obliged to conclude that in animals the blood is driven round a circuit with an unceasing, circular sort of motion, that this is an activity of the heart which it carries out by virtue of its pulsation, and that in sum it constitutes the sole reason for that heart's pulsatile movement.

But even now Harvey is not finished. He draws attention to still further consequences of a lower order of generality—the distribution of certain pathological conditions by the blood stream, the consonance between the development and tone of the heart and other muscles in particular subjects. These gave his theory its maximum extension. Only one link was missing. How does the blood pass from the finest arteries to the finest veins? For the capillaries are invisible, and the microscope was not yet at hand. Nevertheless, Harvey posited their existence, and when that instrument did come into use, his prediction was verified—by Malpighi, who in 1661 identified capillary structures in the lungs of a frog.

There are writers who would make Harvey a heart-worshiper as Copernicus was a sun-worshiper. And indeed there are a few echoes of the debate between Plato and Aristotle over the relative excellence of the liver or the heart:

> We must equally agree with Aristotle's view about the pre-eminence of the heart, and refrain from asking if it receives movement and sensation from the brain and blood from the liver. . . . For those who attempt to refute Aristotle with such questions disregard or do not appreciate the chief point, namely, that the heart is the first part to exist, and that it was the seat of blood, life, sensation and movement before either the brain or the liver had been created, or had appeared clearly, or at least had been able to perform any function. With its special organs designed for movement the heart, like some inner animal, was in place earlier. Then, with the heart created first, Nature wished the animal as a whole to be created, nourished, preserved and perfected by that organ, to be in effect its work and its dwelling place. Just as the king has the first and highest authority in the state, so the heart governs the whole body.

But Harvey is here writing with reference to embryology, not to mysticism, and the admiration he often expresses for Aristotle is no more than the reverence which any honest biologist who has truly studied the classics must feel.

Indeed, it distorts the import of Harvey's approach to assimilate his outlook to anything like the world alive. Its place in the history of physiology is obvious and central. The phenomena ordered by Harvey could not reach beyond that science. Nevertheless, his work was the first, if partial, breach opened by the scientific revolution in the life sciences. His subject is not the ineffability of life. It is a problem in fluid mechanics. The heart is a pump, "a piece of machinery in which though one wheel gives motion to another, yet all the wheels seem to move simultaneously." The veins and arteries are pipes. The blood, so far as the problem is concerned, is simply a liquid, a lubricant to be passed periodically through the air filter of the lungs. No vital spirit, no principles of nourishment intrude into the analysis.

However different his subject from Galileo's, and however different their methods, Harvey's work portended the same conception of science. In him a new science which makes objective measurements superseded an old science which found qualities, humors, purposes, and tendencies indwelling. On yet another front, personality was displaced as a category of scientific thinking and as the model for order. Galileo had excluded the biological metaphor from physics. Harvey went further and introduced mechanistic thinking into organic studies. And by a simple though systematic extension, Descartes would find a machine in man. Dim fears stirred at all these vaguely sensed implications. Harvey's views were not generally accepted in the thirty years or more before Malpighi's discovery forced assent. As Harvey foresaw, he had against him more than force of habit. For his hydraulics of the blood stream destroyed a whole philosophy of the body in order to establish a single phenomenon of nature.

❖

THE misunderstandings between Harvey and Francis Bacon make an ironical commentary on the relationships between those who do science and those who write of its methods. *De Motu Cordis* might be taken as the purest and best Baconianism. In fact, Harvey dismissed Bacon as one who "writes philosophy like a Lord Chancellor," and Bacon in one of his less fortunate pronouncements denied the circulation of the blood. Not only so, but Bacon also rejected Gilbert on magnetism, Copernicus on the sun, and Kepler on the planets. Nor did he understand Galileo. These misjudgments might seem to cast some doubt upon the right that Bacon claimed at great and eloquent length to speak as the philosopher of the new science. Nevertheless, that right has been very generally accorded him. Bacon created no science. But his prophecy—he was nothing if not a prophet—of a world perfected by a reformed science created an image of scientific progress which has been vastly more popular than science itself can ever think to be. With all his scorn for the scholastics, he founded a new school, a new orthodoxy which organizes knowledge, not for understanding, but for use. With all his facile contempt for Aristotelians, he made himself the Aristotle of the philosophical bourgeoisie and held out his method as a royal road to science for the middle-brow mind. For the shadows of the Middle Ages lingered, and Bacon's dream of progress was not altogether glad and confident. He did not rightly know whether modern times are the morning or the evening of history, but he did not suppose that modern man could attain the stature of the ancients. We must, therefore, make the most of our smaller forces: "The course *I* propose for the discovery of the sciences is such as leaves but little to the acuteness and strength of wits, but places all wits and understandings nearly on a level."

This is comfortable democratic doctrine, and it is obvious why Baconianism has always held a special appeal as the way of science in societies which develop a vocation for the betterment of man's estate, and which confide not in aristocracies, whether of birth or brains, but in a wisdom to be elicited from common pursuits—in seventeenth-century England, in eighteenth-century France, in nineteenth-century America, amongst Marxists of all countries. Nevertheless, it would be simply snobbish to deplore Bacon's influence. His was the philosophy that inspired science as an activity, a movement carried on in public and of concern to the public. This aspect of science scarcely existed before the seventeenth century. Since then it has accompanied and at times enveloped intellectual effort. There is bound to be a certain tension in affairs which associate the expert with the layman. The scientist wants of his subject an intellectual adventure which can only remain private. The public wants advantage for itself, and though it does not wish to pay the mental price of understanding, may all too humanly resent exclusion. But tensions bind while they distort, and science and public welfare are not likely to separate out of the complex fashioned by Bacon.

There was more than a gibe in Harvey's remarks, for Bacon was a lawyer and politician. His career is difficult to admire. As defence against the charge of accepting bribes while in high judicial office, he pleaded that he had not allowed them to influence his decisions. This reasoning failed to placate, and he ended in disgrace: "The rising unto Place is laborious; and by pains men come to greater pains; and it is sometimes base; and by indignities men come to dignities. The standing is slippery, and the regress is either a downfall, or at least an eclipse, which is a melancholy thing." Such was the lawyer-like worldliness which invested science with a civic dimension.

The subject matter of Bacon's writings falls into three categories: demonstration of the worth and dignity of learning; analysis of the obstacles which kept it languishing in futility; and prescriptions for its reformation and advancement. It is not, perhaps, necessary to insist much on the first point—indeed, it was not so necessary in the early seventeenth century as Bacon would imply. His pleas for learning generally took the form of a rather scornful repudiation of all that passed for such. As for the hindrances, it was trite enough to blame the sterile habit of reliance on authority and the circularity of scholastic logic. But though no student of science, Bacon was an extremely acute student of human beings, and in his discussion of the obstacles raised by the intellect against itself, he showed his mettle. There is that in the very constitution of our understanding which renders the mind a pesky instrument for innovation. "Idols," Bacon called these innate blinders. The "Idols of the Tribe" are distortions which arise from our common nature: "The human understanding is no dry light, but receives an infusion from the will and affections; whence proceed sciences which may be called 'sciences as one would.' For what a man had rather were true he more readily believes." The "Idols of the Cave" compound this common tendency to error with the favorite prejudices or enthusiasms of the individual man, each of whom "has a cave or den of his own, which refracts and discolours the light of nature."

Third, are "Idols formed by the intercourse and association of men with each other, which I call Idols of the Market-Place on account of the commerce and consort of men there. For it is by discourse that men associate, and words are imposed according to the apprehension of the vulgar. And therefore the ill and unfit choice of words wonderfully obstructs the understanding." This was perhaps the most penetrating and valuable of Bacon's observations. Not much can be done about human nature, after all, any more than about gravity or inertia, even when its disadvantages are recognized. But identification of the error that lurks in words was the first step to correction. The attempt to put precision into scientific language has never since been relaxed. Humanists may complain of the jargon of the specialties, sometimes with justice. But no science can flourish until it has its own language in which words denote things or conditions and not qualities, all loaded with vague residues of human experience.

Finally, Bacon held up to suspicion the "idols of the theater," by which he meant the systematic dogmas of philosophies, "because in my judgement all the received systems are but so many stage plays, representing worlds of their own creation after an unreal and scenic fashion." Many gratuitous suppositions of the ancient philosophies—the perfection of circles, for example, or the notion of purpose in nature—go far to justify this indictment of great flights of system. Nothing, as Descartes was to observe a few years later, is so absurd that it has not been said by one of the philosophers. And this proscription of system, coupled with analysis of language, was to become one of the main motifs of science in the eighteenth-century Enlightenment. But though it brought much sanity, it developed the drawback of too sane an outlook—it discountenanced imagination. Reinforced by a misunderstanding of Newton's famous "I frame no hypothesis," it discouraged theory in favor of accumulation of detail, abstract generalization in favor of natural history. In Baconian science the bird-watcher comes into his own while genius, ever theorizing in far places, is suspect. And this is why Bacon would have none of Kepler or Copernicus or Gilbert or anyone who would extend a few ideas or calculations into a system of the world. But reason had its revenge on Bacon's dogmatic empiricism in the manner of his death:

he took a chill while stuffing a chicken with snow, the most famous experiment he is known to have performed.

So radically inane, in Bacon's view, was the corpus of received philosophy that it must simply be jettisoned. "It is idle to expect any great advancement in science from the super-inducing and grafting of new things upon old. We begin anew from the very foundations, unless we would revolve forever in a circle, with mean and contemptible progress." There must be a New Learning, then, with Bacon as its guide. And the goal? In a word, Progress— progress as it has been understood everywhere in the West since the seventeenth century, progress through technology and the domination of nature. "The true and lawful goal of the sciences is none other than this: that human life be endowed with new discoveries and power." Thus will be turned the last of the barriers to knowledge, at once the most obstructive and the most unnecessary, left standing through the failure to demand practical results from learning. Natural philosophers had been allowed to enter on speculations, not with a view to applying their results, but to please their own vanity, or out of idle curiosity. There is, therefore, no choice to be made in Bacon's philosophy between basic and applied science. On the contrary, applied science is by definition basic: it is the object of the search.

The positive side of Bacon's program for building an infinity of better mousetraps into a better world envisioned three related stages: application of the inductive method, creation of a universal natural history, and the public organization of science. Induction will redress the empty rationalism of science and start anew by reversing its procedures. Scholastic science, thought Bacon, began with the principle and deduced the consequence. He would begin with the particular fact, with all relevant facts, and rise by successive steps to the general principle. Not that he was quite so uncritical as this might seem. He placed great emphasis on comparative analysis and the exclusion of facts or possibilities which do not carry the mind toward general laws. Suppose, for example, one is studying heat. Metals can be heated without producing light. The moon gives light, but no heat. Therefore, in distinguishing the laws of heat, the phenomena of light are to be excluded. This principle of elimination is obviously important in scientific investigation. It is doubtful, however, that it needed Bacon to become established. It is implicit, after all, in Galileo's resolution of motions, explicit in Harvey, and inherent in all workable science.

More important, Bacon's emphasis on experiment did shape the style of science. So strongly did it do so that the term "experimental science" has become practically a synonym for "modern science," and nothing so clearly differentiates post-seventeenth-century science from that of the Renaissance, or of Greece, as the role of experiment. Where is Bacon's new scientist to find the facts from which he will draw out inductions? He will look to observations of nature, of course, but preferably to that artificial reproduction of nature which is experiment. Experimentation practices the principle of elimination of the irrelevant. It puts the inquirer in control of the inquiry. It reveals more than passive observation: "Nature when vexed takes off her mask and reveals her struggles."

The prospects for examining nature by the inductive process obviously depended on having a vast collection of particulars from which to begin. Like any philosopher who mistrusts abstraction, like Aristotle, for example, Bacon was driven back on classification as an instrument for ordering the world—an instance, by the way, of the relationship of "frères-ennemis"—obtaining between Bacon and his largest target. A "natural and experimental history" was, therefore, prerequisite to progress. In *The*

Dignity and Advancement of Learning (1623) he classified all knowledge as history, poesy, philosophy, and theology, and divided history into the two categories of civil and natural. This is the beginning, perhaps, of the separation of science and philosophy at the higher level, and of science and the humanities at the lower. For civil history embraces human affairs and natural history the facts of nature. Nor was Bacon modest. He meant his natural history to range into place nothing less than all the facts. They were to be gathered everywhere, even from accounts of sorcery, witchcraft, and magic, if there be facts therein. But it was to the mechanical arts, to industry, trade, cookery, agriculture, seafaring, in a word to the practical man, that Bacon looked with fondest confidence. In this emphasis he struck a note which resounds very far back in western history, spoke indeed as an echo of his namesake, that other technological seer, friar Roger Bacon, who equally impatient of scholastic subtleties, and even guiltier of them, sought in the thirteenth century to define knowledge as what men can do rather than what definitions they can look up. There is both good sense and a constant flirtation with vulgarity in that practical western instinct which holds that the blacksmith knows metals and the metallurgist only books, that in real life an Edison outweighs an Einstein, and a businessman a scholar. And nothing has so flattered the audience that always welcomes Bacon's philosophy as this central imperative which would send errant philosophy back to the school of industry to learn what's what by experience.

Bacon did not expect to complete the natural history himself. The collection and encyclopaedic ticketing of all the facts would be an enterprise of considerable magnitude, requiring cooperation and subsidy. But it was a finite venture. In principle (again as in Aristotle) a completed science is perfectly feasible. "I take it that all these things are to be held possible and performable, which may be done by some persons, though not by one alone; and which may be done in the succession of the ages; though not in one man's life; and lastly, which may be done by public designation and expense, though not by private means and endeavours." Since science will benefit humanity, it is up to the state to support and organize science. Nowadays, it is even so, of course—except that the work no longer seems likely to be finished. Science has in fact become a cooperative enterprise. There are indeed vast numbers of technicians, worker-bees in the Baconian hive, passing information up through computers to the master theorists. In his last book, *The New Atlantis*, Bacon clothed his dream in that favorite device of the man with a message, a Utopia. He even situates it on a remnant of that lost continent which in the cosmological allegory of the *Timaeus* had carried Plato's final evocation of philosopher-kings down to oblivion. There the shipwrecked Europeans find a land in perfect order, full of untroubled knowledge, empty of discord or jealousy. The explanation, it emerges, lies in the nature of the sovereign, which is no king nor any tyrant, but a community of sages, living apart and devoted to learning and guidance of the land. The very model of a wise government, then, is a society of scientists with leisure to accumulate knowledge and power to apply it to the public weal, and the catalogue of wonders they have wrought reads like one of the crasser paeans to the American standard of living.

After Galileo, science could no longer be humane in the deep, internal sense of its forerunner in classical antiquity. Instead, Bacon dressed it out as humanitarianism. The reasons for his popularity are obvious, therefore. He makes science what it has become in part, and what the public tends to wish it were in its entirety: an innocuous instrument of human betterment which requires of him who would master it, not difficult abstract thought, but only patience and right method. It is less obvious why Baconianism has often enjoyed popularity among scientists themselves. No discovery has ever been made by following his method. Scientific thought itself is bound to be far more abstract, elegant, and intellectually aristocratic than Bacon foresaw or would have approved. But scientists are likely to be humane men who wish to do good and like to be told that they do. And an appreciation—like Pascal's, for example—of the poignancy of science is fairly rare among them: that is to say that the act of understanding is an act of alienation. Pascal took the contemporary attitude to authority wryly, as a commentary on the perversity of human nature. In science, he pointed out, men are so respectful of authority and fearful of innovation that they disbelieve in the vacuum on the word of Aristotle, whereas in theology, where they ought to depend on authority, they go running after novelty. But the wringing of hands has never been an influential posture in western history, and it is the materialistic commitment of a Bacon, at once tough-minded and humanitarian, rather than the delicacy of mind of a Pascal, which has shaped the technical tradition. The public soon grows bored with the man behind the tragic mask.

THE NEW PHILOSOPHY

THE THOUGHT OF RENÉ DESCARTES moved across the gap in the scientific revolution between the physics of Galileo and the prophecies of Bacon. In its success it complemented each. In its failure it announced the need for a scientific declaration of independence from philosophy. Descartes' *Discourse on Method* is often read along with Bacon as propaganda for that seventeenth-century science which so insisted on its newness that it risked reducing novelty to banality. The two spoke with one voice, impatient and intolerant, on the emptiness of philosophy and on brushing aside that antique shell, preparatory to erecting the structure of knowledge by right method. They did, it is true, differ over what right method is. Bacon would rebuild science on utility, and Descartes on clarity. Bacon put his confidence in experiment and induction, and Descartes in reason and deduction. But empiricism and rationalism touch in most important investigations, and science has travelled both roads, often interchangeably, to their convergence in mathematical physics.

Nevertheless, Bacon was only a prophet of science, whereas Descartes was a founder. "Who would know nature," went one of Leonardo's aphorisms, "must know motion." And Descartes stated the principle of inertia correctly: 1) "Every individual body remains in the same state so far as possible and changes its state only by impact with other bodies"; and 2) "Every body tends to continue its motion in a straight line, not a curved line, and all curvilinear motion is motion under some constraint." In this unassuming way Descartes exhibited the courage of Galileo's principles. There was a touch of snobbery in Descartes. He lifted an eyebrow at the large-gesturing theatrics of Galileo's *Dialogue*. But it was Galileo who had taught him that motion is a state to which a body is indifferent, not a process which involves it. Galileo had made motion persist, though in circles. Descartes simply moved its destination out from the cyclic to the infinite.

Nothing is more characteristic of Descartes' intellectual style than that he should thus have quietly introduced as a corollary of inertia the image of an infinite universe— the notion which Galileo had not quite dared draw out of his new physics, the speculation which had led Bruno to the stake, the consequence which Copernicus had more prudently left up to the philosophers. In the opinion of Alexandre Koyré, one of the most searching historians of science, the infinity of the universe worked a deeper philosophical disorientation than any other scientific development. The question is not the size of the world, but whether man can ever again feel that he fits. For there can be no mean between the finite and infinite, no correspondence by science or any route between man and the universe, no place in nature where man specifically belongs because there is no such thing as place in infinity. All a man can do with the world is try to understand something about it. He can no longer think to participate in the cosmic process, and science can never hold comforts.

But however that may be, Newton would make the laws of motion out of the principle of inertia. Beyond this, the legacy of Cartesian philosophy to physics included analytical geometry, its essential mathematical instrument, and the correct law of refraction, the point of departure for a rational optics. Moreover, Descartes housed physics in that Euclidean conception of space which science inhabited until Einstein. It was Descartes, further, who systematically substituted the impersonality of the machine for the purposiveness of the organism as the model of order embracing all nature. But finally, as the crowning defect of the great Cartesian virtues of clarity and simplicity, he fashioned out of metaphysics a deeply erroneous and an extremely interesting physics. One often reads that Newton had to overcome Aristotle. This is not correct. Pressing into the breach opened by Galileo, Descartes had already routed Aristotle. But he overshot the mark, and Newton had to supplant Descartes in order to set physics back on the road mapped out by Galileo.

The reader of the *Discourse on Method* should be warned, therefore, that Descartes wrote it, not as a preface to what science has become since Newton, but to what he meant science to be. In form the essay is an intellectual autobiography composed in a beautifully quiet French. The vein of introspective soliloquy, the praise of tranquility, the retiring manner of Descartes' life (in a Dutch retreat, out of reach of the colleagues, the censors, the critics who made the intellectual life of France the whirlwind that it always is)—these mannerisms seem to assimilate Descartes to the posture of skeptical detachment assumed by Montaigne.

The manner is misleading. The *Discourse* is extremely radical. "Philosophy," he writes, in reviewing his education, "affords the means of discoursing with the appearance of truth on all matters, and commands the admiration of the more simple." But for himself, "I was especially delighted with the mathematics, on account of the certitude and evidence of their reasonings; but I had not as yet a precise knowledge of their true use; and thinking that they but contributed to the advancement of the mechanical arts, I was astonished that foundations, so strong and solid, should have had no loftier superstructure reared on them."

The "method" which the *Discourse* expounds aims at nothing less lofty than raising that superstructure. For Descartes' mind was one of the most daring of those which experience in mathematical demonstration the cynosure of truth. He would embrace the world in the clear and simple, and therefore true, ideas of geometry, not like Plato by denying existence to the merely physical, but by a total criticism which would strip philosophy of the illusion that understanding may be had on vaguer terms. Such was the instrument of systematic doubt, wielded so ruthlessly that it left Descartes in what most people would find an uncomfortable isolation, all alone in the world with his reason. This was the one entity whose existence withstood doubt, since it was itself doing the doubting. "I think, therefore I am," in the famous assertion that existence begins with mind. And now Descartes proposed to reconstruct science and philosophy out of the mathematical concepts of number, motion, and extension. For his criticism of philosophy was a summons to reform, not a repudiation of its mission to comprehend the world in a single rationale.

A passage in the *Discourse* exemplifies the unifying imperative in Cartesian thought. It tells how, in accordance with his own precept, Descartes began his reformation of learning with "objects the simplest and the easiest to know." Descartes thought like a geometer in and about straight lines. He would study, not the objects themselves, but—here speaks the geometer—the

> relations or proportions subsisting among those objects. . . . perceiving further, that in order to understand these relations, I should sometimes have to consider them one by one, and sometimes only to bear them in mind, or embrace them in the aggregate, I thought that, in order the better to consider them individually, I should view them as subsisting between straight lines, than which I could

find no objects more simple, or capable of being distinctly represented to my imagination and senses; and on the other hand, that in order to retain them in the memory, or embrace an aggregate of many, I should express them by certain characters the briefest possible. In this way I believed that I could borrow all that was best both in geometrical analysis and in algebra, and correct all the defects of the one by the help of the other.

There is no happier instance of the contrast in Descartes between moderation in statement and originality of conception. This mild sentence explains why he framed analytical geometry in rectangular (Cartesian) coordinates. It alludes to the notation (a, b, c for knowns; x, y, z for unknowns) still in use. Moreover, applied first to light and then to motion, the rectangular resolution of relations returned Descartes the law of refraction and the principle of inertia, and applied to matter as that which is extended, it assimilated matter to space and geometrized both.

Cartesian geometry is algebra applied to spatial relationships. A moment of reflection will convey its significance to anyone with an elementary knowledge of the calculus. One may write an equation determining a line or circle, and not just draw it. Even more important, it is possible to write an equation of motion for falling bodies, and not just develop its quantity as Galileo's triangle between time and velocity. Unification of algebra and geometry was far more than a convenience. Throughout the entire history of science, the style of analysis may appear in retrospect as derivative from the particular branch of mathematics which is bodied out onto nature. In this sense, classical physics was an application of Euclidean geometry to space, general relativity a spatialization of Riemann's curvilinear geometry, and quantum mechanics a naturalization of statistical probability. Until Descartes, algebra as the mathematics of discrete quantity was suited only to atomistic presuppositions about the structure of reality. Geometry, on the other hand, traditionally the mathematics of the spatial continuum, served only the Platonistic or the Stoic traditions, which sought to contemplate the unity of nature rather than to number her parts.

Descartes abolished that distinction. The possibilities were too immense for his generation, and he himself was too much the geometer to exploit them. But in the act of striking out the world picture of classical physics, Newton was to invest this same step with physical meaning. One element in the Newtonian synthesis would be to unite an abstract and continuous conception of space with a concrete and atomistic conception of matter. Even without that hidden implication, however, the invention of analytical geometry was the most momentous contribution to mathematics since Euclid. Since the treatise which described it was one of three by which Descartes practised the argument of the *Discourse on Method*, it is obvious why that method carried authority.

The second of those treatises dealt with optics, and the third with what Descartes called "meteorology"—actually with the physical environment. Nor is it inconsistent with his temper of mind that he should have moved into physics by way of optics. For the neo-Platonists light was the bearer of truth. Mysticism apart, light has always held a certain primacy in the tradition of Platonic realism, no doubt because of the simple elegance of optical phenomena. It did for Copernicus. It did for Kepler. It would do so in a much deeper way for Einstein, in whose physics light is the bearer of signals. When light passes from air, say, into water, it is bent toward the normal, and the sine of the angle of incidence bears a fixed proportion to the sine of the angle of refraction. This is the law of refraction. On a purely empirical basis it had been anticipated by

Snell, to whom Descartes gave no credit. His attitude to forerunners of humbler powers has been less rare than might be hoped among original minds. Newton and Lavoisier behaved no better. But Descartes went beyond most innovators, perhaps, in implying that to pass a discovery through his method rendered it his own intellectual property.

Nevertheless, it was Descartes who made the law interesting. He defined light as inclination to motion in luminous bodies. The *Dioptrique* then presents the law of refraction by means of a curious comparison to an imaginary ballistic experiment. (Descartes' thought was nothing if not mechanistic.) The reader is to think how a ball would pass obliquely, first, through a cheese-cloth barrier doing duty for a deflecting surface, and then through such a surface into water. The trajectory will be bent toward the horizontal and the velocity decreased. Just suppose for argument, however, that the velocity be increased. Such is the case for refraction of light which unlike ballistic bodies is a pressure transmitted more readily through the optically denser medium. And Descartes made the geometry yield Snell's Law out of the relative proportions of the normal and parallel components before and after refraction.

Descartes has often been reproached for his excessively abstract habit of mind. Justly so, one must agree, and yet this it was which freed him from the scruples that kept inertia circular in Galileo. Infinity held no terrors for Descartes, not that he was braver, but that he was more indifferent. He felt no compunction about turning man loose in an infinite universe, so long as his own ideas were clear and simple. Galileo had had the world to worry about, after all, and the world is round. But Descartes had only clarity, a kind of alarming consistency heedless of consequences. Or rather Descartes would handle consequences by addressing himself, not largely to the public, but fastidiously to the philosophically initiate. He derived the principle of inertia from two premises: the homogeneity of the straight line, and the immutability of God, of which the constant quantity of motion in the world is an expression. In this high way he was able to identify as the foundation of physics that which can never occur in fact—rectilinear inertia, motion persisting in a straight course to the place where parallel lines meet in the never-never land at the end of an infinite universe. And since this concept of inertia is absolutely abstract, it may well be that only a mind capable of greater confidence in its own ideas than in physical appearances could ever have formulated it: "For, in fine, whether awake or asleep, we ought never to allow ourselves to be persuaded of the truth of anything unless on the evidence of our reason. And it must be noted that I say of our *reason*, and not of our imagination or our senses."

Space itself, finally, has got to be an embodiment of the abstract in the hypothetically concrete. Descartes makes of space the physical casing of solid geometry, an endless three-dimensional box in which the straight line is the shortest distance between two points. This is the idea of space which for a long time now has been mistaken for common sense, but which never started as common sense. It was rather the complement to Galileo's (even more difficult) treatment of time. Galileo had turned time from the course of our lives into a dimension, an abstract parameter of that state of motion which is what science measures and numbers. Descartes crossed time at right-angles with the other coordinate of classical physics, not just linear distance, but space in depth.

It has already been hinted how Newton would step from the wings and express physically the combination of numerical and spatial which was Cartesian geometry. And once again, on the even more comprehensive problem of space, Newton's thought moved out from Descartes—but this time in the opposite direction. For with that extraordinary physical intuition (which was precisely the quality that Descartes most signally lacked) Newton abstracted the Cartesian idea of space from its assimilation to matter. For all the rancor Newton developed against Cartesianism, he paid it the supreme compliment of making space the physical embodiment—or rather disembodiment—of a more sophisticated geometry than Euclid's: that of Descartes. In Newton, space becomes the abstract system of Cartesian coordinates to which absolute motion, the stirring of reality, is referred.

This is to anticipate, however, and to complicate. Descartes, in his commitment to simplicity, had a more concrete use for space. His space is not empty. If it were, one could say nothing about it. On the contrary, space is what the world is full of, as it is of matter. Space, indeed, is the same thing as matter. And this takes us back to the fundamental dualism of Cartesian metaphysics, which proposed to make the world of mind and matter. Mind is what thinks. Matter is what is extended. Nor could Descartes have taken any other view. If mind is to think truly about matter, matter has got to be such that it can be handled geometrically. To admit of clear and simple ideas about it, matter has got to be extended in the geometric continuum.

Having thereby filled the world as full of itself as an infinite egg, Descartes laid down the proposition of universal mechanism in cosmology. For like Galileo before him and Newton after him, Descartes, too, had to answer the great question of seventeenth-century science. What keeps the planets in orbit, what holds the world together, in a universe where motion persists? This was the problem on which the scientific revolution turned, the problem which Galileo had met—or failed to meet—with circular inertia. Descartes, for his part, answered with a clear and simple idea. The world is a machine. One often reads in intellectual histories about the "Newtonian World-Machine." The world-machine was no such thing. It was Cartesian. It is only the science of mechanics, a far more restricted topic, which was Newtonian.

Descartes himself never thought to stop with physics, but immediately embraced all science, even biology, in mechanism. One of the few discoverers for whom he admitted admiration was Harvey. Cartesian biology generalized the hydraulics of the circulatory system into a comprehensive discussion of animals as machines. Arms and legs are levers worked by pulleys. Nerves are hollow tubes through which messages are puffed by the nervous fluid as if in some modern pneumatic communications system. Indeed, it has been plausibly argued that Harvey's theory of the circulation served Descartes as pilot for his cosmology, which transposed the cosmic animal that the Greeks made of the world into a cosmic hydraulic system. There can be no void in Cartesian physics. There are only different states of matter. Rigid matter composes solid bodies. Subtle fluid matter coils throughout interplanetary space in great eddies. The luminous matter of the sun and stars, finally, streams in straight lines through the vortex.

Extension gave Descartes a key for many doors. His *Principles of Philosophy* (1644) begins by disarming the Church. There is no reason to think his alarm at the condemnation of Galileo unworthy of a good Catholic. He withheld *The World*, his first essay in cosmology, lest it transgress in some equally unforeseeable fashion. Now, however, he has found the way out of the dilemma posed by his agreement with Copernicus. The earth does go around the sun. At the same time the authorities were quite right. It does not move. For the earth is stationary with respect to its own immediate envelope of space-matter. Motion is in the vortex, not the earth. Some fourteen vortex systems keep the planets and their satellites orbiting like corks in a complex but orderly maelstrom. Centrifugal force throws the bulk of the ethereal matter outward in each vortex. This lowers the density toward the center, and creates a centripetal pressure inward. The line of equilibrium describes the elliptical path along which planets are forever constrained out of the straight way of inertia. Gravity, tides, chemical phenomena, light, heat, sound— all are explicable as manifestations of figure, motion, and extension, and (again following Galileo) qualities are only modes of perception in us. Regularity in nature bespeaks mechanism, not intelligence or will. Descartes was as caustic as Galileo on final cause. God did indeed make the world, but not for us.

Clearly then, the historian of science must reckon with Descartes. But he is not an easy figure to assess. His impetus was not just another push down the right path. It gives pause to think that so subtle a mind should have spun so crude a physics. And on reflection it seems clear (as Descartes, indeed, would wish) that the difficulty was in his idea of nature. Economy in explanation is, of course, a goal of science. But for Descartes what is simple is nature herself, whereas every neat-handed physicist knows that nature is very complex, and that only the laws of nature are simple. Cartesian thought was excessively mathematical, after all. In Galileo, in Newton, in modern physical science, mathematics is a tool, a means of expressing quantity. It is the language of science. Descartes mistook the language for the subject. "It is true," he wrote, though not as an admission, "that my physics is nothing but geometry." His thought went off into clarity and left the world behind. It was not in his character to do otherwise. He was interested in reason, not in nature. "The seeds of the sciences," he writes in one of the *Meditations*, "are *in us*."

Not in nature, in us, and so he examined his own ideas to see if they met the test of truth in clarity, and not the world where there is much less clarity to be seen, whatever there may be of truth.

A further difficulty with Cartesian science is not unconnected with that impression of arrogance which an excessively mathematical temper is likely to convey. It was too ambitious. It overestimated the function of scientific explanation. It tried, not just to describe some set of phenomena, but to explain both the behavior and the reason in things by a single generalization. It is not enough to say that the planets sail around the sun in ellipses. Science must say why they do so, and that is a question which cannot yet be answered in the way Descartes required. No one knows the cause of gravity. One only measures certain effects called that. Descartes looked further. To explain everything, he resorted to a mechanism. But, alas, the mechanism was only a clear and simple idea. And so close was Cartesian reasoning, that with its failure fell not only a physics. There collapsed, too, a metaphysics, a beautiful metaphysics, which made clarity and simplicity signs of truth.

Deep in the structure of the scientific revolution, therefore, Descartes stood at the divide between antique and modern. The science of the Renaissance, like its prototype in Greece, had been largely derivative from culture and philosophy. Since the seventeenth century that primacy has been reversed. Today culture and philosophy—especially philosophy—have become largely derivative from science. The tide turned with Cartesianism. After this, knowledge of nature passed over from philosophy to science, which assumes only uniformity of law and neither unity of truth nor cosmic personality. Descartes was the last of the great systematic philosophers to make integral contributions to science directly out of a metaphysics. For the image of infinite mechanism which he imagined as the object of science, modified though it was in structure and restricted in import by Newton, proved impervious, if only by irrelevance, to further metaphysical essays in revision or replacement. Since the nineteenth century it has yielded, but only to the physical.

❖

THE HISTORIAN finds a dialectic informing successive resolutions of the great dilemma in which science oscillates between the unity of nature and the multiplicity of phenomena, the one and the many. Is the universe a single continuum, to be described in a geometrical physics? Or is it a congeries of discrete entities?—atoms, bodies which, in Clerk Maxwell's straightforward definition, "cannot be cut in two." Is the world, as Bertrand Russell somewhere asks, a bucket of molasses or a pail of sand? The issue divided Einstein from most of his fellow physicists at the end of his life. And since this problem, though ever more fruitful, is no nearer solution after 2,500 years than when it was discovered in Greece, it seems safe to say that its merit lies in the discussion, not in the answers.

Perhaps the answer will be given on that distant day when science is complete. But until then, all experience suggests that neither approach—neither geometrizing nor counting—could yield a complete description. The choice would appear to be a matter of temperament. A mathematical spirit, an Einstein, a Descartes, will speak out the unity of nature in the language of geometry. Descartes was nothing if not consistent. Atoms could never be fundamental. Space-matter is infinite both ways, in extension and division: there is no point below which a straight line may not be further divided. But the investigator whose intuition is physical will seek a definite term to measurement, something numerable for science to come down to, at least in principle, if not in fact. And the whole experience of science bears out a curious paradox. Again and again it has proved the better part of physical wisdom to postulate infinity in extension but to exclude it from division. There seems no reason for this in the logic of things. But history never gives on to vistas of logic, though claims have sometimes been made for it as a storehouse of wisdom.

It is time, therefore, to introduce atomism. For schematically speaking, theoretical physics is a prolongation into science of Platonism, and experimental physics of atomism. Atomism made the opposing jaw of the pincers which was to find its hinge in Newton and which finally squeezed the credibility out of the Aristotelian world picture built by forms and qualities out of common sense data. Four major figures followed each other in the ancient atomic school: Leucippus, of whom we know almost nothing; Democritus, the most powerful thinker, on whom we are not much better informed; Epicurus, who incorporated the doctrine into one of the two leading philosophies (the other was Stoicism) of the Hellenistic world; and Lucretius, a Roman who transmitted the tenets in his poem *On the Nature of Things*. Like the earliest Milesian philosophers, the atomists accepted motion as a fundamental condition and matter as that which is conserved. By making a distinction between mathematical and physical subdivision, their doctrine saves both motion and conservation from the logical traps in which these preconditions of science could otherwise be enmeshed. Without such a distinction, for example, the conception of real motion was exposed to paradoxes like that which makes it impossible for Achilles to catch the tortoise, since to halve the distance always takes a finite time. Even more important, they saved themselves from the necessity to *explain* motion, which is what led Aristotle astray. They put their particles in a void extended to infinity, and thus rendered motion possible in a universe where matter is conserved.

"To weave again at the web," writes Lucretius, "which is the task of my discourse, all nature then, as it is of itself, is built of these two things: for there are bodies and the void." Change and process consist not in flux or penetration by soul or realization of the goal of life, but in the physical rearrangement of varied particles of specific shape and size which do have objective existence. No more than this (though this is a great deal) must be read into it. Atomism was not some kinetic theory of the macrocosm. The atoms simply drift through the void. There is no positive analogy to be made with the atoms of the periodic table of chemistry. There is, of course, a schematic analogy, though judgments differ on its significance. Still, the atomists did compose nature of an imaginary alphabet of atoms, even as language is composed of letters which combine according to laws of spelling and syntax. And S. Sambursky, a physicist who has written on Greek science, esteems most highly of all its achievements this empiricism of the imagination, this mode of inference from the visible to the invisible in concepts which are adopted simply because they make objective reasoning possible.

Nevertheless, encased in the Epicurean philosophy, the atomic doctrine could never be welcome to moral authority. Nor could it ever be popular. However serene, objectivity is an Olympian posture. Indeed, it was more Olympian than the Olympians, for the Epicurean gods neither created the world nor paid it the compliment of attention. "Nature," says Lucretius, "is free and uncontrolled by proud masters and runs the universe by herself without the aid of gods." Only the atomists among the schools of Greek science divorced law from mind and purpose, and theirs was the one view of nature quite incompatible with theology. Like a pair of eighteenth-century *philosophes*, Epicurus and Lucretius introduced atomism as a vehicle of enlightenment. They meant to refute the pretensions of religion "which gives birth to deeds sinful and unholy," and release men from superstition and the undignified fear of capricious gods. Consequently, a hint of Epicureanism came to seem the mark of the beast in Christian Europe. No thinker, unless it is Machiavelli, has been more maligned by misrepresentation. Materialism remains a pejorative word; and to the half-educated, Epicureanism suggests self-indulgence rather than disciplined serenity of taste or that courageous resignation which contemplates the world, not as one would have it for one's spiritual ease, but as it actually appears through the windows of sense. For sensation is contact with Epicurean truth.

It is unlikely that proscription by authority deprived the public of anything it would have welcomed. The order we sense in the world becomes, to borrow the old cliché, a fortuitous concourse of atoms. All that exists in nature is impenetrability, shape, and arrangement of atoms. The secondary qualities of bodies, those we appreciate and judge—color, odor, taste, form, feel—are only modes of perception in us. Truly, as Lucretius admits, this robs the glory and beauty out of nature as the price of understanding, and men resent the loss. To parse the sentences in Milton, to count the frequency of the letter "e" or "a" in *Paradise Lost*, is not to experience the reality of the poem. Not only have the categories of consciousness no existence in nature, but soul and mind themselves are simply arrangements of the finest particles. And Epicurus had to spoil his consistency to make a place for free will. In an exception which justly provoked the ridicule of critics, he let in chance, and therefore choice, by allowing his atoms a tiny but unpredictable swerve in their downward drift through the nothing. Historically, therefore, it may be simply the accident that atomism became the ontology of classical physics which made science seem an enemy sometimes to religion and morality. For atomism offends the most common instincts of humanity in order to achieve understanding and serenity for philosophers.

Not till the seventeenth century was this last of the Greek systems domesticated by science and admitted into polite company. It is a measure of the neglect into which atomism had fallen that it was retrieved, not by a Galileo or a Descartes, but by a thinker of the second rank, Pierre Gassendi, a Provençal, who had minor astronomical observations and physical measurements to his credit. His reconciliation of that system with Catholic dogma does not convince, and is intolerably diffuse. But that is no matter. He paraded Democritus and Epicurus before the Republic of Letters, and his place in the history of science is secure. For atomism spoke with authority to the physical intuition of the seventeenth century, to those who wanted to do science with their hands as well as their heads. One could scarcely hope to demonstrate empirically the existence of atoms which, by definition, lie below the dimensions of sense. But one might, perhaps, demonstrate the existence of the void—not, to be sure, the infinite void, but at least the local vacuum. Just as Galileo and the theoretical physicists founded dynamics in their attack on the Aristotelian concept of local motion and its principle that motion argues a mover, so experimental physicists proposed to disprove the impossibility of the void and to set at naught the principle that nature abhors a vacuum.

Miners had long known that a suction pump will not lift water more than thirty-four feet. The question gathered some importance with Galileo (for schematizing the pedigree of theoretical and experimental physics must not be carried to the point that the one excludes the other). But the Aristotelian *horror vacui* was too alien to Galileo's own habit of mind, too childish perhaps, for him to see the problem as explosive in its possibilities. It figures in the *Discourses* as a digression from the strength of materials. He simply suggests in passing that the column may break of its own weight at thirty-four feet. This is wrong, and the question was picked up, with others of Galileo's loose ends, by Torricelli, one of his most brilliant students and his successor as mathematician to the Grand Duke. Torricelli's bent was experimental. He had the happy idea of reducing the behavior of liquids in closed standpipes to a laboratory scale by substituting for water a column of mercury. This was the invention of the mercury barometer. Torricelli himself was less interested in the device as a metrical instrument than in the implications of the empty space left at the top of the tube when the mercury sank into the sustaining basin of liquid metal until it stood at its normal height of 30 inches. "Many have said that a vacuum cannot be produced. . . . I reasoned in this way: if I were to find a plainly apparent cause for the resistance which is felt when one needs to produce a vacuum, it seems to me that it would be vain to try to attribute that action, which patently derives from some other cause, to the vacuum; indeed, I find that by making certain very easy calculations, the cause I have proposed (which is the weight of the air) should in itself have a greater effect than it does in the attempt to produce a vacuum." This was written in 1644, and Torricelli gave his experiments on air pressure all the generality of which they were capable in respect to the way the world is made: "We live submerged at the bottom of an ocean of the element air, which by unquestioned experiments is known to have weight."

Physicists derive satisfaction from verifying and extending that which can be established. In Germany the Burgomaster Otto von Guericke of Magdeburg dramatized

Torricelli's results in massive Teutonic fashion by causing great brass hemispheres to be machined so perfectly that the space inside could be exhausted, and two mule-teams failed to pull them apart against atmospheric pressure. In Paris, on the other hand, Pascal developed the implications with the finesse and subtlety which were the hallmarks of one of the most disturbingly refined minds in history. Pascal's intellectual career was a kind of agony of penetration. He saw so far that, instead of drawing from the problems of physics that perception of beauty to which his genius entitled him, he ended—so he felt—by seeing right through science itself: "From that time forth," his brother-in-law, François Perier, tells us, "it was his firm belief that religion was the one worthy object of the thoughts of men. . . . He would frequently say on this subject *that all the sciences could not comfort them in the days of affliction, but that the doctrines of Christian truth would comfort them at all times both in affliction and in their ignorance of those sciences.*" Nor is one's discomfort at such tenderness of mind lessened on finding that Pascal in his brilliance and humanity was not always altogether frank and just in his science.

His style was Archimedean like Galileo's, attended by that same malicious capacity for imagining those physical consequences which would place his opponents in the most embarrassing light. For though Pascal wrote in praise of meditation—and believed what he wrote—in action he depended on polemics (even in theology) as an athlete on gymnastics. It is a defect of the present book that no place has been found for the physics of Simon Stevin, a Flemish contemporary of Galileo, his equal some would say, in power of thought if not of propaganda, who cast the Archimedean legacy of statics into the form which served classical mechanics. Pascal brought Stevin's work into the main stream. His *Treatise on the Equilibrium of*

Liquids prepared Stevin's hydrostatics for unification with Torricelli's atmospheric hypothesis. The urbanity with which Pascal described experiments creates the impression that he had actually performed them all—until one reflects a moment about the glass barometer forty-six feet long, the physicist doing experiments twenty feet below the surface, and the fly "which can live in luke-warm water as well as in the air" and experiences no discomfort on submersion because the pressure acts equally from all directions. Pascal's second essay, *Treatise on the Weight of the Mass of the Air*, brings home the analogy for men swimming in the ocean of the atmosphere, of which Pascal calculated the total weight to be 8.28×10^{18} pounds.

In 1648 Pascal imagined and commissioned the famous experiment which caught the imagination of Europe in its demonstration of the decline of atmospheric pressure with altitude. He had Perier carry a barometer from the black lava city of Clermont, where they had been born, to the volcanic peak of the Puy-de-Dôme "some five hundred fathoms above" and compare the readings. The difference was over three inches of mercury. The intervening figures, characteristically enough, are so perfect that they smell of interpolation rather than observation. One may, perhaps, quote the consequences Pascal drew, not only for the structure of nature, but for the structure of scientific explanation:

> Consequently, I now find no difficulty in accepting . . . that nature has no repugnance to a vacuum, and makes no effort to avoid it; that all the effects ascribed to such abhorrence are due to the weight and pressure of the air, which is their only real cause; and that for lack of knowledge, people have purposely invented this imaginary abhorrence of the vacuum in order to account for them. This is far from being the only case in which, when the weakness of men has made them unable to discern true causes, their subtlety has substituted for them imaginary causes to which

they have attached specious names which fill the ears, but not the mind. Thus they say that the sympathy and antipathy of natural bodies are the generic efficient causes of several effects, as if inanimate bodies were capable of sympathy and antipathy.

Truly experimental physics came into its own with Robert Boyle. He spared his reader no detail. No one could doubt that he performed all the experiments he reported, hundreds and hundreds, thousands of them, bringing to his laboratory great ingenuity, incomparable patience, and that simple honesty which makes experiment really a respectful inquiry rather than an overbearing demonstration. His distinctive technique was the application to considerable receivers of an exhaust pump. He confirmed Torricelli and Pascal on the existence of the vacuum, and quietly went on to perform experiments in it and there demonstrate the silence of the ringing bell, the collapse of the puff of smoke, the feathers falling like buckshot, the demise of the inevitable mouse.

Boyle was a gentleman, youngest son of the first "and great" Earl of Cork, founder of one of the families whom the first and great Elizabeth loosed upon Ireland to secure the protestant ascendancy and enrich themselves. Drawn to science while living in Oxford, he read of the pneumatic experiments on the continent. In 1660 he published *New Experiments Physico-Mechanicall Touching the Spring of the Air*. He became in later years a senior member of the generation of genius which Restoration England contributed to science. In Boyle the English character in science is already apparent. He was patient in observation, handy in manipulation, and tedious in exposition. His writing betrays no stylishness of mind—only occasional sallies too heavy and infrequent to lighten the mass. ("I should scarce have ventured to entertain you so long concerning such empty things as bubbles.") He was well-meaning but simple-minded in theology, and would rebut the stigma of atheism attaching to atoms-and-the-void:

> When I speak of the corpuscular or mechanical philosophy, I am far from meaning with the Epicureans, that atoms, meeting together by chance in an infinite vacuum, are able of themselves to produce the world . . . but I plead only for such a philosophy, as reaches but to things purely corporeal, and distinguishing between the first original of things, and the subsequent course of nature, teaches, concerning the former, not only that God gave motion to matter, but that in the beginning he so guided the various motions of the parts of it, as to contrive them into the world he designed they should compose.

Boyle endowed a series of lectures on physics as the study of God in His works which continued to be given into the eighteenth century. He had no mathematics. And yet in a certain dogged way, he not only got the texture of matter right; he got the point, the only point, perhaps, on which he could agree with Aristotle: that atomism contemplates a world of number, not abstract number like Plato's or geometrical form like Galileo's, but a world of numerable things.

Many have lost the thread of Boyle's thought in the vast mass of pneumatic evidence. For he was a thinker about nature. Boyle's Law—that in a confined gas pressure times volume is a constant—was a by-product, not the object of his inquiry. Neither was he the simple empiricist lacking in strategy which students of his chemistry have sometimes made him. It is true that *The Skeptical Chymist*, his most reprinted book, is more successful as a destructive criticism of the essences and principles of alchemists and spagyrites than as a constructive guide to the groping science of chemistry. More than a chemist, Boyle will be better understood as the atomic physicist he meant to be. He exhausted his reader along with his receiver in exhibiting the void. But Torricelli, Pascal, and others had already done that, as Boyle well knew. He thought to go beyond them, beyond the assertion of the negative to the demonstration of the positive, beyond—or into—the void to the atoms. What interested him in his very first pneumatic experiments was less the vacuum produced than the action of his pump itself, and the reaction—the "spring"—of the air, which anyone feels who pumps up a tire by hand:

> By which . . . spring of the air, that which I mean is this; that our air either consists of, or at least abounds with, parts of such a nature, that in case they be bent or compressed by the weight of the incumbent part of the atmosphere, or by any other body, they do endeavour, as much as in them lieth, to free themselves from that pressure. . . .
> This notion may perhaps be somewhat further explained by conceiving the air near the earth to be such a heap of little bodies, lying one upon another, as may be resembled to a fleece of wool. For this . . . consists of many slender and flexible hairs; each of which may indeed, like a little spring, be easily bent or rolled up; but will also, like a spring, be still endeavouring to stretch itself out again.

For Boyle designed his experiments as material essays in the "corpuscular philosophy." While still an undergraduate, Newton probably read *The Spring of the Air*, and certainly Boyle was the immediate source of Newtonian views on the structure of matter. Beyond this, Boyle etched certain lines of deeper perspective into the world picture of classical physics. His work expressed a more cautious phenomenalism than the assumptions of mathematically minded theorists of the continent. He would not say that the air is atoms—only that the atomic model makes its phenomena "intelligible," by which he meant accessible to science. But as for the essence of matter,

> I shall decline meddling with a subject, which is much more hard to be explicated than necessary to be so by him,

whose business it is not, in this letter, to assign the adequate cause of the spring of the air, but only to manifest, that the air hath a spring, and to relate some of its effects.

And again,

I consider, that the chief thing, that inquisitive natural- ists should look after in the explicating of difficult phae- nomena, is not so much what the agent is or does, as, what *changes* are made in the patient, to bring it to exhibit the phaenomena, that are proposed; and by what means, and after what manner, those changes are effected.

Hence the interest in chemistry, the attempt to redeem the science of *combining* matter from the mists of alchemy and the receipts of medicine and "to beget a good under- standing between the chymists and the mechanical phi- losophers," the latter all scornful of the former. Boyle was the first important physicist to take chemistry seriously, as a means to the end of establishing the corpuscular phi- losophy. This was a profound ambition, a major step into a science which objectifies change. Galileo had laid the foundations in the positions of matter by turning motion— translational change—from process to state. Boyle would assimilate the constitution of matter to science by con- sidering substantial change in the same way. Substantial change is no penetration by active qualities—heat, color, life—no reshuffling of the stuff of the world amongst categories of form. Instead, the "mechanical philosopher being satisfied, that one part of matter can act upon another but by virtue of local motion, or the effects and conse- quences of local motions, he considers, that as if the pro- posed agent be not intelligible and physical, it can never physically explain the phaenomena"; and he assigns all changes to "these two grand and most catholick principles of bodies, matter and motion." Change, in short, is re-ar- rangement in the parts of an objective world. For if science is to be possible, it *must* be like this. Otherwise, everything blends into everything and the world is the way Goethe will want it to be, not to be embraced by measurement but to be penetrated by sympathy, where Faust will take his shortcut to knowledge, and power, not through science, but through magic.

Boyle's was the common sense of science. And yet there is an element, if not of failure, at least of inconclusiveness in his career. For he never in fact established the corpuscular philosophy as more than an inference, an assertion about matter which—even like Descartes'—was really an assertion about method. He never made chemistry quantitative. The reason is simple. He found the physical properties of air, but not the chemical properties of gases, the identity of which as chemical substances eluded him. Consequently every chemical agent which acts in the gaseous state escaped his control. Not till Dalton, over a century later, would the "corpuscular philosophy" take on the positive meaning that comes of being clothed in numbers. In Boyle, therefore, as in Democritus, and indeed throughout the seventeenth and eighteenth centuries, atomism remained rather a precondition of an objective science than a finding of an experimental science. Matter "being a finite body, its dimensions must be terminated and measurable: and though it may change its figure, yet for the same reason it must necessarily have some figure or other." And it was Boyle's devotion which, more than his success, lent the dignity of practice to Baconian experimentalism as a way of scientific life.

Boyle's publisher wrote a preface to his *Origin of Forms and Qualities*. For once, a publisher's was a just estimate:

And though the most noble author hath herein, for the main, espoused the atomical philosophy, . . . I may not scruple to call it a new hypothesis, peculiar to the author, made out by daily observations, familiar proofs and experiments, and by exact and easily practicable chymical processes; whereby one of the most abstruse parts of natural philosophy, the origin of forms and qualities, which so much vexed and puzzled the antients, and which, I would speak with the leave of the *Cartesians*, their ingenious master durst scarce venture upon, or at least was unwilling to handle at large, is now fully cleared and become manifest: so that from this very essay we may well take hope, and joyfully expect to see the noble project of the famous *Verulam* [i.e. Bacon] (hitherto reckoned among the *Desiderata*) receive its full and perfect accomplishment; I mean a real, useful, and experimental physiology [i.e. physics], established and bottomed upon easy, true, and generally received principles.

❖

ENTHUSIASM FOR EXPERIMENT expanded through the scientific literature of the seventeenth century until it reached the proportions of a moral cause. "This noble Design of Experiments," Bishop Sprat calls the program of the Royal Society in the apologia which he wrote as a contemporary history of its foundation. In Bacon, who never performed an experiment worthy of the name, such insistence may arouse impatience as a piece of pretentiousness. But there is no denying the admiration which Boyle and the charter members of the Royal Society expressed for Bacon. It is not too much to say that, acting through Boyle, Bacon's inspiration produced atomic physics—out of the void. In Sprat's words again, "there should have been no other Preface to the *History* of the *Royal Society*, but some of his Writings."

It is not just a question of method. That may be found earlier in Harvey, and even more perfectly in the optical researches of Newton, who held aloof from the Royal Society in his creative years. The question is of scientific style, of taste; and it may, perhaps, be permissible to suggest that what is vulgarity in Bacon, who only wrote about experiments as the easy alternative to the hard, abstract thought which orders in concepts, is a seemly humility in men who actually did experiment to find out how the world is made by taking bits of it apart:

> It is enough (writes Sprat), that we gather from hence; that by bringing *Philosophy* down again to men's sight and practice, from whence it was flown away so high: the Royal Society has put it into a condition of standing out, against the invasions of *Time*, or even *Barbarism* itself: that by establishing it on a firmer foundation, than the *original Notions* of men alone, upon all the *works of Nature*; by turning it into one of the *Arts of Life*, of which men may see there is daily need; they have provided, that it cannot hereafter be extinguish'd . . . but that men must lose their *eyes* and *hands*, and must leave off desiring to make their *Lives* convenient, or pleasant; before they can be willing to *destroy* it.

Experimenters were the craftsmen of science—Malpighi, Snell, Robert Hooke, even Boyle. There is, after all, an arrogance in mathematics, originating in the mind and not in nature. A kind of justice thus attends the chastening which every theory must undergo at the hands of what has been called the terrible experimental method. Descartes needed to be redressed by the serious, modest seeker after fact, however naïve and undirected the early experimental programs were. And what redeems them from the vulgar anti-intellectualism into which Bacon fell was that they never opposed the order to be won by accumulation and classification of fact to that higher order won by abstraction and mathematical formulation. Confronted with a Newton, theirs was a genuine humility.

Science developed its social character out of the necessity for cooperation, communication, and patronage. Historically the two most eminent scientific bodies are the Royal Society of London (1662) and the French Academy of Sciences (1666). Ephemeral literary and cultural academies (the word comes from Plato's circle) abounded in Renaissance Italy. They were called into being as the ornament of some court, the pastime of some prince, and vanished as easily as formed. The first to set itself a scientific object was the *Accademia dei Lincei* (lynx-eyed) founded in 1603 in Rome. The patron was Prince Federigo Cesi, who expended his youthful enthusiasm—he was eighteen when he began his academy—on natural history. Galileo was a member, but the group did not long survive his disgrace or Cesi's death in 1630. A far more solid undertaking was the Florentine *Accademia del Cimento* (experiments) founded in 1657. It was the embryo of project research. Problems would be propounded by members or correspondents. The experiments would be performed in the Academy's rooms in the Pitti palace by Giovanni Borelli or Vincenzio Viviani, the best of Galileo's students, or by another of the nine members. But the decision as to what problem to pursue belonged to Prince Leopold, the brother of the Grand Duke Ferdinand II, who acted as director. The Academy published its researches on atmospheric pressure, thermometry, barometry. Instrumentation was perhaps their finest contribution. They found that freezing occurs at constant temperature. For one final decade of brilliance, Florence was again the home of the most portentous branch of culture. It was the swan-song of Italian cultural leadership. In 1667 the support of these latter-day Medici faltered in a thickening climate of clerical animosity, and the Academy collapsed.

In London and Paris at the same time the climate of culture did conduce to that continuing growth which is the sign of vigor in learned disciplines. In science as in other realms the French were quicker to feel the need for communication than the English and slower to stabilize institutions. Already in the first half of the seventeenth century, the intelligentsia of Paris formed that shifting salon where generations of men of letters have succeeded each other, ever contending over the French conscience for which they take special responsibility. Their relations with colleagues in the provinces were by letter. The center of the group, the abbé Mersenne, made a career as a scientific gossip, telling the latest news of Galileo's experiments on dynamics, of Pascal's demonstration of the vacuum, of Descartes' views on light. From 1620 to 1648 his correspondents depended on his devotion for the information their successors would find in learned publications. Gallic ingratitude made a joke of him as the "letter-box to the learned world," but he was in fact an influential exponent of mechanism.

More substantial arrangements had to await the rationalizing regime of Colbert in the first constructive years of the great monarchy of Louis XIV, by which time the Royal Society had stolen a march. But despite the stimulus of the English example, the *Académie royale des sciences* was conceived less in the image of Bacon's New Atlantis than in the tradition of French statism. A faith has always animated its great officials that by taking thought they could add a cubit to the stature of the body politic. Accordingly, the *Académie des sciences*, unlike the Royal Society, had statutory responsibilities for technological supervision and improvement of French industry, though whether it accomplished much in this line is less clear than what was expected of it. Fortunately the question was seldom pressed, and academicians were pensioned by the crown and enjoyed honorific distinctions.

The spirit differed from the honest amateurishness of the Royal Society, which exacted no qualifications of its fellows beyond a personal undertaking to be interested. Places in the *Académie des sciences* were limited. By the modified constitution of 1699 three "pensioners" and three "students" were to sit in each of the six sections—geometry, astronomy, mechanics, anatomy, chemistry, and botany. Ever since 1635 the *Académie française*, founded by Richelieu, has been ruling over letters (not without opposition) with the mission of purifying and guarding the French language, that vehicle of civilization. The *Académie des sciences* was given the same responsibility for scientific standards. The result was a more professional institutionalization of science in France than in England. But this admirable design was frustrated by the waywardness of genetics. The scientific minds which matured in France under Louis XIV were far less fertile than those of the generation of Descartes and Pascal, far less productive than the cluster of English genius which crowded in upon the Royal Society in Newton's time. France had to await the Enlightenment of the eighteenth century for the pre-eminence intended by Colbert.

In the seventeenth century it was rather the Royal Society which set the tone and style of science, arising as the spontaneous answer to the need for a scientific public. The creation of such a public was a condition though not a cause of scientific culture. It lent body to what would otherwise have continued, on too rarefied a plane for social vitality, as the exchange of high concepts between successors and peers of Galileo and Descartes.

The Royal Society issued not from the discoveries of great minds, but from the serious discussions of earnest minds in the effort to comprehend and further those discoveries in their bearing on godliness, learning, and humanity. In part it was a refuge from the civil wars— the "Invisible College," Boyle called their association during Cromwell's regime. He was nineteen when he fell in with the "virtuosi" in 1646. Doctors were the most numerous group: Jonathan Goddard, George Ent, Francis Glisson, Christopher Merret, Thomas Willis. One of the first social statisticians, William Petty, was of the circle. Two Germans, Theodore Haak and later Samuel Oldenburg, played the part of Mersenne in Paris as "intelligencers." There were divines who tended to be of a Puritan persuasion: John Wallis and Seth Ward, whose interests were mathematical and astronomical, and John Wilkins, who married Cromwell's sister and became Warden of Wadham College.

The commitment of the group appears to its best advantage, perhaps, in the career of Wilkins. In 1648 he published works expounding the new mechanics and cosmology. His terms foretell with surprising insight the accommodation to be reached between Galileo's mathematicization and Bacon's socialization of science. He was a moving spirit. Some members—particularly Boyle—followed him from London to Oxford in the 1650's, and back again to London after the Restoration dispossessed Puritan appointees from their college livings. Most interesting of all, he perceived that one consequence of science is the possibility of definitive communication by means of symbols which represent things and not opinions. He devised, therefore, a "philosophical language," which anticipates John Locke's psychology in important ways and seeks to exorcise in practise Bacon's idols of the market place.

In the more settled atmosphere of the Restoration, the group moved for royal favor and a permanent organization. Charles II, who easily forgave the Puritan past, granted his approval. A preliminary charter was issued in 1662, and in 1663 letters patent authorized the group to be known as *The Royal Society of London for Improving Natural Knowledge*. The Society's *Philosophical Transactions* began from 1665 the long series of scientific memoirs never since interrupted. But the epithet "Royal" signified only the indulgence, not the support of the crown. The Royal Society was a voluntary body in the English pattern wherein private enterprise undertook what on the continent would have been civic functions. It associated public-spirited patrons like Lord Brouncker and literate men about town like Samuel Pepys and John Evelyn—"gentlemen free and unconfin'd," Sprat called them—with those who, like Boyle, Robert Hooke, and Edmund Halley, actually engaged in "improving natural knowledge" in their private laboratories or in that—never very well endowed—of the Society. Thus the Royal Society embodied a concerted movement of culture under the Restoration.

The movement amounted to an English rehearsal for the European Enlightenment of the eighteenth century. Only one feature was missing. There was no hostility to Christianity. It must not be supposed from the discredit of atomism or the indignities suffered by Galileo that religion and science are always in conflict. On the contrary, Puritan dedication and earnestness passed over from the ethos of religion to that of science. The career of Benjamin Franklin serves Americans as an evocative instance of the Puritan ethic secularized in the practical man of science and affairs. In Boyle, the ethic was not even secularized. His peculiarly English school of natural theology rested with all the insecurity of great sincerity on the evidence for Nature as the art of God.

The correlation of Calvinist behavior patterns—hostility to tradition, utilitarianism, calculating self-denial, a calling to work in this world, rationality and the individual interpretation of experience—the correlation of these qualities with practical business and science (it is less notable when it comes to speculative or theoretical science) is a very general feature of Western cultural history. There can simply be no doubt that protestant and bourgeois milieux have encouraged talent and ambition to rise through science, and that catholic and aristocratic milieux have inhibited the development of scientists.

Scotsmen and Dutchmen flock through the history of science; Irishmen and Spaniards are scarcely to be found.

But the forces are sociological, not doctrinal. One can discern them at work not only as between Protestant and Catholic countries, but inside both. In France, for example, an undue proportion of scientists were of the Jansenist persuasion, which is psychologically akin to Puritanism within the Catholic fold. In England, the immense majority of scientists have been Nonconformists from the plainer social classes, and not from the Anglican gentry. A recent survey of the provenance of American scientists finds them coming in large proportion from the small denominational colleges of the Middle West, from the Corn Belt, and not from the South with its aristocracy of ghosts, nor from the Ivy League, whose graduates move typically toward the law, diplomacy, or affairs. And who knows, finally, what avatars of Puritan purposiveness, what imperatives to progress in this world, work in the great Russian mass toward the ultimate socialization of science?

A stereotype of the lonely scientist, isolated by his knowledge, sometimes rises before the layman's eyes. Nothing could be further from the true social nature of the modern scientific enterprise. One humanist, for example, has the impression that his scientific colleagues, with an altogether charming gregariousness and an enviable access to funds, travel all over the world to meetings where their inability to speak one another's languages seems no barrier to fruitful discourse. All speak the language of their science. In *Science and the Common Understanding* Robert Oppenheimer, whose country failed so discreditably to understand the meaning of his career, reflects movingly on the true community which lives and has its being in science. This it was that the Royal Society achieved at the very outset, at a time of deep civil disturbance when men of good will could take little heart in the state of the world, nor find causes to support which did not offend against moderation and educated taste. Sprat tells how their constitution forbade discussion of religion and politics:

> To have been always tossing about some *Theological question*, would have been, to have made that their private diversion, the excess of which they themselves disliked in the publick: To have been eternally musing on CIVIL BUSINESS, and the distresses of their country, was too melancholy a reflexion: It was Nature alone, which could pleasantly entertain them in that estate. The contemplation of that, draws our minds off from the past, or present misfortunes, and makes them conquerors over things, in the greatest publick unhappiness: while the consideration of Men, and *humane affairs*, may affect us with a thousand various disquiets: *that* never separates us into mortal *Factions*; that gives us room to differ, without animosity; and permits us, to raise contrary imaginations upon it, without any danger of a *Civil War*.

CHAPTER IV

NEWTON WITH HIS PRISM
AND SILENT FACE

THE MIND OF SIR ISAAC NEWTON was one of the glories of the human race, and one of its mysteries. "How did you make your discoveries?" an admirer is said to have asked. "By always thinking unto them," replied Newton, but did not then say what is even more daunting, that he did most of the creative work in two periods of about eighteen months each, in 1665-66 and 1685-86. In those three years of intensive application, interspersed by twenty years of study and reflection, Newton united knowledge of heaven and earth in the mathematical structure of classical physics. For over two centuries that structure contained the thinking of a science which, no longer struggling to be born, grew exponentially in vigor as in volume. "There could be only one Newton," Lagrange is supposed to have said to Napoleon (who was fishing for a comparison and resented the remark), "there was only one world to discover." Contemporary physics has transcended Newtonian in the reaches of the very small and the very fast. But our own century has only hastened the pace of science. It has not altered the rules or the nature of the enterprise. And surely it will always repay effort to study the mind and personality which founded science in generality and once for all. Fellow beings have the right to share in that triumph, and the duty to respect it. It enhances all humanity.

Born in 1642, the year of Galileo's death, Isaac Newton was a posthumous child in a family of minor Lincolnshire gentry. His mother remarried, and his childhood was not happy. A girl of the neighborhood remembered him as "a sober, silent, thinking lad," who "was never known scarce to play with the boys at their silly amusements." When he was fourteen his stepfather died. His mother set him to farming the manor. This was not a success. She wisely put him back in school, and in 1660 sent him up to Cambridge, where he matriculated in Trinity College. There he worked under the Master, Isaac Barrow, a classicist, astronomer, and authority on optics. "In learning Mathematicks," wrote Fontenelle, Newton's first biographer, "he did not study Euclid, who seemed to him too plain and simple, and not worthy of taking up his time; he understood him almost before he read him, and a cast of his eye upon the contents of the Theorems was sufficient to make him master of them. He advanced at once to the Geometry of Des Cartes, Kepler's Opticks, &c., so that we may apply to him what Lucan said of the Nile, whose head was not known by the Ancients,

> Nature conceals thy infant Stream with care
> Nor lets thee, but in Majesty appear.

Barrow did perceive the quality of that stream. He knew in extraordinary measure the finest of a teacher's joys, a fine student. In 1669 he resigned his Lucasian Chair of Mathematics that Newton might have it. This gracious precedent must alarm any professor who becomes aware that his student is abler than he is. But Barrow himself did not then know the portent of what Newton had secretly begun. In the same year he published a book which was obsolete before the type was set, in consequence of his former student's optical experiments. Newton was not ready to communicate these, or other musings. But in preliminary studies at the age of twenty-three he had sketched the world picture of classical physics.

Athletes of the intellect, theoretical physicists build careers upon the innovations of their youth. The plague was in Cambridge in 1665. To escape it, Newton went down to his mother's manor of Woolsthorpe. It is pleasant to be able for once to record the truth of a legend. As he sat in the garden, a falling apple did indeed set his mind

> into a speculation on the power of gravity: that as this power is not found sensibly diminished at the remotest distance from the center of the earth, to which we can rise, neither at the tops of the loftiest buildings, nor even on the summits of the highest mountains; it appeared to him reasonable to conclude, that this power must extend much farther than is usually thought; why not as high as the moon, said he to himself? and if so, her motion must be influenced by it; perhaps she is retained in her orbit thereby.

The account is Henry Pemberton's, who was much with Newton in old age, and wrote one of the first and best explanations of his system. But Newton's retirement was no desultory meditation at the end of college. He himself left a fragmentary memoir of these months of discovery:

> I found the Method [of fluxions—i.e. the calculus] by degrees in the years 1665 and 1666. In the beginning of the year 1665 I found the method of approximating Series and the Rule for reducing any dignity of any Binomial into such a series [i.e. he had formulated the Binomial Theorem]. The same year in May I found the method of tangents of Gregory and Slusius, and in November had the direct method of fluxions [the differential calculus], and the next year in January had the Theory of colours, and in May following I had entrance into ye inverse method of fluxions [integral calculus]. And the same year I began to think of gravity extending to ye orb of the Moon, and having found out how to estimate the force with w^{ch} [a] globe revolving within a sphere presses the surface of the sphere, from Kepler's

Rule of the periodical times of the Planets being in a sesquialterate proportion of their distances from the centers of their Orbs I deduced that the forces wch keep the Planets in their Orbs must [be] reciprocally as the squares of their distances from the centers about wch they revolve: and thereby compared the force requisite to keep the Moon in her Orb with the force of gravity at the surface of the earth, and found them answer pretty nearly. All this was in the two plague years of 1665 and 1666, for in those days I was in the prime of my age for invention, and minded Mathematicks and Philosophy more than at any time since.

The calculus, the composition of light, the law of gravity—the first two were fundamental, the last both fundamental and strategic. As if by instinct, Newton asked not what the forces are that keep the planets in orbit, but what the proportions of those forces are. In part Newton's was a winnowing genius. He took the planetary laws from Kepler. (Kepler had made them serve the tangential drag of sympathetic attractions.) He took from Descartes the argument that curvilinear motion argues a constraint against inertia. (Instead of formulating the quantity of that constraint, Descartes had imagined a mechanism.) He took from Galileo the perception that, though motion is the object of science, the handle to grasp in numbers is change in motion. (Because Galileo was a purist about any hint of animistic or occult qualities, he had made falling the source of motion and had never asked the questions which would relate acceleration to a force law. Galileo remains the founder of kinematics, therefore, and left Newton to found dynamics.)

The writings of Christiaan Huygens contain the missing piece. "What Mr. Hugens has published since about centrifugal forces I suppose he had before me," wrote Newton reluctantly (for there is a kind of avarice about discovery which may be one of its springs of action). A Dutchman, Huygens made his career in Paris. He combined his native experimental tradition with Cartesian rationalism, often in criticism against the more naïve physical propositions of the master. The pendulum clock owes its design to his studies, which he addressed rather to specific problems—the laws of impact, conservation of momentum, a wave theory of light—than to establishing some world view. There he remained faithful to the Cartesian conception of science as the mechanistic rationale of material reality.

His analysis of centrifugal force (later objections to the term do not diminish the historical value of the argument) considers circular motion as inertial and centrally accelerated. His reader is to imagine a man—a physicist, let us say, for Huygens is an early example of the physicists' genre of instrumental playfulness—attached to the rim of a wheel and holding a plumb bob on a wire. The wheel rotates, and the physicist experiences a tension in the wire indistinguishable from the pull of gravity when it is still. Now let him release his hold, and by a very elegant geometric proof, Huygens showed that the distance from the plumb bob sailing out along the tangent and the physicist on the rim increases as the square of the time of rotation. Let him hold on to it, therefore, and appreciate that the formalism of its angular motion is identical with the law of falling bodies, and that the concept of acceleration includes change in direction as well as velocity. It appears that Newton worked out the same result in ignorance of Huygens' demonstration. But he does not need the credit. For he saw in it what Huygens did not: that by this argument the moon is forever falling around its orbit even as the apple falls, that any acceleration supposes a force, and that if moon and apple move under the same force, then celestial mechanics becomes a sublime instance of inertial motion under a universal force law.

This was the comparison that Newton found "to answer pretty nearly." Nevertheless, he did not then press on to formulate the universal law of gravity. Nor did he generalize the measurement of force by acceleration into the laws of motion. Instead, he kept all these things to himself, laid the work aside, and did not return to it for thirteen years. Various explanations have been advanced for the delay. He was working away from books, and had the wrong figure for the size of the earth—60 miles to the degree instead of 69½. It is said, too, that Newton thought the discrepancy—he says "pretty nearly," not "exactly"— might be caused by other forces at work concurrently with gravity—Descartes' vortices, perhaps, for he was not yet ready to introduce the void as the arena for gravity. What was more important, an essential proof eluded him. He had treated the earth and moon as points, all mass concentrated at the center. The intuition does not compel assent. Nor could Newton then prove the theorem which justifies it. It is a most difficult problem in integration, which he resolved only in time to write the *Principia*. And though posterity is fascinated by the divining power of his intuition, he could hardly come before his contemporaries except in the full force of geometric demonstration.

❖

MEANWHILE, in his later twenties, Newton's mind, and now his hands too, were full of optics and of chemistry— alchemy some commentators say, but wrongly, for his chemistry was in the spirit of Boyle's corpuscular philosophy. In 1672 he sent the Royal Society an account of the "oddest if not the most considerable detection, which hath hitherto been made in the operations of nature."

I procured me (he began) a Triangular glass-Prisme, to try therewith the celebrated *Phaenomena of Colours*. And in order thereto having darkened my chamber, and made a small hole in my windowshuts, to let in a convenient quantity of the Suns light, I place my Prisme at his entrance, that it might thereby be refracted to the opposite wall. It was at first a very pleasing divertisement, to view the vivid and intense colours produced thereby; but after a while applying myself to consider them more circumspectly, I became surprised to see them in an *oblong* form, which, according to the received laws of Refraction, I expected should have been *circular*.

Newton was the first to analyze the spectral band rather than the first to see it. Having ruled out accidents like imperfections in the glass or curving rays, he performed his "Experimentum Crucis." He refracted a ray of each color through a second prism and determined that refrangibility was a constant quantity, specific to the color, greater toward the violet and less toward the red. It follows that white light is composite, "a confused aggregate of rays indued with all sorts of Colours, as they are promiscuously darted from the various parts of luminous bodies." And this was verified by experiments in combining colors:

> These things being so, it can be no longer disputed, whether there be colours in the dark, nor whether they be the qualities of the objects we see, no nor perhaps whether Light be a Body. For, since Colours are the *qualities* of Light, having its Rays for their intire and immediate subject, how can we think those Rays qualities also, unless one quality may be the subject of and sustain another; which in effect is to call it *Substance*. We should not know Bodies for substances, were it not for their sensible qualities, and the Principal of those being now found due to something else, we have as good reason to believe that to be a Substance also.
>
> Besides, whoever thought any quality to be a heterogeneous aggregate, such as Light is discovered to be. But, to determine more absolutely, what Light is, after what manner refracted, and by what modes or actions it produceth in our minds the Phantasms of Colours is not so easie. And I shall not mingle conjectures with certainties.

No summary can do justice to the cogency of Newton's experimental practise, in execution as in design. "When we are for prying into Nature," wrote Fontenelle, "we ought to examine her like Sir Isaac, that is, in as accurate and importunate a manner." His first paper is the simplest and most straightforward piece he ever wrote. The vein is frank and youthful, almost innocent. He seems confident that everybody will be as pleased to find out about light and colors as he was. Discovery is exciting. He awaited with confidence the recognition that is one of its rewards.

He proved right about the oddity of his discovery. It went against the instinct of centuries, so deep as to be axiomatic, that light is simple and primary. This made sense of light, and nothing in Newton's own experience forewarned him of the tenacity of intellectual habit. He was not prepared for opposition. Neither was he yet aware of the seamy side of scholarship—though his own ungenerosity to rivals was to become its most illustrious example—which is that reputation accrues at the expense of someone else's status. The scholarly community has developed norms to repress such unworthy chagrins. They were not then strong. The young Newton was a David confronting no Goliath, who would win to the top by force of superiority, in ways not altogether fair, at the cost of growing secretiveness of mind and bitterness of soul. For Newton was a most complicated personality, not at all innocent really, his disillusionment excessive, his dismay extreme, when confronted with what was only human reality and not unjust treatment. "Newton was a nice man to deal with," wrote John Locke (meaning touchy), "and a little too apt to raise in himself suspicions where there is no ground." And John Flamsteed, the Astronomer Royal, with whom he broke, found him "insidious, ambitious, and excessively covetous of praise, and impatient of contradiction . . . a good man at the bottom; but, through his natural temper, suspicious."

The incomprehension which greeted his theory of colors was the more frustrating that it raised objections among inferior minds whose applause he craved and who truly could not understand what he meant, so deep and novel was his insight, so new and different his conception of science. Newton undertook to answer each of the criticisms communicated to the Royal Society—from Paris by Adrien Auzout and Father Ignatius Pardies, from Liége by Franciscus Linus, an English Jesuit in exile, from Paris again by no less a person than Huygens, from London and the heart of the Royal Society itself by Robert Hooke, its great experimentalist and author of *Micrographia*, a Baconian cornucopia of observations and experiments. Newton succeeded only with Pardies, who thereby earned the distinction of having understood an argument and changed his mind. As to the rest, the confusion reached deeper than the evidence, right into the question of what science does. For they insisted on seeing colors as modifications of light—the "acts and sufferings of light" Goethe would call colors a century later in a last romantic fling against Newton's "anatomy of light"—and for them optics was not just the science of its behavior, but also the explanation of its nature.

Patiently (at first) Newton tried to explain himself. And the effort was worthwhile. Besides converting the amiable Pardies, he made explicit that limitation of science and that conception of scientific method on which his physics always acted, even when he himself did not. It was in defining what he was saying about light that Newton first laid down the standpoint "Hypotheses non fingo," which seems an almost Baconian repudiation of theory and has so puzzled critics, coming as the phrase does at the end of the *Principia*, that most elegant and comprehensive work of theoretical science in all literature. In one of his replies he addressed himself to the supposition of his critics "in which light is supposed to be a power, action, quality, or certain substance emitted every way from luminous bodies."

> In answer to this, it is to be observed that the doctrine which I explained concerning refraction and colours, consists only in certain properties of light, without regarding any hypotheses, by which those properties might be explained. For the best and safest method of philosophizing seems to be, first to inquire diligently into the properties of things, and establishing those properties by experiments and then to proceed more slowly to hypotheses for the explanation of them. For hypotheses should be subservient only in explaining the properties of things, but not assumed in determining them; unless so far as they may furnish experiments. For if the possibility of hypotheses is to be the test of the truth and reality of things, I see not how certainty can be obtained in any science; since numerous hypotheses may be devised, which shall seem to overcome new difficulties. Hence it has been here thought necessary to lay aside all hypotheses, as foreign to the purpose, that the force of the objection should be abstractedly considered, and receive a more full and general answer.

Hooke's objections are the most interesting for the grammar of assent in science. For there has frequently been a stage at which the precepts of science itself—economy, for example, mechanism, realism—have been introduced so literally and at so low a level of abstraction that they have blocked sophistication instead of advancing theory. "*Whiteness* and *blackness*," wrote Hooke, "are nothing but the plenty or scarcity of the undisturbed rays of light" and those "two colours (than the which there are not more compounded in nature) are nothing but the effects of a compounded pulse." He likened Newton's theory that colors "should be originally in the simple rays of light" to saying that the sounds which issue from a musical instrument were originally in the bellows of the organ or the strings of the fiddle. And he criticized the "indefinite variety of primary or original colours" as an inadmissible multiplication of entities. There is, indeed, no better way to summarize the issue than to juxtapose their two definitions of light. Hooke's view was that

> Light is nothing but a simple and uniform motion, or pulse of a homogeneous and adopted (that is a transparent) medium, propagated from the luminous body in orbem, to all imaginable distances in a moment of time, and that that motion is first begun by some other kind of motion in the luminous body; such as by the dissolution of sulphureous bodies by the air, or by the working of the air, or the several component parts one upon another, in rotten wood, or putrifying filth, or by an external stroke, as in diamond, sugar, the seawater, or two flints or crystal rubbed together; and that this motion is propagated through all bodies susceptible thereof, but is blended or mixt with other adventitious motions, generated by the obliquity of the stroke upon a refracting body . . . I believe MR. NEWTON will think it no difficult matter, by my hypothesis, to solve all the phaenomena, not only of the prism, tinged liquors, and solid bodies, but of the colours of plated bodies, which seem to have the greatest difficulty.

But Newton found this meaningless. For his definition of light was less capacious: "By light therefore I understand, any being or power of a being, (whether a substance or any power, action, or quality of it) which proceeding directly from a lucid body, is apt to excite vision."

Four years of controversy left Newton bleakly confronting that failure in communication to which his successors have become habituated in the progress of science and specialization. He was not the man to resign himself to this predicament. But his reaction was ambivalent. On the one hand, he affected renunciation: "I was so persecuted with discussions arising from the publication of my theory of light," he wrote to Leibniz, "that I blamed my own imprudence for parting with so substantial a blessing as my quiet to run after a shadow." And to Oldenburg: "I see I have made myself a slave to philosophy, but if I get free of Mr. Linus's business, I will resolutely bid adieu to it eternally, excepting what I do for my private satisfaction, or leave to come out after me; for I see a man must either resolve to put out nothing new, or to become a slave to defend it." And he did refuse to make a treatise of his optical researches until after Hooke's death. So it happened that, though the work was done first, the *Opticks* itself, Newton's most approachable and appealing work, was published last, in 1704.

On the other hand, provoked beyond endurance, he threw off the mask of cautious phenomenalism, violated his own privacy, and, from the inner springs of his being, revealed quite another scientific personality, not the correct empiricist whose theories must just embrace the evidence, not Newton the scientist who may be assimilated to positivism, but Newton the man and the discoverer, the rhapsodist who studied the mystical works of Jakob Boehme even as he studied the mysterious works of God, that secret Newton who was the most daringly speculative thinker about nature known to history, and the most fertile framer of hypotheses. This Newton *must* communicate, even if he has to give his critics what they want:

> And therefore, because I have observed the heads of some great virtuosos to run much upon hypotheses, as if my discourses wanted an hypothesis to explain them by, and found, that some, when I could not make them take my meaning, when I spake of the nature of light and colours abstractedly, have readily apprehended it, when I illustrated my discourse by an hypothesis; for this reason I have here thought fit to send you a description of the circumstances of this hypothesis as much tending to the illustration of the papers I herewith send you.

Yet he haughtily makes it clear that he is talking down to them:

> I shall not assume either this or any other hypothesis, not thinking it necessary to concern myself, whether the properties of light, discovered by me, be explained by this, or Mr. HOOKE'S, or any other hypothesis. . . . This I thought fit to express, that no man may confound this with my other discourses, or measure the certainty of the one by the other, or think me obliged to answer objections against this script: for I desire to decline being involved in such troublesome and insignificant disputes.

Thus Newton opened his Second Paper on Light and Colours in 1675. The change in tone is distressing. The difference in content is striking. The paper consists of two parts. In the second, Newton shifts his ground—not for the only time—to obviate, and denigrate, certain of Hooke's experimental objections. In the opening part, he proceeds to the hypothesis: "First, it is to be supposed therein, that there is an aethereal medium much of the same constitution with air, but far rarer, subtler, and more strongly elastic."

With this, Newton introduces the aether, not precisely, nor into the structure of physics, but ambiguously, and as a condition for the intelligibility of physics. Having failed with demonstration, he appeals to imagination and gives his own fancy full license:

> Perhaps the whole frame of nature may be nothing but various contextures of some certain aethereal spirits, or vapours, condensed as it were by precipitation, much after the manner, that vapours are condensed into water, or exhalations into grosser substances, though not so easily condensible; and after condensation wrought into various forms; at first by the immediate hand of the Creator; and ever since by the power of nature; which, by virtue of the command, increase and multiply, became a complete imitator of the copies set her by the protoplast. Thus perhaps may all things be originated from aether.

Perhaps it is this subtle aether which kicks the motes about in electrostatic situations. "It is to be supposed that the aether is a vibrating medium like air, only the vibrations far more swift and minute." Like water rising in capillary tubes, the aether permeates the pores of solid bodies, "yet it stands at a greater degree of rarity in those pores, than in the free aethereal spaces." It may be the aether—and to this Newton devotes some pages—which will resolve "that puzzling problem" how soul acts on body: "Thus may therefore the soul, by determining this aethereal animal spirit or wind into this or that nerve, perhaps with as much ease as air is moved in open spaces, cause all the motions we see in animals."

Now, this must not be read as animism if one wishes to understand Newton's thought. Aether is not the same thing as soul. It is not some world-spirit creating unity by blending everything into everything. It is not activity taking ontological precedence over matter and motion. It does not permeate matter to unite it with space. On the contrary, the universal impermeability of matter is a cornerstone of Newtonian doctrine, and the aether permeates only the pores between the particles. "That Nature may be lasting," Newton will say much later (even more clearly than Boyle), "the Changes of corporeal Things are to be placed in the various Separations and new Associations and Motions of these permanent Particles." Aether, in other words, is not the Stoic *pneuma*, and not an ineffable refuge of consciousness. It is a subtle fluid, itself particulate in structure. For the fancy Newton is indulging is a scientific fancy, an enrichment but no escape from science. In the same way, to come back from soul to optics, "light is neither aether, nor its vibrating motion, but something of a different kind propagated from lucid bodies." Aether is the medium for light:

It is to be supposed, that light and aether mutually act upon one another, aether in refracting light, and light in warming aether; and that the densest aether acts most strongly. When a ray therefore moves through aether of uneven density, I suppose it most pressed, urged, or acted upon by the medium on that side towards the denser aether, and receives a continual impulse or ply from that side to recede towards the rarer, and so is accelerated, if it move that way, or retarded, if the contrary.

From the aether itself, Newton moved on in the second part of this, his last reply to Hooke on optics, to its role in the explanation of colors. But now he was concerned less with the prismatic spectrum than with the rings which appear shiftingly in very thin translucent bodies like sheets of mica or soap bubbles. These interference phenomena (as they have since been called) had been described roughly by Hooke in his *Micrographia*. He had objected that neither they nor other instances of diffraction were accounted for in Newton's theory of colors. So clearly was he right that Newton extended the knowledge of the phenomena by a very precise and beautiful series of experiments with a "thin-plate" of air between two optical surfaces, one ground slightly convex, so that by turning one upon the other the rings might be varied and observed from different angles.

The phenomena, Newton saw, argue an element of periodicity in light. In the case of monochromatic light, the rings were alternately light and dark: "If light be incident on a thin skin or plate of any transparent body, the waves, excited by its passage through the first superficies, overtaking it one after another, till it arrive at the second superficies, will cause it to be there reflected or refracted accordingly as the condensed or expanded part of the wave overtakes it there." But when the rings are colored, it is because in compound light the rays "which exhibit red and yellow" excite "larger pulses in the aether than those, which make blue and violet." By measuring the separation of those rings Newton computed the thickness of the air film corresponding to each ring and color. This was a pesky task, for the boundaries were shadings. Over a century later Thomas Young, employing his new principle of transverse interference, used Newton's measurements to compute the wavelengths of the visible spectrum. His results agreed closely with the figures now accepted.

A cluster of conflicting interpretations rose up in later years to obscure Newton's reasoning. Eighteenth-century atomism committed itself to the corpuscular model of light, and nineteenth-century physics to the wave theory. From both points of view Newton seems inconsistent as between prismatic and thin-plate colors. In fact, however, this is a false problem. Newton did not himself adopt a crude optical atomism—it was fathered on him. It is true that his phraseology does sometimes give occasion for uncritical successors to represent the stream of particles as the Newtonian theory of light. But this was only a manner of speaking. The heart of Newton's theory is the composite nature of light rather than its corpuscular texture. Its parts are rays, not corpuscles. It is the *rays* which differ from each other "like as the sands on the shore." What led him to his theory was its structural congruence with philosophical atomism, rather than a literal analogy between the parts of light and the parts of matter. It was, therefore, no inconsistency, but an enlargement of his views, adopted to meet different facts from those encountered in his first paper, when he introduced vibrations as the physical basis of interference phenomena.

The argument has also been represented as a concession to Hooke's modification theory of color. That was Hooke's view. "After reading this discourse," runs the closing note in the minutes of the Royal Society for 16 December 1675, "Mr. HOOKE said, that the main of it was contained in his *Micrographia*, which Mr. NEWTON had only carried farther in some particulars." Newton's reply was categorical, and delivered only five days later: "I have nothing common with him, but the supposition, that aether is a susceptible medium of vibrations, of which supposition I make a very different use; he supposing it a light itself, which I suppose it is not." And properly appreciated, this distinction should clarify all the ambiguity. For it is the fundamental distinction, that which brings some category of phenomena within the scope of objective science—the same which Galileo established between motion and the moving body, the same which Boyle tried to introduce between substance and change, the same (to go back to the beginnings of objectivity) which Democritus established between atoms and the void. In Newton's work the advancing front of objectivity moves through optics. As always, numbers spelled success. Hooke had indeed observed

> plated bodies exhibiting colours, a phaenomenon, for the notice of which I thank him. But he left me to find out and make such experiments about it, as might inform me of the manner of the production of those colours, to ground an hypothesis on; he having given no further insight to it than this, that the colour depended on some certain thickness of the plate; though what that thickness was at every colour, he confesses in his Micrography, he had attempted in vain to learn; and therefore, seeing I was left to measure it myself, I suppose he will allow me to make use of what I took the pains to find out.

And all the mistake has been to read Newton's optical atomism literally instead of strategically. In a sense, Newton is saying that in some situations it is helpful to consider light as particles, in others as waves, and always as a composite of colored rays each of specific properties—but only in a sense, for before too much prescience is attached to this wisdom, it should be remembered that Newton's waves are longitudinal pulses, not transverse undulations.

✧

AFTER 1676 Newton gave over contending for his theory of colors and withdrew into his alternate posture of renunciation. "I had for some years past," he wrote in 1679, "been endeavouring to bend myself from philosophy to other studies in so much that I have long grutched the time spent in that study unless it be perhaps at idle hours sometimes for a diversion." It is not known in detail how he spent those years. On theology and biblical antiquities certainly, on mathematics probably, on chemistry and on perfecting his optics perhaps, for it is in character that he should have nursed his disenchantment in public and continued his work in private. In 1679 he was recalled to science, but to dynamics this time, by a further letter from Hooke, now become Secretary of the Royal Society. Hooke approached him on two levels. Privately, the letter was an olive branch. Officially, it was the new secretary bespeaking the renewed collaboration of the most potent of his younger colleagues, sulking in his tent.

Newton answered, correctly enough in form, but not very frankly, not at all cordially, affecting ignorance of an "hypothesis of springynesse" (Hooke's law of elasticity) on which Hooke had invited his opinion. So as to disguise without taking the edge off his snub, he threw in as a crumb "a fancy of my own," the solution of a curious problem he had toyed with in one of those idle hours. It concerned the trajectory of a body falling freely from a high tower, supposing the earth permeable and considering only the diurnal rotation. This was in fact a famous puzzle suggested by the Copernican theory, the same problem which Galileo had so curiously and erroneously answered with a semi-circle to the center of the earth. Since then it had been much discussed in obscure and learned places. And having brought it up himself, as if to flex a mental muscle in Hooke's face, Newton gave an answer as wrong as Galileo's. The trajectory, he casually said and drew it, will be a spiral to the center of the earth.

Now, Hooke did not know the right answer. The forces are in fact complex: the force of gravity increases by the inverse square relationship as far as the surface of the earth and thereafter as the first power of the distance. Hooke, along with many others, surmised the former (though he was too feeble a mathematician to handle gravity other than as constant) but was ignorant—as Newton then was—of the latter fact. He did have the happy thought of eliminating Coriolis forces by putting his tower on the equator. But Hooke did not need to solve the problem correctly to perceive that the initial tangential component of motion will not only, as Newton pointed out with an air of correcting vulgar errors, carry the body east of the foot of the tower, but by the same reasoning will insure that one point which the body can never traverse, either on a spiral or on any other path, is the center of the earth. Hooke was not the man to resist this opportunity. He had invited Newton to a private correspondence. He communicated Newton's reply to the Royal Society, and corrected his error publicly.

It would be tedious to follow the ensuing correspondence: the outward forms of courtesy, the philosophical tributes to truth as the goal, the underlying venom, the angry jottings in the margin. Newton "grutched" admitting error far more than the time spent on philosophy. He never did solve the problem. But he left it as the most important unsolved problem in the history of science. For it drew his mind back to dynamics and gravity, back to where he had left those questions thirteen years before. And in the course of these geometrical investigations, he solved the force law of planetary motion: "I found the Proposition that by a centrifugal force reciprocally as the square of the distance a Planet must revolve in an Ellipsis about the center of the force placed in the lower umbilicus of the Ellipsis and with a radius drawn to that center describe areas proportional to the times." He would prove the point mass theorem only after 1685. But he had proved the law of gravity on the celestial scale, not just approximately for circular orbits as in 1666, but as a rigorous geometric deduction combining Kepler's laws with Huygens' law of centrifugal force. And he told no one, "but threw the calculations by, being upon other studies."

It is one of the ironies attending the genesis of Newton's *Principia* that no one knew beforehand of his work on celestial mechanics. In inviting Newton's correspondence, Hooke may even have thought that he was taking his rival onto his own ground. For the problem of gravity was constantly under discussion. Hooke had certainly surmised that a gravitating force of attraction was involved in the celestial motions, and that it varied in power inversely as the square of the distance. So, too, had Christopher Wren, then one of the most active of the virtuosi, and the young astronomer, Edmund Halley. But none of them was mathematician enough to deduce the planetary motions from a force law.

Far more than Boyle, Hooke was the complete Baconian. The only plausible explanation of his later conduct is that he truly did not understand the necessity for mathematical demonstration. He relied uniquely upon experiment to sort out the good from the bad ideas that crowded out of his fertile imagination. He seems to have been prepared to build even celestial mechanics out of experiments on falling bodies like those improvised to test out

Newton's spiral. Nor could he see that the rigorous geometrical demonstrations of the *Principia* added anything to his own idea. They gave the same result. Once again, thought Hooke on seeing the manuscript, Newton had wrapped his intellectual property in figures and stolen it away.

Halley was more sophisticated. He was also an attractive and sympathetic young man. In August 1684 he went up from London to consult Newton. An account of this visit by John Conduitt, who later married Newton's niece, is generally accepted.

> Without mentioning either his own speculations, or those of Hooke and Wren, he at once indicated the object of his visit by asking Newton what would be the curve described by the planets on the supposition that gravity diminished as the square of the distance. Newton immediately answered, *an Ellipse*. Struck with joy and amazement, Halley asked him how he knew it? Why, replied he, I have calculated it; and being asked for the calculation, he could not find it, but promised to send it to him.

While others were looking for the law of gravity, Newton had lost it. And yielding to Halley's urging, Newton sat down to rework his calculations and to relate them to certain propositions *On Motion* (actually Newton's laws) on which he was lecturing that term. He had at first no notion of the magnitude of what he was beginning. But as he warmed to the task, the materials which he had been turning over in his mind in his twenty-five years at Cambridge moved into place in an array as orderly and planned as some perfect dance of figures. Besides proving Halley's theorem for him, he wrote the *Mathematical Principles of Natural Philosophy*. The *Principia*, it is always called, as if there were no other principles. And in a sense there are none. For that book contains all that is classical in classical physics. There is no work in science with which it may be compared.

"I wrote it," said Newton, "in seventeen or eighteen months." He employed an amanuensis who has left an account of his working habits.

> I never knew him to take any recreation or pasttime either in riding out to take the air, walking, bowling, or any other exercise whatever, thinking all hours lost that was not spent in his studies, to which he kept so close that he seldom left his chamber except at term time, when he read in the schools as being Lucasianus Professor. . . . He very rarely went to dine in the hall, except on some public days, and then if he has not been minded, would go very carelessly, with shoes down at heels, stockings untied, surplice on, and his head scarcely combed. At some seldom times when he designed to dine in the hall, [he] would turn to the left hand and go out into the street, when making a stop when he found his mistake, would hastily turn back, and then sometimes instead of going into the hall, would return to his chamber again.

Mostly Newton would have meals sent to his rooms and forget them. His secretary would ask whether he had eaten. "Have I?" Newton would reply.

The Royal Society accepted the dedication, undertook to print the work, and like a true learned organization found itself without funds. The expense, therefore, as well as the editing came upon Halley. He was not a rich man, but he bore both burdens cheerfully, with devotion and tact. He had the disagreeable task of informing Newton that upon receipt of the manuscript Hooke had said of the inverse square law, "you had the notion from him," and demanded acknowledgment in a preface. Upon this Newton threatened to suppress the third book, the climax of the argument, which applied the laws of motion to the system of the world. He was dissuaded, as no doubt he meant to be, but one can understand how his feeling for Hooke turned from irritable dislike to scornful hatred:

Now is not this very fine? Mathematicians, that find out, settle, and do all the business, must content themselves with being nothing but dry calculators and drudges; and another that does nothing but pretend and grasp at all things, must carry away all the invention, as well of those that were to follow him, as of those that went before. Much after the same manner were his letters writ to me, telling me that gravity, in descent from hence to the centre of the earth, was reciprocally in a duplicate ratio of the altitude, that the figure described by projectiles in this region would be an ellipsis, and that all the motions of the heavens were thus to be accounted for; and this he did in such a way, as if he had found out all, and knew it most certainly. And, upon this information, I must now acknowledge, in print, I had all from him, and so did nothing myself but drudge in calculating, demonstrating, and writing, upon the inventions of this great man. And yet, after all, the first of those three things he told me of is false, and very unphilosophical; the second is as false; and the third was more than he knew, or could affirm me ignorant of by any thing that past between us in our letters.

The provocation was great, as was the strain under which it was given. A few years after completing the *Principia* Newton suffered a nervous collapse. He wrote very strange letters. One of them accused Locke of trying to embroil him with women—Newton, who was as oblivious to women as if they were occult qualities. Alarmed, his friends had arranged a move to London, to bring him more into company. He gave up solitude in Cambridge with no regrets, became after a few years Master of the Mint, then President of the Royal Society which once he had held at such a haughty distance. Knighted in 1705 he lived out his years until 1727, the incarnation of science in the eyes of his countrymen, a legend in his own lifetime.

But he did very little more science.

❖

THE *Principia* is an intractable book. It is doubtful whether any work of comparable influence can ever have been read by so few persons. The scientific community itself required forty years of discussion, rising at times to controversy, to grasp the implications of Newton's achievement and to assume the stance of classical physics. Thereafter the *Principia* scarcely needed to be read. It was enough that it existed. Up to 1900, mechanics, now including celestial mechanics, was a formal development of Newton's laws by more sophisticated and rigorous mathematical techniques. Though of first importance to the technical history of science, classical mechanics had made its contribution to the intellectual history of science in Newton, its founder. Even the other domains of physics, electro-magnetism, heat, optics, were conceived with varying success as extensions of Newtonian principles and practice to new ranges of phenomena.

Indeed, no sooner was this development under way than the *Principia* became, if not impossible, at least impracticable to read. For it is expressed in an archaic formalism, not in the new analytical mathematics of the seventeenth century, but in the synthetic geometry of the Greeks. In his mathematical taste, Newton, like Pascal and Galileo, was a purist. He must first have satisfied himself about crucial theorems by his own "fluxions," or calculus. But he demonstrated them as theorems in classical geometry.

The ancients, wrote Newton in the preface, had distinguished between geometry and mechanics, the one rational and abstract, the other having to do with manual arts. As theory, geometry deals with magnitude. As practice, "the manual arts are chiefly conversant in the moving bodies," and mechanics, therefore, is commonly referred to the motion of things. He proposes to unit the two, "and therefore we offer this work as the mathematical principles of natural philosophy. For all the difficulty of philosophy seems to consist in this, from the phenomena of motions to investigate the forces of nature, and then from these forces to demonstrate the other phenomena."

Next Newton defined his terms. They are the basic quantities of classical physics, made explicit for the first time—mass, momentum, and force, the latter from several points of view with special attention to centrally directed forces. His language alone establishes that physics is fundamentally an affair of metrics. Thus for mass: "The quantity of matter is the measure of the same, arising from its density and bulk conjunctly." And of momentum: "The quantity of motion is the measure of the same, arising from the velocity and quantity of matter conjunctly." The definition of force closes with an important qualification:

> I likewise call attractions and impulses, in the same sense, accelerative and motive; and use the words attraction, impulse, or propensity of any sort towards a centre, promiscuously, and indifferently, one for another; considering those forces not physically, but mathematically: wherefore the reader is not to imagine that by those words I anywhere take upon me to define the kind, or the manner of any action, the causes or the physical reason thereof.

An important scholium to the last definition distinguished between absolute and relative time, absolute and relative space. This, of course, was the metaphysical chink into which criticism would bore as it had done into the Aristotelian doctrine of motion. But rather than anticipate, let us leave it for this chapter in Newton's own words at the end of his definitions:

> I do not define time, space, place, and motion, as being well known to all. Only I must observe, that the common people conceive those quantities under no other notions but from the relation they bear to sensible objects. And thence arise certain prejudices, for the removing of which

it will be convenient to distinguish them into absolute and relative, true and apparent, mathematical and common.

I. Absolute, true, and mathematical time, of itself and from its own nature, flows equably without relation to anything external, and by another name is called duration; relative, apparent, and common time, is some sensible and external (whether accurate or unequable) measure of duration by the means of motion.

II. Absolute space, in its own nature, without relation to anything external, remains always similar and immovable. Relative space is some movable dimension or measure of the absolute spaces.

Finally, he completes his premises by stating the "Axioms, or Laws of Motion": inertia, the force law, the equivalence of action and reaction.

The *Principia* consists of three books. Book I develops the motion of bodies in unresisting mediums. It is a set of geometric theorems, on the method of limits and exhaustions, on problems of the center of force, on motion in conic sections, on the determination of orbits, on the attraction exerted by spherical bodies, on the motions of mutually attracting bodies, and on other topics. Book II is on the motion of bodies in resisting mediums. Much of it has to do with hydrodynamics, and it might seem a digression. Neither is the discussion always correct. But it was included because, like Galileo before him, only Newton's treatment was austere and mathematical. His purpose was philosophical, not to say polemical. He proposed to refute Cartesianism with its bodies swirling through spatial fluids and show the vortex system to be untenable on strictly mechanical grounds.

Throughout Book II, however, Newton left this an implication to be drawn. So far the structure of his book (like the structure of his space) is Euclidean, a set of mathematical deductions following from a few fundamental definitions and three axioms. But with Book III it becomes rather Archimedean, and the argument is applied to the physical information supplied by astronomy. For he means to compel agreement about universal cosmology, not by metaphysical reasoning, but to compel it with all the force of geometric demonstration. Applying the laws of motion to the solar system, he showed that they contain Kepler's orbits as a celestial consequence. In eternal unpropelled inertial motion, the moon and planets are constrained in their orbits by the universal force of attraction which every body—every particle—in the universe exerts over every other in an amount proportional to the product of the masses divided by the square of the distances. Weight is simply gravity acting on mass. And Newton included a vast array of calculations on fine points of the motion of the moon and tides as illustrations of gravity at work. Like Galileo, he turned to the tides for earthly evidence of his cosmological theory. But he had the principle that Galileo lacked, the answer to the more general, indeed the fundamental, question of what holds the world together? What will unify our science in an infinite universe?

The answer was the law of gravity.

Such was the book which formed the picture of the world in which everyone now alive was brought up. For it is safe to say that relativity and quantum physics have not yet been taken for granted as are Newton's notions of time, space, place, motion, force, and mass. It is easy to summarize the *Principia*. It is less easy to see how it affects our consciousness, though to have been brought up in the Newtonian world certainly does shape that consciousness, as it does to have been brought up an American rather than a Frenchman, a Christian rather than a Moslem. It is an element of culture, and to exist in a culture with no notion whence it came is to invite the anthropologist's inquiry rather than to live as an educated man, aware and in that measure free.

❖

"I AM ALWAYS READING ABOUT THE NEWTONIAN SYNTHESIS," an English professor once said irritably. "What did Newton synthesize?" It is a fair question. On the most immediate level, theory met experiment on equal terms for the first time in Newton. In practise as in principle, Newton achieved the correct relationship between physics as the science of metrics and mathematics as the language of quantity. The problem had bedevilled science ever since Plato and Aristotle had separated the two in opposing but equally defeatist ontologies. Galileo, it is true, had had it right, but in insufficient generality, and Descartes had confused the issue once again. Newton, therefore, had to redistinguish mathematics from physics, and with it space from matter. Thus he was able to unite physics and astronomy in a single science of matter in motion. Finally, by flinging gravity across the void, he reconciled the continuity of space with the discontinuity of matter. This was his resolution of the last of the great Greek philosophical problems which Europe clothed in science, whether the world is a continuum or a concourse of atoms? It is both. In force and motion it is one, in matter the other. And that unites the Platonic-Archimedean tradition with atomism.

People accustomed to think in these separate channels could not easily lose themselves in the great stream of science. But more than habit blocked assent. Newton's science did not answer all the traditional questions. It did not even ask them, and one is tempted to attribute its ready acceptance in England to national pride rather than to superior culture. For to the most refined and subtlest minds on the continent, it seemed that Newton committed two mortal sins in metaphysics. First, the void introduced the existence of the nothing. Second, gravity supposed action at a distance, bodies affecting each other through a mystery rather than a medium. Indeed, the specific complaint which united all Newton's critics was that gravity as attraction was an occult force no better than it should be, a reversion to the innate tendencies which Aristotle put in bodies. It could hardly be expected that Newton should have been understood at once. Science would have to live with these difficulties for a time, after which it would forget them in its own success rather than resolve them—both those which were trivial and those which were profound.

The first objection was only a misunderstanding. The void as Newton used it was not the metaphysical nothing. It was the complement of the aether, that which motion occurs *in*, translational motion in the void and vibrational motion in the aether. The void was introduced for the same reason, not as a positive physical hypothesis, but as a condition for the possibility of physics. The second point, the "cause" of gravity, is more interesting. For it turned on the problem of what it is that science explains. In the lesser person of Hooke, on the lesser issues of optics, Baconianism had already failed to understand the import of Newtonian science. Now it was the turn of Leibniz and the Cartesian school to miss his meaning on the universal plane of gravity.

The web of metaphysical resistance to Newton was complex. On theology he had to survive a cross-fire. It is well known that Newton casually allowed God a hand in the solar system to repair certain irregularities that he thought cumulative. Among people who know little else about Newton, this is, indeed, altogether too well-known, considering what a trivial point it was, and how irrelevant to the structure of his physics. It is more interesting that Newton was a profoundly religious man. Like many later rationalists, he could not credit the Trinity. He was a Unitarian before this position had become respectable. He did certainly believe in the free creation of the world by God and its government under Providence. His was a personal belief, not a principle of physics, any more than was the occasional repair of the solar system. But he was criticized for holding these views (particularly the latter) by the Cartesians, who regarded any finalism as childish. And he was criticized for failing to make providential destiny part of physics by Leibniz, who had united his own system of the world, not by a physical principle like Cartesian extension or Newtonian gravity, but by the metaphysical principle of pre-established harmony. And it was Leibniz who turned the odium traditionally incurred by atomism against Newton, and accused his science of a tendency to lead down the path already trodden by Hobbes to a self-sufficient materialism destructive of natural religion.

Newton's critics, in short, wanted more out of science than he found there. In the Cartesian view, for all its hostility to scholasticism, science moves through nature from definition to rationale; in that of Leibniz, it moves rather from principles to values; and in that of Newton, from descriptions and measurements to abstract generalizations. Strictly speaking, therefore, Newtonian science could never get outside itself, and might be said to be a tautology, or at least to accomplish nothing of human interest or value. The trouble was not in the evidence. No one complained of the mathematics. But taken as an explanation of the universe, the system failed—or rather it was no explanation at all, since no cause could be assigned and no mechanism imagined for its central principle, the principle of attraction. For the concrete, working, mechanical picture of the Cartesian universe, it substituted a set of geometrical theorems.

There is an irony in all this. Countless intellectual historians have followed Leibniz in describing Newton's theory as responsible for the picture of a soulless, deterministic world-machine, that same theory which at the time was rejected by men as discriminating as Huygens and Fontenelle for being overly abstract, insufficiently mechanistic, and subservient to natural theology. Indeed Newton has never been able to give critics what they wanted, a system which saw nature steadily and saw it whole, which accounted at once for the behavior and the cause of phenomena, the "how" and the "why" of nature. He did not know the cause of gravity. Gravity in Newton was a mathematical, not a mechanical force. Nor did he, in fact, believe in action at a distance, or gravity as an innate tendency: "You sometimes speak of gravity as essential and inherent to Matter," he wrote to Bentley. "Pray do not ascribe that Notion to me; for the Cause of Gravity is what I do not pretend to know, and therefore would take more Time to consider of it." But ignorance of the cause is not to deny the effect. "To us it is enough"—so he says in the General Scholium at the end of the *Principia*—"that gravity does really exist, and acts according to the laws which we have explained, and abundantly serves to account for all the motions of the celestial bodies, and of our seas."

To the Cartesians, however, it was not enough.

Growing old, but never mellow, Newton responded to incomprehension in the pattern of his youthful optics. He wrote this "General Scholium" for the second edition (1713) of the *Principia*. The penultimate paragraph works up to the austere rebuke: "But hitherto I have not been able to discover the cause of those properties of gravity from phenomena, and. [in the translation newly established by Koyré] I feign no hypotheses. For whatever is not deduced from the phenomena is to be called an hypothesis; and hypotheses, whether metaphysical or physical, whether of occult qualities or mechanical, have no place in experimental philosophy." And then, to make interpretation as difficult as science, the next and last paragraph begins:

And now we might add something concerning a certain most subtle spirit which pervades and lies hid in all gross bodies; by the force and action of which spirit the particles of bodies attract one another at near distances, and cohere, if contiguous; and electric bodies operate to greater distances, as well repelling as attracting the neighbouring corpuscles; and light is emitted, reflected, refracted, inflected, and heats bodies; and all sensation is excited, and the members of animal bodies move at the command of the will, namely, by the vibrations of this spirit, mutually propagated along the solid filaments of the nerves, from the outward organs of sense to the brain, and from the brain into the muscles.

Again, it is as if there were two Newtons speaking in turn. Once again, the frustrations of the empiricist release the affirmations of the visionary. It must not be supposed that Newton's life in London was only what it seemed, the ceremonial existence, all passion spent, of the elder statesman of science. Behind the scenes, he looked to his polemical interests with an undimmed eye. Hooke died in 1703. In 1704 Newton published the *Opticks*, writing in English now, and very well, as if he meant to be read. The early experiments had been refined and extended. And the book closed with the famous "Queries," that moving and beautiful series of rhetorical speculations about light, heat, and electricity, the aether, the atoms, and God, which Newton left as his legacy of unsolved problems, and to which he added in later editions. (It is seldom noted that in 1672, his very first attempt to explain himself after the mixed reception of his paper on prisms had taken the form of "Queres" addressed to the Royal Society.)

Except in the *Opticks*, Newton chose to retire behind the advocacy of disciples, whom he probably coached. Roger Cotes wrote the preface for the second edition of the *Principia*. Samuel Clarke published a philosophical debate with Leibniz. This discussion developed out of an ignoble squabble over the invention of the calculus, in which the Royal Society acted as umpire in no very just or impartial spirit. Nor can it be said, any more than of the earlier polemics, that all these unworthy quarrels served no higher purpose. They brought the issues before the Republic of Letters as perusal of the theorems of the *Principia* would never have done, if only because those theorems so coldly discourage perusal.

Newton himself spoke out again in the General Scholium. It closes with the renewal of that aethereal hint just given. But what had wounded Newton most deeply was the attribution of infidelity. And his views on divinity do, and should, carry more interest than his hypothesis— for so he had called it himself when first he brought it in— of the aether. God is neither hypothesis nor object of science. He is certainty:

> He endures forever, and is everywhere present; and, by existing always and everywhere, he constitutes duration and space. Since every particle of space is always, and every indivisible moment of duration is everywhere, certainly the Maker and Lord of all things cannot be never and nowhere. . . . Whence also he is all similar, all eye, all ear, all brain, all arm, all power to perceive, to understand, and to act; but in a manner not at all human, in a manner not at all corporeal, in a manner utterly unknown to us. . . . We have ideas of his attributes, but what the real substance of anything is we know not. In bodies, we see only their figures and colours, we hear only the sounds, we touch only their outward surfaces, we smell only the smells, and taste the savours; but their inward substances are not to be known either by our senses, or by any reflex act of our minds; much less, then, have we any idea of the substance of God. We know him only by his most wise and excellent contrivances of things, and final causes; we admire him for his perfections; but reverence and adore him on account of his dominion: for we adore him as his servants; and a god without dominion, providence, and final causes is nothing else but Fate and Nature. . . . And thus much concerning

God; to discourse of whom from the appearances of things, does certainly belong to Natural Philosophy.

To discourse, but not to prescribe, nor to presume. For all this speculation on the aether, all this reverence for God, these considerations are interesting for the inspiration of Newton's science, but irrelevant to its validity. Its validity is to be judged—did not Newton say so?—in relation not to Newton, but to nature. Indeed, one of the most elementary though disregarded of distinctions is that between the scientist and his science. Science is created by the scientist, but about nature, not about himself. Once it is created, it has the independence of any work of art. One sometimes reads of the arrogance of science. And Newton was subject to unseemly spells of haughtiness when crossed. But surely—to insist upon the distinction— his science is rather an expression of modesty. That limitation of *allowable* theories to the evidence was no positivist skepticism about truth in the world of things. Rather, it was modesty. Descartes was the one who presumed to prescribe what the world must be. Newton only said how it is, and how it works. And it is right, therefore, to let Newton the scientist, rather than Newton the controversialist, or Newton the theologian, have the last word. It comes from the closing sentence of the final definition at the start of the *Principia*: "But how we are to obtain the true motions from their causes, effects, and apparent differences, and *vice versa*, how from their motions either true or apparent, we may come to the knowledge of their causes and effects, shall be explained more at large in the following treatise. For to this end it was that I composed it."

CHAPTER V

SCIENCE AND
THE ENLIGHTENMENT

THERE IS A STORY that only once was Newton known to laugh, when someone asked him what use he saw in the *Elements of Euclid*? But how much more incongruous is the place he has come to occupy in the history of ideas as godfather to the eighteenth-century Enlightenment. For, when one thinks of it baldly, how can the *Principia* and the *Opticks* have inspired Jefferson's proposition that under the law of nature a broken contract authorized Americans to rebel against George III? Or the constitutional device of checks and balances? Or the atomic model of society, individualism in Europe replacing organic feudal corporatism? Or the instinct of Jean Jacques Rousseau that the state of nature is a state of virtue? Or the feeling of romantics that society should be based, therefore, on virtue, and of rationalists that it should be based (on the contrary) on talent? Or the campaign of sly sardonic japing which Parisian skeptics levied against the Christian religion? Or Jeremy Bentham's vocation as the Newton of legislation, his identification of utility, the greatest-happiness principle, as the social law of gravity, and his gentle determination to make the nations observe it? What, indeed, can Newton, devout and secret man, or his sublime, impersonal science have had to do with any of these humane and public preoccupations of the Enlightenment?

This is no simple question. It immediately raises the disparity between what science is and does, and what people say it is and use it for, between its content and its exploitation, whether by moral, social, or logical philosophy. The stereotype of rationalism conceiving the Enlightenment on the body of science may be quickly rehearsed. Armed with Newton's principles, heartened by his success, the *philosophes* would make over society and culture, in the image of nature, by the use of reason. What did Newton reveal in nature? Harmony, order, things that fit, a world well made. He found cosmic law as an objective fact. What, on the contrary, does the enlightened man observe when he compares society to nature as the Newton-given norm? Conflict, disorder, anachronistic institutions, priests and noblemen who foster ignorant superstition and exploit it for power. Philosophy, therefore, become the science of humanity, will find the principles of order in men and affairs, state these laws of human nature, and a world of reasonable beings will read and conform.

Now, there can be no doubt that the prestige of science derived from Newton's triumph. The force of the mind seemed equal to the power of nature in the act of understanding. But prestige is one thing. Genuine comprehension is quite another, and on this score the lines begin to blur and run together. The physical science of the eighteenth century could only be Newtonian, of course. But the scientific ideology of the Enlightenment was something less or more. Its authors read their Newton—if, indeed, they always did read him—in the light now of Bacon and again of Descartes. When it was a question of technology and progress, or of descriptive science, optimism was tinged with a certain sentimentality about practical artisans, experiments, and collecting of facts. What chateau lacked its cabinet of natural history?

But the confidence in Baconian classification as the route to knowledge found reinforcement in the Cartesian faith in reason. The most striking illustration is the rapid commitment to just that philosophic mechanism which Newton had expressly repudiated. Newtonian mechanics was universalized in thought, not because of any evidence that the world is nothing but a machine, but because of its rationality. Just so would the Enlightenment bring the Cartesian method from metaphysics to social philosophy (whence, by the way, Descartes had expressly excluded it). Doubt and criticism would purge away obscurity and error. Then reason would rebuild the world of humanity, arming herself with the prestige of science, which we know from Newton cannot err.

Thus the rationalists set about to condemn the structure and institutions of the old Europe as absurd and unjust because contrary to nature. It is not for the historian to complain of the sleight of hand which put this message in the mouth of science. He should only be clear that Newton's name, elevated now into a symbol, was conjured with in all senses of the verb. The leaders of the Enlightenment practised rational criticism. They adopted the utilitarian idea of progress. And they invested both with the high authority of Newton. Their success testifies to the force of their humanitarian commitment. A liberal is bound to applaud the results. He is bound to approve that movement of thought, radiating from critical France, which carried all the West from the religious and dynastic preoccupations of the seventeenth century, through the Enlightenment and the revolutionary struggle for popular sovereignty, into the democratic preoccupations of the nineteenth century and of our own. Nevertheless, one is equally bound, however liberal, to recognize that the authority which social science (for the thing antedates the name) drew from natural science was vitiated from the outset by the characteristic determination of social scientists to do good. For Bentham's relationship to the principle of utility is what Newton's would have been to the law of gravity, had Newton established that law by persuading the planets to obey the inverse square relationship in their own interest. That is to say, the noble eighteenth-century faith in natural law involved a fundamental confusion between the declarative and the normative senses of law, between "is" and "ought."

It is best to be explicit about this at the very beginning of a discussion of the eighteenth century when science, no longer simply drawing strength or problems from culture, began to shape the world. Science may be admirable. This book is written in the conviction that it is the distinctive achievement of our history, and that nothing less momentous than the preservation of our culture hangs on understanding its growth and bearing. But the influence of science is not simply comfortable. For neither in public nor in private life can science establish an ethic. It tells what we can do, never what we should. Its absolute incompetence in the realm of values is a necessary consequence of the objective posture. That necessity has never been more narrowly identified than by Henri Poincaré, in an essay of 1913 on "Morality and Science":

> It is not possible to have a scientific ethic, but it is no more possible to have an immoral science. And the reason is simple; it is, how shall I put it? for purely grammatical reasons.
>
> If the premises of a syllogism are both in the indicative, the conclusion will equally be in the indicative. In order for the conclusion to be put in the imperative, it would be necessary for at least one of the premises to be in the imperative. Now, the principles of science, the postulates of geometry, are and can only be in the indicative; experimental truths are also in this same mode, and at the founda-

tions of science there is not, cannot be, anything else. Moreover the most subtle dialectician can juggle with these principles as he wishes, combine them, pile them up one on the other; all that he can derive from them will be in the indicative. He will never obtain a proposition which says: do this, or do not do that; that is to say a proposition which confirms or contradicts ethics.

"Ye shall know the Truth and the Truth will make you free"—it is one of the great themes of the Enlightenment. But for most people emancipation was not enough. It never has been. They are not content to take science for what it is intellectually, a great creation, a description of how the physical world works, beautiful and admirable in itself, but empty of morals and lessons. They want more. They want reassurance about the existence of God from the design of nature, and about His loving care from His continuing to repair its imperfections. They want a license to rugged individualism in the theory of evolution. They want a crumb of comfort about free will in the unpredictability of the electron. They want studies of the mind considered as a digital computer. In short, they want science to give us a world we can fit, as Greek science did, and not just a world like any external object that we can first measure, and then destroy.

In moments of discouragement, therefore, one is tempted to think that the history of the influence of science in culture is bound to be the history of a misunderstanding, in which what changes is the way in which the import of science is misunderstood. But this would be too despairing a view, too priggish an approach, to be adopted by the historian. Whatever else it may be, science is an instrument. It is bound to be used in ways to be judged pragmatically, and not only by their fidelity to the intent of the maker. Science, after all, has no higher claim to innocence than theology or philosophy or literature. It does, however, confront criticism with an essential difference from the other elements of culture. They are about man or God, personality or affairs. Science is about nature. It is about things. And this means that it can scarcely enjoy direct and valid influences as do literary and philosophical schools. So it is that a Marxist regime can permit its physicists but not its writers to receive alien honors. Necessarily, the permeation of culture by science must be a problem in accommodation rather than a study in validity.

The problem is great. Since the eighteenth century, the necessity to work an accommodation for science is what has polarized culture into its two great moods of romanticism and rationalism. Resisting the alienation of nature, seeing Newtonian science only as an impoverishment, the romantic tradition flings out against objectivity, and either declares war against science and all its works, or else (and this continued to be possible right through the nineteenth century) proposes to substitute a subjective science, a science which will fit nature to man by returning from mechanism to organism as the metaphor for order. Nor, though ultimately self-defeating, was this sterile. Biological studies received a tremendous impulse from the romantic attempt to replace physics by biology as the ordering science.

The rationalist tradition, on the other hand, coupled by science with empiricism, proposed rather to accommodate science by fitting man to nature. The instrument was an objectified psychology, which in turn conceived the function of scientific explanation to be a kind of cosmic education of humanity in the order of nature. The most notable scientists of the Enlightenment—as distinct from men of letters and *philosophes*—participated in this development. Ultimately, it led to positivism, which in its extreme form assimilates all philosophy to the study of scientific method. But before that, its most notable feedback into science itself occurred in the chemical revolution, which styled chemistry into a peculiarly eighteenth-century science.

✧

AMONG the philosophic and literary Newtonians of the French Enlightenment, it is the first and foremost, Voltaire, who comes closest to surviving Poincaré's criticism. Voltaire explained Newtonian science to the educated public more successfully than any other writer, perhaps because he took more pains to understand it. Nor did Voltaire turn to Newton as to one who had subjected the world to a mechanistic determinism. On the contrary, he turned to Newton as to a liberator, and devoted to physics his best efforts of the 1730's, the years between his first reputation in *belles-lettres* and the philosophic crusade, between a deep personal humiliation at the hands of the princely Rohan, and his systematic, life-long vindication of human dignity and civilized intelligence looking out from experience at the world as it was.

"Droit au fait—Let facts prevail," was Voltaire's motto. He first became aware of Newton, along with other fruits of liberal thought and practice in England, during his retreat there between 1726 and 1729. He attended Newton's funeral in Westminster Abbey, and presented the first general sketch of his ideas to the French public in three of his *Philosophical Letters* of 1734. The balance is held even between Newton and Descartes, but the very existence of rival systems of science suggests the absurdity of dogma. One may choose, but how?

A Frenchman arriving in London finds philosophy changed as well as everything else. He left the world full; he finds it empty. In Paris the universe is composed of vortices of subtle matter; in London there is nothing of the sort. At home it is the pressure of the moon which causes the tides; in England it is the sea which gravitates toward the moon.

According to the Cartesians everything happens by virtue of an incomprehensible impulse; according to Newton it is by an attraction of which the cause may not be understood. . . . Here, then, are ferocious contradictions.

Only later, however, did Voltaire see all the way down the vista opened by Newton. From 1736 until 1741, he and his companion, the Marquise du Châtelet, "immortal Emilie," delivered themselves over to physics. Her chateau became a learned institute. A laboratory was installed. Researches were put in hand, into chemistry and into fire. Madame du Châtelet translated Newton into the language of Racine, and became an adherent of Leibniz, to Voltaire's annoyance. With her help he set himself to master Newton by study and correspondence in order to "reduce this giant to the measure of the nincompoops who are my colleagues." Voltaire's *Elements of the Philosophy of Newton* appeared in 1738, a labor of dedication though not, perhaps, of love, or not of love of Newton. Like most people, Voltaire really disliked physics. Nevertheless, it was eminently worth five years of his best efforts and those of his mistress. For however ungrateful the subject, it freed the mind from dogma.

The appeal physics did hold for Voltaire emerges in the very arrangement of his book. It begins—and this is the first of the surprises reserved for those who would trace religious skepticism to science—where the *Principia* had ended, with the liberty of God to make the world as He, rather than Descartes, had seen fit. But Voltaire was more interested in human liberty. Because of Newtonian science, thought is free, not just of the censor and his fatuities, but free of the more damaging tyranny of metaphysics. About metaphysics Voltaire expressed that scorn on which modern science has acted. For to Voltaire, the externalization of nature was no tragedy of deprivation. On the contrary, it worked a liberation of thought.

Cartesian science, as distinct from the Cartesian critical imperative, had proved amenable to the uses of orthodoxy. Père Malebranche, a devoted and intelligent priest, took upon himself to be the Aquinas of Cartesianism. Cartesian physics was received into the Catholic Church. And there, under the dome of the Sorbonne, the spirit of doubt was laid to rest. Except for that, the theologians perfectly understood Descartes' *a priori* manner of thought. They, too, habitually defined doctrine and (human) nature instead of accepting men as they are. And just here lay the distinctive advantage of Newtonian science. Once science became a creation of the mind, instead of a projection of concepts, it could no longer serve as a lens for thought, focusing down onto man as dogma all the necessity inherent in the cosmos.

Voltaire found the humane counterpart to Newton's physics in the psychology of John Locke. "Just as a skilled anatomist explains the workings of the human body," he wrote, "so does Locke's *Essay on the Human Understanding* give the natural history of consciousness." And here Voltaire, too, welcomed in empiricism that which wilts presumptuous speculations: "So many philosophers having written the romance of the soul, a sage has arrived who has modestly written its history."

This (almost universal) coupling of Locke and Newton associated science with liberalism in a nexus of English empiricism. No comfort must be sought therein about science depending on liberal polity. The connection lay in coincidence and circumstance, in history and not in some necessity. Nevertheless, Locke was a loose and seminal thinker, and the coincidence embraced much of the Enlightenment, which was nothing if not eclectic. In politics, the *Essay on Civil Government* put the case for government by consent on the classic basis of the social contract under natural law. Though written earlier, it was published as justification for the Revolution of 1688, which secured the supremacy of representative assemblies in the world governed by English institutions. The *Letter on Toleration* is a pragmatic demonstration that civic evils flowed from religious bigotry while profits accrued to the Dutch. The *Reasonableness of Christianity*, not quite a deist tract, is a lawyer's brief for a religion cured of zeal by experience. Atheists pretend to be shocked at Jehovah's vengeful spirit. Locke gave the answer. Condemning all mankind for Adam's sin was no unjust act, for immortality had been a privilege, not a right, and there is injustice only when rights are transgressed. Such was the temper of Locke's mind, sensible, moderate, literal, more than a touch philistine, professorial—or rather donnish, for your Oxford don of Locke's type tends to be a safe man and a bit of a schoolmaster. And what would be the plight of a schoolmaster forbidden by ethics to take away privileges?

The *Essay on the Human Understanding* develops the proposition that all ideas are records of sensations. Man, accordingly, is what he makes of his experience. That the way to improve him is to give him a better one was the obvious lesson which liberal reformers drew from the sensationalist psychology. Locke based himself on the example of science, which he interpreted as total empiricism. He had no mathematics. Too realistic to try the theorems of the *Principia*, he asked Huygens whether the geometry was correct. Assured that he might depend upon it, he ignored Huygens' other reservations and mastered the physical argument and rules of reasoning. In effect, therefore—in this, too, he was prophetic of the Enlightenment—Locke selected from the Newtonian corpus what he could handle.

Going behind Newton, the sensationalist psychology drew inspiration from the whole scientific movement. Locke was a member of the Royal Society in his own right, and a friend of Boyle when Newton was an unknown mathematics don. He had even provided himself with a somewhat ambiguous training in medicine, still the only profession which joined onto natural philosophy. The *Essay* was jotted down after informal discussions among certain "virtuosi" beginning in 1670. It was published only in 1690, and remained more notebook than treatise. "I will not deny," Locke disarmingly admitted, "but possibly it might be reduced to a narrower compass than it is: . . . the way it has been writ in, by catches, and many long intervals of interruption, being apt to cause some repetitions. But to confess the truth, I am now too lazy or too busy to make it shorter."

The method was "to search out the *bounds* between Opinion and Knowledge," for "When we know our own strength, we shall the better know what to undertake with hopes of success." Immediately Locke confronted that same obstacle which in the Cartesian philosophy had obstructed Newton, the doctrine of innate ideas. If science is introspection, we will never understand nature outside. And lacking those external reference points, neither may we aspire beyond science to its complement, an objective description of how the mind works which creates it. Innate ideas are to be excluded, therefore, in principle and as a condition of method. Every death erases the blackboard of the intellect. Thus, every idea is new, the echo in every life of something heard, something felt, something seen, something touched, something tasted. All the fancies and all the dreams that flutter through consciousness and lodge in thought, all the hopes and all the beliefs we ever hold, the highest theories like the lowest superstitions, all these elements of error and knowledge come to mind from experience.

They come at two levels. Sensation prints direct impressions. But Locke knew that he knew something. The faculty, or rather experience, of reflection comprises a second and higher category of ideas. Through self-knowledge we are aware of thinking, doubting, believing, reasoning, and willing. "This source of ideas every man has wholly within himself, and though it be not sense, as having nothing to do with external objects, yet is it very like it, and might properly be called internal sense."

This reduction of consciousness to internal perception led Locke's empiricism toward paradox. For he left the fanning of Baconian optimism about science and progress to successors. He himself pointed rather toward positivism. Only self-knowledge is firsthand and certain. With objects it is even as Newton said. We smell the odors, we taste the flavors, we feel the surfaces, but things themselves we never know. From the very conditions on which science succeeds, therefore, it unexpectedly appears that the proper study of mankind is man. Empiricism walls up truth in humanity, and leaves an educated guess as the best knowledge we can have of nature. In philosophy this skepticism issued in Berkeley's idealism and Hume's criticism of causality. But though it is interesting that science introduced skepticism rather than dogmatism into epistemology, the main concern of the history of science is the congruence between Locke's formulation of the inquiry and Newton's own conception of scientific explanation. Compare, for example, Newton's strictures on hypotheses to Locke's summary on ideas:

> My present purpose being only to inquire into the knowledge the mind has of things, by those ideas and appearances, which God has fitted it to receive from them, and how the mind comes by that knowledge, rather than into their causes, or manner of production: I shall not, contrary to the design of this essay, set myself to inquire philosophically into the peculiar constitution of bodies, and the configuration of parts, whereby they have the power to produce in

us the ideas of their sensible qualities: I shall not enter any further into that disquisition, it sufficing to my purpose to observe, that gold or saffron has a power to produce in us the idea of yellow, and snow or milk the idea of white, which we can only have by our sight, without examining the texture of the parts of those bodies, or the particular figures or motion of the particles which rebound from them, to cause in us that particular sensation: though when we go beyond the bare ideas in our minds, and would inquire into their causes, we cannot conceive any thing else to be in any sensible object, whereby it produces different ideas in us, but the different bulk, figure, number, texture, and motion of its sensible parts.

Or, as Condillac will say in his lapidary fashion, "Our first object, which we must never lose sight of, is to study the human mind, not to discover its nature, but to know its operations."

Taking physics as its model, Locke's science of human nature treated mind on the analogy of Newtonian matter. It was bound, therefore, to become atomistic. Elementary ideas, identical in origin with elementary sensations, made the term of that analysis which would know only operations. These particulate ideas drop into the mind through the five funnels of sense. There they bound, rebound, and combine like the corpuscles of which they are counterparts. In this kinetic theory of the intellect, the association of ideas is the counterpart of the law of universal attraction. And the conception of the mind as an instrument for sorting out discrete impulses becomes, therefore, less novel than the contemporary theorists of communication imagine. For only rarely are technicians correct in their supposition that they were born yesterday.

Will it strain the parallel to carry it one step further? The new psychology followed the new physics in a final respect. It was permitted by a conditional assumption rather than a positive discovery. There was no evidence

that all ideas derive from experience, any more than that
the texture of matter is atomic. But atomism and—true to
the Epicurean prototype—sensationalism allow objectivity. Unless ideas might be related to experience, they
would escape exact and scientific study. No science of
human nature would then be possible. And that great
subject, that closest of subjects, would fall back into the
domain of religion and superstition, back to where Jung
in fact has taken it, into the collective subconscious whence
Freud himself may have emerged rather in word than deed.
For if Locke made a fundamental mistake, it was that he
over-compensated for antique science in treating objectively a branch of knowledge only to be experienced subjectively.

❖

IT WOULD BE DIFFICULT to exaggerate the eighteenth-
century enthusiasm for the proposition that our minds only
organize information, messages fed in from experience.
Versions of the favorite psychological puzzle were imagined
by almost all the famous writers. Voltaire, Berkeley, Condillac, Diderot, all treat the question of how the world
appears to a man deprived of one of his senses. Diderot,
ever the master of the telling example, pointed to a blind
mathematician, one Saunderson, who could conceive geometry to perfection. He had a post at Cambridge. But he
had only a partial grasp on the real world. Each sense is
a dimension, and he was missing one. The idea of beauty,
the idea of duty, even the idea of God, everything is
bound to take different shape in the flatland of a man
who cannot see.

A real case supported this lesson. In 1728 a Dr. Cheselden of Chelsea Hospital operated successfully for congen-
ital cataracts sealing the eyes of a boy of fourteen. The patient was observed with passionate interest. To the delight

of the associationist psychologists, he did not at once understand his eyesight. He had to experience the difference between a cube and a sphere, red and yellow. He thought, said Voltaire—the case figures prominently in the *Elements of the Philosophy of Newton*—that everything he saw was resting on his eyeballs. For the more fortunate relate touch to vision only by a habit of association which that boy had never formed.

Thus they argued, who would develop the associationist psychology into a school of human nature in the analytical spirit of the abbé de Condillac. Time was, when French science shone in brilliance, that several generations of scientists, led first by d'Alembert and then by Laplace, Condorcet, and Lavoisier, admired Condillac as their authority on method. An intimate of Rousseau and Diderot, and even of the materialist d'Holbach, Condillac was one of the intellectual priests whose scholarship adorned the Enlightenment. Like many *philosophes*, he was educated by the Jesuits. In Paris he was much invited by the great hostesses, Mlle de l'Espinasse, Mme d'Epinay, Mme de Tencin, in whose salons the commitments of the Enlightenment started wittily as conversation. He kept his priesthood private. The Essay on the *Origin of Human Knowledge* opens with a passage attributing the occasion for psychology to the fall of man, which deprived us of the divine capacity to know essences. A realistic analysis of the imperfect understanding that survives will permit us to make the most of it. But only there in the preface does Condillac's vocation intrude into his empiricism.

Condillac took Newton for his model, so he tells us, but went to school to Locke. To turn from master to pupil is a relief. Where Locke was discursive, repetitious, and long, Condillac was incisive, analytical, and brief. "A thing said once, where it ought to be said, is clearer than if repeated here and there, now and again." There is no mystery about the principles that organize knowledge. They are those that order the correct composition. For Condillac assimilated effective knowledge to communication. Toward this end, he gathered up Locke's scattered threads of criticism against Cartesian innate ideas and pursued the argument with a truly Cartesian instinct for unity, simplicity, and universality. He took exception, for example, to Locke's passing remark that, for all we know, God might have conferred the power to think on any odd heap of matter. Condillac is short and severe on this confusion of subject with object: "It suffices to remark that the subject who thinks must be *one*. Now, some heap of matter is not *one*; it is a miscellany."

In such rigorous spirit did Condillac address himself to the difference Locke had admitted between sensation and reflection. "If, as Locke urges, the soul experiences no perception of which it is not conscious, so that an unconscious perception is a contradiction in terms, then perception and consciousness ought to be taken for only a single operation. If the contrary be true, then the two operations are distinct, and it would be in consciousness, and not as I have supposed in perception, that our knowledge originates." Locke had never faced up to this damaging consequence. Condillac simply abolished the distinction. He described the most advanced actions of the mind without allowing it any initiative of its own. His instrument was an elaborate and a highly original theory of language.

Though not the most inevitable, Condillac's assimilation of science to language was the most fertile conjunction of empiricism with analytical rationalism in the Enlightenment. The seeds were in Bacon, of course, who had been at his best on the errors imported by words, and on the imperative that names must signify things. But having repudiated abstraction, Baconian natural history could never transcend classification to appreciate the sense in which mathematics is said to be the language of science. Nor, indeed, could this be said with full confidence in its generality before Newton's procedures were understood. Condillac had those procedures before him, an important advantage for a philosopher of science. He took mathematics, not synthetic geometry (which he thought a siren to tempt the mind into metaphysical chimeras), but rather analytical mathematics, to be the exemplar of language. "Algebra, which is adapted to its purpose in every species of expression, in the most simple, most exact, and best manner possible, is at the same time a language and an analytical method." Algebra is a set of exact symbols to be manipulated by conventions. By agreement, they always mean the same. No admixture of our own judgment is in them, no adulteration by wishes, no weakening by fears.

On the analogy of the analytical role of mathematics in physics, Condillac would study to sharpen and improve ordinary language as the instinctive analyst of experience. Rude and imperfect though language is, every mental act which we take to be the expression of innate intelligence is in reality an effort at communication. A baby translates every sensation directly into action, cooing and crying, muling and puking. What with observation and training, this uncouth language of action soon gives way to French or English, to conventional signs and symbols. A child who can talk is no longer at the mercy of random events. He can guide himself among realities with a success proportional to his vocabulary, and draw on the common experience and the common wisdom, which are the same thing. The existence of a symbol for what one needs creates the possibility of securing it. Even if all experience is from outside, therefore, to dispose of a language is to be the master of one's thought. It is to command the instrument which orders the impressions of happenstance into the pattern of civilized existence. The point is to be sure that our words express determinate ideas about things and not figments. Else we err like old philosophers.

Language, then, the conventionalization of symbols from experience, is the cause—in the Newtonian sense in which gravity is both effect and cause, which is to say expression—of the highest and most complex operations of the mind. This was a profound concept, capable of great generality, and highly original. It is odd that it has been so little emphasized in the histories of the Enlightenment, which have preferred to trace to science the vague, ubiquitous preoccupations about deism, naturalism, and social or political reform. Modern literary criticism and modern communications theory sometimes give the impression that they invented the analysis of language, though independently, of course; for one of the findings of that theory, and one of the preoccupations of that criticism, is the isolation which arises from the difficulty of sharing experience. Of that difficulty Condillac was unaware. But surely the assertion is his that "We think only through the medium of words." It remains a principle of the psychology of art that no dragon, be it never so horrid, can be composed except from the organs of known animals. It remains a curiosity of science fiction that the imagination is dependent on actual scientific dreamings. Do with them what it will, it invents nothing.

After centuries of sophistry and superstition, ordinary language is, unfortunately, a dull and rusty tool. To compare it with the precision instrument of algebra is to confront the difference between science and life. And the main object of rationalism in the Enlightenment was to reduce that difference by right education. Thus it came about that eighteenth-century psychology first built its science of human nature on Newton's conception of scientific explanation, and then reciprocated by returning to science a conception of method oriented toward education, education taken in the very largest sense of the progressive improvement of the whole of the human understanding.

By the canons of this method, analysis first identifies the elements of a complex subject. These once clear, it ranges and classifies them according to the logical, which is to say the natural, connections that subsist beneath all the welter of phenomena, all the chaos of irrelevant experience. Finally, analysis finds the science its proper language, a systematic nomenclature designed to fix the thing in the name, assimilate the idea consciously to its object, and cement the memory to nature. Thus, to anticipate, salt of Venus becomes copper sulfate. One small stroke of naming destroys a superstition and teaches us a compound.

Indeed, once the nomenclature is established, analysis and naming become a single act, for languages, like algebra, are instruments of analysis, differing only in degrees of precision and self-consciousness. With this procedure, Condillac meant to succeed where Bacon had miscarried for lack of science, and Descartes for lack of humility. He would bring the whole reformation of learning down to a linguistic reform, redesignating words where necessary to make them speak facts, recombining them in a syntax of experience, lending reality to the expression used of the ancient atomists that theirs was an alphabet of nature. "The art of reasoning is nothing more than a language well arranged." Scientific explanation, then, would consist in resolving a subject into its elements in the objective world, in order that it might be reassembled in the mind by the association of ideas. Thus might the human understanding itself be reassembled in ever closer approximation to the grammar of nature.

This was a philosophy of science well suited to a situation where attention was shifting from the abstract to the experimental and descriptive sciences, from astronomy and theoretical physics to experimental physics, natural history, and chemistry. Newton had carried high theory beyond the capacity of the eighteenth century to do more than formalize the mathematics and perfect the treatment in detail. The humbler sciences, however, did present to method just such a mass of uncoordinated fact as needed ordering by right classification. And in practise Condillac's favorite analysis came to be less the algebraic resolution of abstract relations of quantity than the simple pigeon-holing of the naturalist. For if his inspiration was algebra, his model was botany.

Taxonomy little tempts the historian of scientific ideas. The problems were fussy and practical, but the question whether classifications are natural or artificial did not ultimately prove interesting. In botany the systematics of Carl von Linné prevailed—or Linnaeus, for he universalized his own name in Latin as he did those of plants. Linnaeus held beliefs rather than ideas about nature. He explained the form of species by the most literal and complacent providentialism. This variety of Lapland moss grows just here and not there because God made it to grow here and not there. For "God has suffered me to peep into his secret cabinet." Nevertheless, Linnaeus was a patient and accurate Swede with the soul of a fundamentalist librarian whose system improved on its many forerunners, beginning with Aristotle's, in the universal applicability of the criteria.

That plants have sexuality and that flowers (as a rule) are the sexual organs had been established in the seventeenth century. The Linnaean classification took advantage of their form. It distributed plants into twenty-four classes according to the number, proportion, and arrangement of the stamens; each class into orders according to the number of styles; each order into genera according to the method of fructification; and each genus into species by whatever characteristics distinguished it. The system was definite and easy. Rousseau tells how, in his search for consolation, he learned to identify plants in a few weeks, and "All of a sudden, there I was, as much of a naturalist as anyone needs to be who studies nature only to find new reasons for loving her." Nomenclature was the crux, the binomial system, soon to be extended to zoology, by which the substantive denotes genus and the adjective species. By this means a plant once identified stayed identified. The forget-me-not of England, the *oreille-de-souris* of France, the Vergissmeinnicht of Germany, become the *Myosotis palustris* of science, and naturalists of all countries could understand one another.

"These methods," wrote Condorcet of natural history, "are a kind of real language in which every object is designated by some of its more constant qualities and by means of which, knowing these qualities, we may find the name of the object in the conventional language." Condorcet was a mathematician, an able mathematician, but at his finest as a statesman of science and a patrician liberal. He expressed an admiration for Condillac's analysis of method which was shared, not only by the taxonomists whom it helped, but by his colleagues in the mathematical sciences. Retrospect may suggest that it served those sciences less as tool than as reassurance about their human worth. But even so, the fashion of interpretation was one which lent a dignity to the construction of value which scientists put upon their work. This was the first sophisticated conception of method to be drawn from an analysis of science as we still know it. And that makes the comparison with the twentieth century an interesting one. Our own day sees science as action and approximates a proposition to the scientific in the measure that it is operational. It abandons truth as the goal in favor of economy. Not so the forerunners of positivism in the Enlightenment, who saw science as education of the human understanding, and approximated a proposition to the scientific in the measure that it was "analytical" or "philosophical." Those attributes meant that it lodged in the mind that portion of truth about things which may be found out from their relations and behavior in nature. For the final reward of Condillac's theory of language was that it redeemed science from Locke's skepticism about our capacity to know any objective truths. To name is to know, not essences, nor totalities, but what we can know.

Condorcet's *Sketch for a Historical Picture of the Progress of the Human Mind* movingly records that faith in science as the educator of humanity. The essay is a cardinal document of the Enlightenment. By the invention of language man first steps from savagery into community. There the sharing of experience improves upon it. But the machinations of those with selfish interests in ignorance, superstition, and fanaticism artificially prolonged the childhood of the race and kept men dependent on kings, priests, and the philosophers who served them by multiplying ignorance. All history overcomes this ignorance. All history is the curve, almost horizontal for ages but turning now and rising toward the vertical, along which the human mind asymptotically approaches the order—the ordinate to sustain the figure—of nature. History is the education of humanity, then, as science is the educator:

> If one studies this development as it manifests itself in the inhabitants of a certain area at a certain period of time and then traces it on from generation to generation, one has the picture of the progress of the human mind. This progress is subject to the same general laws that can be observed in the development of the faculties of the individual, and it is indeed no more than the sum of that development realized in a large number of individuals joined together in society.

Moreover, now that science has found its method, there is no limit to the perfectibility of man. For, "All errors in politics and morals are based on philosophical errors, and these in turn are connected with scientific errors. There is not a religious system nor a supernatural extravagance that is not founded on ignorance of the laws of nature."

Nor was this a faith without works. It was in the Enlightenment that technology first began to feel that rationalizing touch of science which in more recent times has transformed the world. The historian may at first be somewhat puzzled by the contemporary testimony to the renovation that science was bringing to industry. For there is very little of science in the complex of technological innovations which made the industrial revolution. Deep plowing and crop rotation, smelting with coke, the spinning jenny and water frame and cotton factory, the puddling process for the conversion of iron, even the improvement of the steam engine by the separate condenser and the sun-and-planet linkage—these owe nothing to the contemporary achievements of basic science, to taxonomy in natural history, to the theory of combustion, to the foundation of metrical crystallography, to the discovery of the electric current, to the extension of the inverse square relationship to magnetic and electrostatic forces, to the formulation of analytical mechanics by Lagrange, or to the resolution of the planetary inequalities and the vindication of Newton by Laplace.

Clearly theoretical science had very little to offer industry in the eighteenth century. What was applied was scientific method. When scientists turned to industry, it was to describe the trades, to study the processes, and to classify the principles. Diderot's *Encyclopedia*, so much the most famous venture of the Enlightenment that the words "Encyclopedist" and "philosophe" became almost interchangeable, was itself a natural history of industry. Its subtitle was "Analytical Dictionary of the Sciences, Arts and Trades." In Diderot's definition a good dictionary should have "the character of changing the general way of thinking." Through extraordinary vicissitudes with the censor, the seventeen volumes appeared before the subscribers in the complementary guises of ideology and technology.

Diderot's masterstroke was to make the technology carry the ideology. It is the latter which has naturally enough monopolized the attention of intellectual historians, for it is the ideology of progress and liberalism which, conceived in the flirtation of the French Enlightenment with Locke and the English Constitution, burst passionately into life in the French Revolution and matured into the verities of modern democratic government. But the articles on politics and religion had to present all this indirectly, in a vein of innuendo and irony. Sarcasm disturbs and wounds and is never popular, even when justified. At the time, therefore, it was not the liberalism of the *Encyclopedia* which was popular. It was the technology, taking seriously the way people made things and got their livings, dignifying common pursuits by the attention of science.

For the *Encyclopedia*, and with it a multitude of industrial studies by the foremost scientists of the Academy, were attempts, Baconian in inspiration but informed by sophisticated method, to lift the arts and trades out of the slough of ignorant tradition and by rational description and classification to find them their rightful place in the great unity of human knowledge. The metal industries, to take an example, were not at first much changed by the development of metallurgy. They simply began to be understood. But that processes will be altered for the better if their principles are understood, that artisans will improve their manipulations if they know the reasons for them, are illustrations of the faith in progress through classification, industrial examples of the belief in scientific enlightenment as a kind of cosmic education. As a result, it was less often necessary to complain of the obstruction of rational procedures by the ignorance and traditionalism of the artisan. Popular superstition was always the dragon for the rational critic to slay, and whether he looked to religion or to technology, he found it flourishing in ignorance and secrecy. And this movement of publicity conjures up the contrast between the enterprising manufacturer of the nineteenth century, the bold engineer, on the one hand, and on the other, the Gothic master-craftsman of olden times, protecting his secrets and his mysteries, bending over his cauldron and stirring some traditional receipt, some confidential brew. "To the tableau of the sciences," writes Condorcet, "must be joined that of the arts, which leaning on the sciences, have made great strides and broken the bonds of routine."

It vindicates the reality of ideas to find them embodied in institutions. A tragedy attended Condorcet's faith in science and reason. He wrote in hiding from the guillotine. Nor did he survive the Jacobin Terror of the Revolution, which struck down, too, the scientific institutions of France, partly as survivals of the old regime, partly, too, in a fit of vulgar, sentimental petulance against the hauteur of abstract science, the impersonal tyranny of mathematics, the superiority of the scientist over the artisan. But liquidation was only one aspect of science in the Revolution. For the views of Condillac and Condorcet came into their own after the fall of Robespierre, when the rationalist tradition of the Enlightenment was institutionalized in the educational foundations of the French Republic.

Upon the *tabula rasa* left by the Jacobins, the Directory erected a new set of scientific institutions: the first *École normale*, the *École polytechnique*, new medical faculties in Paris, Strasbourg, and Montpellier, the *Conservatoire des arts et métiers*. Only the *Muséum d'histoire naturelle* emerged flourishing from the Terror, favored by the romantic enthusiasm for botany and Rousseau's nature. Other schools were revived, the *École des mines*, the *École des ponts et chaussées*, the *Collège de France*. Finally, at the summit was created the *Institut de France*. Thus, France was endowed at one stroke with her scientific institutions, and the first generation who taught and studied in them assured the restoration of her scientific leadership and its enlargement in the early nineteenth century.

It was a remarkable effort, animated by a consistent philosophy, which was nothing less than to unify the sciences through a common conception of scientific method, and in so doing to link them both institutionally and philosophically to realizing the idea of progress. So for a time, science was organized as a function of its educational mission. *Polytechnique* assembled the first scientific faculty, as distinguished a faculty, man for man, as has ever existed. For the first time, students were offered systematic technical instruction, directed toward engineering to be sure, but under the foremost men. The students, able and eager, chosen by competition, were immensely exhilarated by the sense of being conducted at once to the very forefront of scientific conquest, and being told that the future of the Republic, which was to say mankind, depended on how they acquitted themselves in so exposed a situation.

But *Polytechnique* had an equal influence on its teachers. If one were to read only the research memoirs of the Institute in its first ten years after 1795, one would conclude that French science had collapsed with its Academy. Quite erroneously, for the explanation is that scientists were communicating, not primarily with their colleagues, but with their students. *Polytechnique* made scientists into professors—again for the first time. It brought Lagrange back to mathematics from a preoccupation which had enveloped him since completing the *Analytical Mechanics* ten years before. Monge drew descriptive geometry together for his course. So, too, did Laplace come to write the *System of the World* and the *Essay on Probability*. Cuvier's *Lessons on Comparative Anatomy* were given at the Collège de France. Lamarck first presented the idea of evolution as the framework for his lectures at the Museum of Natural History. In short, the systematic treatise displaced the research memoir for a time.

This necessity to reorganize for teaching intervened in the philosophy of science between the Enlightenment and positivism, introducing a displacement toward action, a great enrichment of detail, and a certain access of rigor. It involved the scientists themselves in the preoccupation with method, with classification, with nomenclature. The author of a treatise, at once investigator and professor, would address himself to his entire science. This he would expound according to whatever principles resolved it into a rational body of knowledge. He would present them, not just as an authority, but argumentatively, as an advocate. His claim to originality lay, less in this or that discovery, than in having discerned the principles. So Cuvier founded comparative anatomy in the principle of the subordination of parts. So Monge gave credit to Linnaeus for his idea of grouping surfaces into families. So Bichat brought anatomy to the instruction of physiology by founding histology in the classification of tissues. So Berthollet looked to statics for the idea of chemical masses in equilibrium.

The force and range of the work were remarkable, then.

It would be no exaggeration to call this French essay in rationalization the last thrust of the Enlightenment, by which the Enlightenment returned whence it had originated and repaid the debt it had incurred to scientific culture in the time of Voltaire. A unity was imparted to scientific effort which it was not to know again. No doubt the universality of analytic method rested on just the kind of semantic illusion which that method set itself against. What is an algebraic process in Lagrange remains simple taxonomy in Linnaeus. But the difficulty was concealed by a very worthy commitment to that kind of tolerance and mutual respect which rescues communication from specialization by means of the comparative method. For as a matter of principle, no one of these French savants limited his vision by his science: "It is certainly," wrote Cabanis, who founded his moral philosophy in physiology, "a magnificent and beautiful conception, to consider all the arts and all the sciences as forming a community, an indivisible whole, limbs from the same great trunk, united by a common origin, and still more by the fruit they are destined to bear: the progress and happiness of mankind."

❖

MEANWHILE, what of romanticism? That Protean mood, as everyone knows, occupied the side opposite reason on the coin of the Enlightenment. But definitions have led armies of scholars through Serbonian bogs into dumps of Ajalon. How evident it is that Burke and Shelley are both romantics, yet one is the apostle of aristocracy and the other of humanity. Similar ambiguities couple Blake with Carlyle, Buonarotti with Maistre, Schelling with Fichte. Always a seer, Coleridge observed the French Revolution first through the enthusiastic eyes of radicalism, and then, passing a hand before his gaze, through the horror-struck eyes of reaction. Which is the true romantic, Napoleon or Robespierre, the eagle or the doctrinaire? Such questions are not to be answered by multiplying examples. And it may be that the difficulty of going beyond this in the histories of politics, philosophy, and literature, is that those areas of expression contain only the fruits of romanticism. Its roots go deeper into man's consciousness of nature. Perhaps, therefore, it is the history of science which will permit us to be explicit, dealing as it does with ideas about nature. These are ideas about which it is possible to be definite precisely because they are about things instead of values. Moreover, the general problem presents itself to the historian of science in an inescapable particular case, that of the greatest of all romantic minds. What is to be said of Goethe as a scientific thinker? And as a man of the Enlightenment?

For on the one hand, Goethe made an undeniable discovery in comparative anatomy, the intermaxillary bone in man. His studies of the metamorphosis of plants were meant seriously, and have been taken seriously by biologists. On the other hand, his polemic against Newton, going beyond the theory of colors to Newton's entire philosophy of nature, appears a dismaying anachronism, not only in itself, but in coming at the end of that century supposed to have been inspired by the Newtonian spirit, and on the part of one taken to be the quintessence of enlightenment, or at least of *Aufklärung*. "As for what I have done as a poet," he said to Eckerman in 1829, shortly before he died, "I take no pride in it whatever. Excellent poets have lived at the same time as myself; greater poets have lived before me; and others will come afterward. But that I am the only person in my century who knows the truth in the difficult science of colors—of that, I admit, I am not a little proud, and here I am conscious of superiority to others."

Nor will we find the clue to an interpretation in the rebellion of Jean Jacques Rousseau against the Encyclopedic spirit. We must look for the origins of romanticism into experiences more cosmic than that hot walk to Vincennes in the summer of 1749, when Rousseau, glancing through the *Mercure de France* on his dusty way to comfort the imprisoned Diderot, came upon the essay question set by the Academy of Dijon: "Has the progress of the sciences and arts done more to corrupt morals or improve them?"—and arrived "in a state of agitation bordering on delirium," self-converted to his crusade against the cultivated intelligence. With Goethe we are concerned with something more serious than the heart turning against the head. He is not simply enlarging Rousseau's revolt against Voltaire into a counter-revolution against Voltaire's master, Newton. For one thing, Rousseau was not himself systematically hostile to the spirit of objective science in the eighteenth century. His was rather the sporadic petulance of the paranoid than the vision, or the memory, of some more grateful science. It is significant, for example, that Rousseau as amateur botanist welcomed Linnaean technique in taxonomy. With all his touchiness in personal matters, it never offended him to range the specific productions of nature into fixed categories of form. Not so casual was Goethe, whom that process insulted deeply.

To appreciate the burden of that insult, it is Rousseau's involuntary host whom we must study, Denis Diderot, a spirit as various in expression as romanticism itself, and with Condillac perhaps the most significant of the *philosophes*. Nor need this prove confusing if we do not insist on assigning every figure and every venture categorically to rationalism or romanticism, but let them rather participate in each in some appropriate measure. Thus, for example, through Diderot's *Encyclopedia*, a work, certainly, of rationalism and positive method, run those counter-connotations of populism and sentimentality, secularized echoes of the Christian assertion of the dignity of labor, which assign humility and truth to common pursuits, to technology and not to science, and which in the manner of Bacon flatter the artisan that he is the one who knows in practice what the scientist obscures by theory. So it is throughout Diderot's writings. Each served as a vehicle in praise of some aspect of the natural man. Nor must we look for any more superficial consistency. Diderot cannot be assigned to either the rationalist or the romantic moods. He belonged to both and to neither. Intelligent, amorous, sensuous, cultivated, critical, sentimental, humane, hard-working, humorous—he was in his own eyes above all else virtuous, in that his life and writings affirmed the innocence of nature, of nature in general, of his own nature in particular, and of the opportunities afforded by their congruence.

"The distinction," wrote Diderot of the picaresque title character in *Jacques the Fatalist*, "between a physical and a moral world seemed to him empty of meaning." That remark contains a theme. It is the theme of Diderot's natural philosophy. More than that, it is the basic theme of romanticism, composed by Diderot in the Enlightenment when modern science first confronted culture. For the romantic response to physics in the Enlightenment is perhaps the most important expression of the tension which must exist between science and the aspirations of humanity to participate morally and through consciousness in the cosmic process. Those aspirations require a nature different from that described by post-Galilean science—not the nature of atomism, where science observes motion and measures quantity; nor even the nature of the Aristotelians, where science classifies forms and defines goals; but rather the nature of the Stoics, where science discerns activity, where its object is virtue rising out of nature, and where cosmic personality is the source of order.

Newton's world offered virtue no purchase. And just as in Hellenistic times, when the Stoics spoke for the dynamic unity of nature against the Epicurean death-sentence of atoms-and-the-void, so the romantics of the Enlightenment seem to have recalled as against the Newtonian "corpuscular philosophy" that ancient retort upon atomizing kinetics. It is probable that the eighteenth century knew the school through the Roman Stoics, and particularly Seneca, whose emphasis on civic virtue and duty appealed to its high sense of *res publica*. Senator and patrician, Montesquieu breathed stoicism into the *Spirit of the Laws*. And before considering Diderot's philosophy of nature, it may be worthwhile to recall, for a moment, the main notes struck by this last of the Greek schools to inform the consciousness of nature which we have inherited, clothed in science, and transformed here in our own world of the West.

The concepts of that school had descended from the prehistoric representations of natural forces preserved in myth and legend. Stoic physics was an attempt to elevate this legacy into science and philosophy, and to combine it with the cosmology of Heraclitus, seeing the world as flux and fire, conflagration and return. For the Stoics it is always the activity, and not the matter, in things which has ontological significance. And the principle of activity in the cosmos, what tenses it into unity, is the breath of the spirit—*pneuma*, which ties the world into one dynamic whole, which does for the world what life does for the animal, and which, therefore, is the life of the world. Strictly speaking, however, we can never say what *pneuma* is. Stoic metaphysics treats of becoming, not of being. Stoic logic puts reason in the verb, not the substantive. We can only say what *pneuma* does. And what it does is, first of all, to bind together substances which would otherwise be passive, undifferentiated stuff, to lend them permanence and stability. It acts as the erethitic agent in matter, which it permeates as a sexual stimulus penetrates erectile tissue in animals. It is the flex in the muscle, the stretch in the rubber band. But in order to create unity in nature, *pneuma* has to abandon simplicity for itself. Not only does it bind, it differentiates. Those properties of bodies which the Epicureans attributed to the shape and arrangement of atoms, the Stoics read back to its state of permeation by *pneuma*. For them there are no boundaries in nature. Combination can never arise from juxtaposition, but only from the blending of principles, intimately melting into that perfect union which lies beyond junction in identity. And change is a metamorphosis throughout the whole. A simple rearrangement of the parts is, indeed, no change at all.

The Stoic world, in short, is a dynamic continuum where causality reigns hand in hand with sympathy. Strict causality is implicit in the unity of nature. In Stoic doctrine it assumes the guise of destiny or fate. And this permitted the reconciliation, or the identification (for the Stoic drive is always integral) of causality with Providence—not a capricious Providence like that into which the conflict of subordinate Aristotelian purposes might too easily degenerate (and in Christian natural theology did degenerate), but a lofty Providence which knows its own mind utterly. Perception, in this doctrine, is penetration of the senses by the real qualities that run through things—participation, therefore, of consciousness in the flux and process of the world's great life. By the same token, knowledge is illumination of the soul by the truth in nature, by the truth and by the good. But there is withal a democratic implication, more subtle in its appeal to the Enlightenment, perhaps, than hard lessons of civic virtue, but more popular, too, and much easier. For it is the common understanding which holds the truth. The sage is teacher, not researcher, who discerns the notions common to all men, lays down rules of conduct from nature, and finds the correspondences and sympathies between cosmos and personality, the great world and the little.

✧

THE FIRST THRUST back from the eighteenth century to this more intimate sense of nature came from chemistry, in an attempt to deepen the concept of matter, to restore body to what physics deprives of every attribute but location and dimension. This, it must at once be said, was not our sort of chemistry, not the recognizable chemistry with which we will be concerned in discussing Lavoisier and Priestley. Rather was it an operational mode of communion with nature, an archaic, sympathetic chemistry addressed to a notion of the physical now quite forgotten, and so far as science is concerned, quite rightly forgotten. For in speaking of physics as concerned with bodies in motion, we forget that the word "body" originally implied organization, internal material organization, with which chemistry might as properly concern itself as biology. Indeed more properly—the word "biology" had yet to be invented, and the concept of organism thereby confined to that which lives as an individual instead of as a world.

The article on chemistry in Diderot's *Encyclopedia* is extremely curious. It is by Venel. He invokes a new Paracelsus, who will make chemistry the science that understands nature and displaces geometry from that pretension. He will be gifted, this neo-Paracelsus, with the sheer technical insight to penetrate beyond physics. But he will have a spirit like the pre-Newtonian philosophers.

They "saw nature better as it is," because "a sympathy, a correspondence was only a phenomenon for them, while for us it is a paradox as soon as we fail to bring it down to our pretended laws of motion." Physics, to pursue the comparison, is superficial. Chemistry is profound. Physics measures the gross, external characteristics of bodies. Chemistry penetrates their essence. Physics confounds abstractions with truths. One asks for a fact. It replies with a theorem. The physicist calculates in rigor to arrive at those exact theories, which experiments then confirm— "approximately." The chemist never deludes himself by calculation. He apprehends his theories rather by an "experimental instinct." In his case it is the theory which is approximate. But as the reward of his humility, the fit with nature is exact.

The question, then, is nothing less than the structure of being in nature. Mechanics will never bear the chemist into the heart of things, for the texture of reality is not corpuscular. And the essential merit of chemistry is that it takes the sting out of atomism. For it allows the masses in which atomism resides no ontological interest. Thus chemistry provides the empirical way out of Newton's unreal abstractions. The qualities in things are what impress our senses, our windows on reality, and this reality inheres, not in mass, but in the principles which run through the world as activities, as bearers of quality and agents of perceived effects. The physicist, therefore, who denies existence to entities like yellowness and life and fire is simply presumptuous. They do not fall in his field.

It is not for the physicist to study quality, nor for the chemist to study quantity. Venel dismisses Boyle as a physicist. The chemist's operations will be different. His laboratory will offer no scene of weighing and measuring. The combinations and separations of matter are what concern him in its state of interpenetration, and masses do not combine. They only aggregate. Principles are what combine, and the chemist, therefore, will catch glimpses of "the life of nature" coursing through his laboratory in phenomena which run all through, around, and under mass: in effervescences and distillations, evaporations and condensations, rarefactions and expansions, elasticity, ductility, malleability, and fluidity. The image is of stretchings and blendings in depth, rich with the Faustian sense that nature has an inside. There is nothing transcendental about it. Nor is it the chemistry of a spiritual world. Rather, it is the chemistry of a world alive. Venel is always moving in the mind's eye from fermentations in his laboratory through digestions in the animal, deep into the mineral gestations of the earth, whose cosmic womb is the home of unity in nature. And in his laboratory, your chemist is almost Hippocratic. He wields his implements with art. His hands are gentle. It is the physicist who brutally pulverizes, ignites, and destroys. The chemist does not analyze. He divines.

The chemist's world, then, is a palpable continuum. His science is Cartesianism stripped of geometry with its clear ideas. To replenish the Newtonian destitution of nature, it sees down into a world pulsing with activity. In place of universal attraction between particles, Venel has discovered that the fundamental property of matter is universal miscibility. But chemistry is more than intimacy with nature. It has the common touch. It is everybody's science, the poor man's manual metaphysics, whereby that artisan in whose skills true wisdom lies manipulates reality, not in the humiliating abstractions of mathematics, but with his own hands: "Chemistry speaks a dual language, the popular and the scientific." And all this seems harmless enough, until suddenly, out of the *Encyclopedia*, comes in one startling sentence the authentic voice of vulgarity. "Parlez plus bas," the mathematical physicist is told, "Pipe down! A coal heaver would die laughing if he heard you."

In *Thoughts on the Interpretation of Nature*, Diderot alludes to this science of the chemists as one example to be emulated in handling nature with the surety of experimental art. But sentience and organism weave a more grateful veil, and though Diderot drew his conception of palpable reality from the continuum of activity, he transposed it out of chemistry into the far more plausible terms of natural history, launched it in the flow of time, and created that idealistic outlook which dominated much biological thinking until the import of Darwinism was fully appreciated, which is to say until very recently. The opening paragraphs of that essay prophesy the imminent decline of mathematics. It is an observation which has been frequently forgiven as the momentary enthusiasm of one who would restore ordinary sight to eyes dazzled by the glamor of Newtonianism, or as a passing slip in a prescient vision of the biological shape of things to come. But as often when Diderot spoke lightly, he meant this seriously as a complete repudiation of abstract conceptualization.

His rejection of mathematics was fundamental. He objected to its claim to be the true language of science on all grounds, metaphysical, mechanical, and moral. It is not just that mathematics idealizes. It falsifies, by depriving bodies of the perceptible qualities in which alone they have existence for an empirical, sympathetic science. Mathematics has turned mechanics into trivial description by mistaking the measurement of bodies for understanding the activity which animates them. Worst of all, the mathematical spirit is a blight. Fortunate but rare the mathematician whose own humane sensibilities are not blunted by his subject, which has fallen into aridity and circularity. So must any science do which ceases to "instruct and please." Once idle curiosity is satisfied and novelty wears off, only its power to edify will keep a science living. "I do not except even natural history."

More ominously, Diderot gives back to the Enlightenment, perhaps from his chemical studies, an angrier note, which echoes down the whole romantic movement. Mathematics is worse than inhumane. It is arrogant. In a sense, no doubt, everyone who has felt himself reach his mathematical frontier, whether at long division or out somewhere beyond the calculus, must know something of the helpless resentment engendered by the hidden beauty of the abstract. But Diderot's own competence in mathematics was by no means contemptible. He fully appreciated its value as an instrument of precision in subordinate matters. His indictment is curious and interesting and not mere petulance. Mathematics is the science by which a finite intelligence purports to plumb the infinite. Now, man aspiring above himself incurs the classic guilt of hubris, the Christian guilt of pride. The prospect of an infinite universe has always disconcerted those who would render science humane. But Diderot was no Pascal to agonize over infinity. We are in the eighteenth century, and he responds with admirable nonchalance. He simply dismisses infinity as uninteresting. Since we shall need some criterion to establish bounds between knowledge and the infinite unknown, why let it be our interests. "It will be utility which, in a few centuries, will establish boundaries for experimental science, as it is about to do for mathematics." And Diderot restores the mind, in a sense, to a finite cosmos, by wrapping science tight around humanity.

Nor in form are Diderot's writings on nature an artless collection of *aperçus*. In the *Dream of D'Alembert,* that distinguished mathematical colleague is put into a trance, almost a delirium, out of which he speaks truths instantly recognized as such, and easily anticipated, by whom? His interlocutor is a doctor, the universal doctor, bending over the bedside of us all, who sees nature across the perspective of human nature, and who knew the answers all the time: "There is no difference between a doctor keeping watch and a philosopher dreaming." And the apparent formlessness of the *Interpretation of Nature* is skillfully adapted to convey the congruence between man and nature. For it is written as a stream of consciousness, a reverie on the Experimental Art, the true route to a Science of Nature, moving out toward three objects: Existence, Qualities, Use. A threefold object, but a single purpose—what is the young man to look for in Diderot's natural philosophy? "An abler than I will teach thee to know the forces of nature; it is enough for me to have made thee try thy own."

Thus Diderot reverses Descartes, who studied himself to know nature. Diderot attends courses on chemistry. He reads Buffon, he studies nature—to know himself. But communication is direct, experiential. It does not lie through mathematics. It lies, instead, through craftsmanship. And this complicates the interpretation which makes the technology of the *Encyclopedia* an expression of the Newtonian spirit, which is to say of science. For Diderot, as for the chemists, truth opens to the common touch, and—as in Bacon—the importance of right method is that it dispenses the ordinary man from the need for genius. Genius, in its pride, is inclined to draw a mathematical shroud of obscurity between nature and the people.

Those are in error who say that some truths can never be put "at everyone's disposal." Certainly, ordinary people will never see merit in what cannot be proven useful. But in this, they see aright—or rather, they are aright to fail to see. Only experiential philosophy is an "innocent study," in that it supposes no prior preparation of the mind. The habit of actually handling materials in dumb, untutored experiments, develops in him who performs the coarsest operations an intuition which has the character of inspiration. Manual facility gives a power of divination, the ability to "smell out—*subodorer*" how it must be with nature. But how do you know you have this power? How do you know you are right? It is—if the analogy is permissible—like awareness of Grace. It is like Virtue. It is participation in the Truth. You recognize it in yourself, in your own intimacy and more than intimacy, your solidarity with nature. In such a breast, science and nature are one, the reality of the great organism suffusing for the moment the material consciousness of the little. Not mathematical abstraction from nature then, but moral insight into nature, is the arm of science. Presuming to prescribe as rules for nature his own formulations, the scientist in his pride conceals from himself and others that it is not his laws of nature which are simple, but nature itself, in its essential unity.

For nature is the combination of its elements and not just an aggregate. Otherwise there is no philosophy— "Without the idea of the whole, philosophy is no more." And Diderot, therefore, is bound to interest himself in continuity and not in divisibility: "Acknowledge that division is incompatible with the essence of forms, since it destroys them." When he writes of molecules, it is of their transience, not their existence. In genetics, he rejects the notion (*emboîtement*) that each animal is contained full-formed in an infinite regression of *ova* reaching back through all the generations to the first creation. But it is the atomistic as well as the theological implications which Diderot finds unacceptable. For nature knows no limits. The male exists in the female, and vice versa (hence the curious fascination with hermaphroditism which runs through his writings). Mineral blends into mineral. The qualities of one living species penetrate in some degree the others. Minerals are themselves fused into living matter through the nexus of the plant which feeds on minerals and aliments the animals. Individual animals are real eddies of tighter organization, the ultimate but impermanent units, borne along a stream of seminal fluid flowing down through time and out from the matrix womb of nature herself. Even the physicist will do better to devote attention to what endures and spreads—to resonance, for example, to fire and electricity, to sulphurous exhalation, and to standing waves. Diderot, too, has a substitute for the universal attraction of corpuscular physicists: it is universal elasticity.

"Tout change; tout passe; il n'y a que le tout qui reste—Everything changes; everything passes; nothing remains but the whole." And Diderot uses two figures to express this unity. The second is the more familiar, the universe as a cosmic polyp—time, its life unfolding; space, its habitation; gradience, its structure—for this embodies the twin ideas of universal sensibility and of what has often been called evolution. Diderot treats the continuous development of the universe as a consequence of the indivisibility of cosmic time. But his time—like his whole natural philosophy—is that of biological subjectivity, and in no way dimensional. And although this is consonant with historicism, there is (as will appear) no serious sense in which it foreshadows Darwinism. Not this, therefore, but rather the first of Diderot's metaphors is the more significant. In it he evokes the swarm of bees. For the solidarity of the universe is social. On a cosmic scale, it is that oneness which the social insects know, among whom the laws of community are laws of nature. "Only the bad man lives alone," he told Rousseau at their sad and angry parting. And in this social naturalism there is a more prescient (and to the liberal a more alarming) concordance between the whole and the parts, the one and the many, than in reversion to an antique hylozoism.

❖

INDEED, Diderot was the Spinoza of biology before ever the science had its name—or its Newton. His was no feminine dislike of precision, no soulful sense of God in nature, but a philosophy of necessitarian organism. He would reach into the heart of science to turn it into moral philosophy. And now, looking back on that old sense of nature through Diderot's clear eyes, we are prepared to grasp the meaning of Goethe's more Delphic, his essentially poetic response to the world picture of Newtonian science. Being the man he was, Goethe would not simply lament the deprivation wrought by that tale of mournful numbers. He would do something about it. He would replenish the dream, not only with the Faust legend, and magic, "this one Book of Mystery/From Nostradamus' very hand," but also—"Thee, boundless Nature, how make thee my own?"—with the intermaxillary bone, the metamorphosis of plants, and a theory of colors as the "deeds and sufferings of light."

Botany is the most accessible and consoling of the sciences, and (following in this Rousseau's example) Goethe began with plants. He took up his Linnaeus to learn how to classify those he might encounter in his rambles about Weimar. But not for him the ticketing of static forms, not for him taxonomy—Linnaeus, he later said, had a greater influence on him than any thinker other than Shakespeare and Spinoza (for Goethe, too, would find destiny and necessity indwelling in flux and process, in the organismic and not the mechanistic universe). Though profound, Linnaean influence was repulsive. "For, even while I sought to assimilate myself to those sharp separations, . . . a conflict developed in my breast:

That which he tried to separate by force of distinctions, tended by the most intimate necessity of my whole being to unite together." There was in Linnaean botany nothing but nomenclature, nothing for the intelligence nor the imagination, no place for loveliness of form and flower. Counting stamens and pistils, it founded itself on numbers. Like some inhuman anatomist, Linnaeus could study a plant as well dead as alive. It was in this state that he habitually did study it, turning the stream of life into a mosaic of skeletons.

A friend of Goethe, a naturalist called Batsch, had developed an alternative system, one of the many "natural systems" of the eighteenth century. His practice ranged the plants by form according as they progressed along the chain of being. Goethe seized on the notion of progressive form. It was an ancient notion. On the principle that everything which can exist must exist, a perfect continuum of form rises from the lowest to the most high. Or must continually be rising, for it was Goethe's view that no plant is to be taken simply as an object, ready made from the hand of the Creator. Growing things are to be studied in their cyclic living. One plants the seed. It burgeons. One follows the unfolding, day by day. Here is slow motion. At first imperceptible, the organs—never say when —appear. Now they differentiate themselves from the embryonic matrix of leaf-form. As the glory and the climax comes the end, the orgasm of the flower, dying into being, life's withdrawal into death to return from the seed. And this is thrilling to observe, an affirmation and release after the pedantic, weary, and disgusting chore of classification, shuffling among minutiae as empty of meaning as the books littered about in Faust's study.

Goethe's *Metamorphosis of Plants* treats the plant as the epitome of all becoming. Unlike rootless organisms, the plant wears its inwardness outside for all to see. All are variations on the archetypal theme of an *Urpflanz*, deep down in being, way back in time: "All have a similar form, yet none is the same as the other. So this chorus of growth shows a mysterious law." It is a true chorus. All the parts of a plant are variations on the single aboriginal organ, the Leaf, "the true Proteus," out of which seed and form, stem and flower, successively permute:

> Always changing, firm persisting
> Near and far, far and near,
> Thus in forming and transforming
> To your wonder I am here.

And when in the 1790's Goethe extended his interests to zoology, it was in the same idealistic vein. All animals are variations on an archetype; in the individual animal, the whole body is a succession of vertebral permutations. The skull is a metamorphosed vertebra, and this symmetrical theme further unfolds in the symmetry of the body. Hence Goethe's joy in his discovery of the inter-maxillary bone in man, all overgrown and merging into the maxillae, but *there*, as in progression from the great apes it should be, there and identifiable, an anatomical bond linking man morphologically yet a little more intimately into the rest of nature.

It is on turning to Goethe's *Teaching on Colors* (for the literal translation of *Farbenlehre* conveys the spirit better than the more usual "Theory") that one finds how inevitably this attempt to embrace all science in morphology implied repudiation of physics. "My *Teaching on Colors*," he wrote, "is as old as the world, and in the long run it cannot be denied and thrust aside. These gentlemen may behave as they will, but at least they will not get this book out of the history of physics." He wrote it only in 1810. But his passion for the subject was aroused in the Italian journey of 1786-87. Like many Germans he experienced the brilliance of that landscape as a revelation. Its pure, strong colors enchanted him. He watched the artists working with those colors. It came to him that "one must first get at colors as physical phenomena, from the standpoint of Nature, if one would gain control over them for purposes of Art." Once again he turned to science and to the standard work. He took up Newton's *Opticks*.

And again he was repelled by Newton as formerly by Linnaeus, and for the same reason. Anatomizing activity, Newtonian analysis takes what is whole to bits. It turns acts into husks. "As by an instinct," he knew that Newton's theory of colors was false. Then he borrowed a prism, and he *saw* ("Zum sehen geboren—to see is my birthright") where Newton went wrong. The white wall opposite remained white though seen through the prism. Colors appeared only where the white was bounded by dark, as at the window frame. And how evident it is that edges bruise the light into colors, and that color manifests the tension between the primal polarities of light and dark. Light (for this is a syncretism between neo-Platonic and Stoic residues) is a streaming of reality, the manifestation of the immanent divine, as indivisible as soul. Darkness is suffering, not-being, death. Nor do colors simply register. Their perception is an optical act. So, for example, the eye may evoke them even when closed if it be struck a blow. But in normal circumstances it simply selects from the polarity what it needs to be at ease, demanding brightness when confronted by darkness, and for every color requiring its complement. Nor, though he looked through a prism, did Goethe believe in experiment. On the contrary, Newton's errors were the price he paid for his methods, mathematicizing nature into abstraction, torturing her with instruments, with telescopes, prisms and mirrors, until she expires like a butterfly on a pin. Goethe would observe phenomena as they really are, under the open sky, without complicated and artificial arrangements, in a lifetime of sympathetic perception—

> Friends, avoid the darkened prisons
> Where they pinch and tweak the light
> And in pitiful decisions
> Bow to rays distorted quite,
> Worshippers most superstitious
> Thronged in plenty down the year.
> Leave in hands of teachers vicious
> Spectres, madness, cheats, and leers!

It is impossible to read the *Farbenlehre* without an acute sense of embarrassment at the painful spectacle of the author, a great man, making a fool of himself. Even the most rapt devotees have preferred to dwell upon his biological philosophy. And yet is it any better? Is not the advantage the biology seems to enjoy an illusion created by the lag between that science and physics? For really Goethe's science is all of a piece with his personality, and with his poetry, as Bayard Taylor, perhaps the best of his translators, noticed many years ago. "His intellect had succeeded in uniting Man and Nature, the individual, the race, and the planet, in one consistent and harmonious scheme, wherein the poem and the mountain, the flower and the statue obeyed the same laws of growth." Neither for the poet nor the scientist did the goal ever change:

> That I may detect the inmost force
> Which binds the world, and guides its course
> Its germs, productive powers explore
> And rummage in empty words no more.

In Germany, however, Goethe, the scientist, has always been taken seriously. The spate of solemn symposia shows no sign of slackening with the passing of the bicentennial. Nor is there any denying the importance of his influence in the nineteenth century. It was consonant with the idealist bent of philosophy in Germany. It eloquently rein-forced that bent with the prestige of Germany's greatest man of letters, her latter-day Leonardo, her universal man. A whole school of *Naturphilosophie* sprang up, or un-folded, to carry forward, or inward, the study of archetypal biology. It gave a great impetus to morphology and em-bryology. It would, of course, be an exaggeration to at-tribute the primacy of Germany in those studies to Goethe's influence. Few important scientists were among the sec-tarians. Nevertheless, even fewer remained untouched by biological romanticism. It set the style of German science, as Cartesianism did (and still does) in France, and Baconi-anism in England. And Goethe's influence may be taken as an instance, not just of cultural nationalism, but also of the anxiety of cultured scientists not to be divorced from culture by their own creation of science. If Goethe was right about enriching science with his insights, or even partly right, why science partakes of humanism in just that measure.

Nevertheless, the historian is bound to represent this Goethean intrusion as profoundly hostile to science, hos-tile to physical science and misleading, even if stimulating, to biology. Whatever the incidental contributions, it was biological romanticism which created the image of biology as a different kind of science. This was the origin of the distinction that makes life or organism, form or goal, the object of biology. This is what divided biology from physics on the supposition (or defense) that the biologist must characteristically study the nature and the wisdom of the whole rather than the structure or arrangement of the parts. In a sense, indeed, Goethe's attack, and Diderot's, upon Newtonian science may be seen as a wrong turning in that long dialogue which science has conducted through-out its history between the unity of nature and the multi-plicity of phenomena. Against the wrecking into atoms, or fixed entities of any sort, whether light rays or Linnaean species, Goethe asserted the biological continuum, the stream of life.

Now, the continuum is by no means always the wrong side. Who studies to perceive the unity of nature necessarily adopts it (though at the price, perhaps, of unity of science). But the mathematical expression of the continuum was geometry, as in Einstein, as in the Newtonian void, as in Descartes, and ultimately back in Platonic mathematical realism. This is rational. It is objective, or may be. Not so the refuge

> In the tides of Life,
> In Action's storm
> A fluctuant wave
> A Shuttle free
> Birth and the Grave
> An eternal sea
> A weaving, flowing
> Life, all-glowing.

Goethe's nature is not objectively analyzed. It is subjectively penetrated. His is the continuum, not of geometry, but of sentience, not to say sentimentality. Nor does this vision of flux and process lead on to evolution in the proper sense. On the contrary, the unity of nature triumphs over the diversity of experience in universal metamorphosis. Man is neither product nor observer of nature. Instead, he is participant. He is communicant. For through the ashes between the surrender of Cartesian science and the rising of the Romantic phoenix, there ran this bond of dissatisfaction with a poverty in the Newtonian conception of scientific explanation. To the Cartesians, nature was the seat of rationality, and Newton's laws appeared intellectually trivial. To the romantics, nature was the seat of virtue, and Newton's laws were morally unedifying. The work of the romantics, therefore, had to be to preserve the continuity of man and nature by opening the personality to reality rather than the intellect. For if nature is congruent with man, if science is the correspondence, the universe has to be a continuum, a whole, a *tout*, as d'Alembert is made to see in his dream. But it is the whole personality which communicates, and not just the heart. Until Rousseau's traumatic conversion, there is no question of irrationality.

Now, therefore, we may be in a position to venture that consistent account of romanticism which eludes us in the history of politics, philosophy, and the arts. Romanticism began as a moral revolt against physics, expressed in moving, sad, and sometimes angry attempts to defend a qualitative science, in which nature can be congruent with man, against a measuring, numbering science which alienates the creator of science from his own creation by total objectification of nature. For physics romanticism would substitute biology at the heart of science. For mechanism as the model of order, romanticism would substitute organism, some unitary emanation of intelligence or will, or else identical with intelligence or will. Romanticism might take any form in politics, art, or letters. But in natural philosophy there is an infallible touchstone of romantic tendencies. Its metaphysics treats becoming rather than being. Its ontology lies in metamorphosis rather than atomism. And always it wants more out of nature than science finds there.

Indeed, the renewals of this subjective approach to nature make a pathetic theme. Its ruins lie strewn like good intentions all along the ground traversed by science, until it survives only in strange corners like Lysenkoism and anthroposophy, where nature is socialized or moralized. Such survivals are relics of the perpetual attempt to escape the consequences of western man's most characteristic and successful campaign, which must doom to conquer. So like any thrust in the face of the inevitable, romantic natural philosophy has induced every nuance of mood from desperation to heroism. At the ugliest, it is sentimental or vulgar hostility to intellect. At the noblest, it inspired Diderot's naturalistic and moralizing science, Goethe's personification of nature, the poetry of Wordsworth, and the philosophy of Alfred North Whitehead, or of any other who would find a place in science for our qualitative and aesthetic appreciation of nature. It is the science of those who would make botany of blossoms and meteorology of sunsets. And perhaps the humanist attempt to understand nature through self-knowledge, though never again to be the way of science, will always be the way of art. Not only of art, but of history, or rather historicism, for Herder's seminal philosophy of history presupposes the same idea of nature as Diderot and Goethe held. Its reality is process and unfolding. Its laws are universal extensions of those which govern the birth, growth, and life course of the single organism. It saves the correspondence of microcosm and macrocosm by transposing it from space to time.

In a book on conservatism, one comes upon the following statement, appreciatively underlined by a succession of student readers, fortunate young men of the American élite: "Without the creative principle of voluntary action and a healthy degree of self-organization the organic life of society perishes in the arms of an efficient despotism, even though it takes unto itself the sacred name of democracy. The purpose of government is not to concentrate but to diffuse power. Diffusion of power is the characteristic of organic life, just as the concentration of power is the characteristic of mechanism." And if one follows in imagination this lead from pure romanticism to the political realm, where the nexus of authority running between the one and the many really matters, it is almost alarming to think for a moment of the vast structures of reasoning about the state and society which depend upon substituting the metaphor of organism for atomism and mechanism. Where is Burke left without it? Where is the whole conservative apologia of the nineteenth century if it read the wrong science when it assimilated the ancient notion of the body politic to naturalism? What becomes of socialism if the idea of society as a collectivity, which it lifted from romantic conservatism, crumbles into atomistic (or individualistic) dust? Nothing, of course, would happen to political realities. But at least political apologists would be deprived of the right to draw dogmas from the nature of things and thrown upon their own resources—though since that is where Voltaire threw them two centuries ago, there is no reason to expect them to remain any more content with their own resources than other people do.

All of which is only to say that deep interests have been bound up with the romantic view of nature, deep interests and deep feelings, and also that it is the wrong view for science.

CHAPTER VI

THE RATIONALIZATION
OF MATTER

THE *Memoirs of the Royal Academy of Sciences* for the year 1783 (printed, however, in 1785) contain a paper, "Reflections on Phlogiston, to Serve as a Development of the Theory of Combustion and Calcination Published in 1777, by M. Lavoisier." It opens with this paragraph:

> In the sequence of Memoirs which I have communicated to the Academy, I have passed the principal phenomena of chemistry in review. I have insisted on those which accompany combustion, the calcination of metals, and in general all the operations in which there is absorption and fixation of air. I have deduced all the explanations from a simple principle. It is that pure air, or vital air, is composed of a particular principle peculiar to it, which forms the basis of it, and which I have named the *oxygenic principle*, combined with the matter of fire and heat. Once this principle is admitted, the principal difficulties of chemistry have seemed to dissipate themselves and to vanish, and all the phenomena may be explained in astonishing simplicity.

The "matter of fire and heat"—caloric, he later called it—has disappeared from the memory of modern chemistry, which has further simplified these views in recalling Lavoisier as the creator of the correct theory of combustion. In his own belief, his work had a more general significance, a deeper interest. He meant it to be the reform of a whole science. Every science has its orderer in the structure of history, one who first framed objective concepts widely enough to reorient its posture: Galileo for kinematics,

Newton for physics, Darwin for biology. That high place Lavoisier claimed for himself in chemistry. And rightly, for the science owes him far more than its grasp on oxygen as what combines in burnings. It owes him, too, its form, that characteristic combination of material algebra and nomenclature in a language of which the syntax preserves, even now, an eighteenth-century analysis perpetually in act. Its equations balance. Its names say compounds. For Lavoisier read Condillac's educational philosophy of science into laboratory investigations of combustion, heat, and gases, and fashioned his chemistry out of the combination.

The spirit of the *philosophe* re-entered science in the person of Lavoisier, and chemistry found its rationale in a treatise which is both synthesis and primer. The *Elements of Chemistry* has none of the inaccessibility of Newton's *Principia*. Published in 1789, it was written to start the science and its students off together on the right foot of method. Lavoisier remembered with impatience the confusion that had presided over his own chemical education: the preliminary Newtonian pieties in no way borne out by the chaos of ingredients and recipes, the contrast between these verbal thickets and courses on mathematics and mechanics wherein consequences really did open out of postulates and definitions, the professor's tacitly sharing his bewilderment by supposing that his pupils already knew what he was unsure how to teach. "These inconveniences are occasioned not so much by the nature of the subject as by the method of teaching it; and, to avoid them, I was chiefly induced to adopt a new arrangement of chemistry." Given his philosophical background, given his generation, Lavoisier's rearrangement of his science could scarcely have taken any guise other than a naturalistic pedagogy.

Nevertheless, the *Elements of Chemistry* is no mere discourse of method. It contains, too, the food on which that method fed, an account of the great experimental discoveries of Lavoisier's lifetime, repeated and refined in his own incomparable laboratory, where if they were not all or any of them originated, there alone were they understood. For the incoherence of his chemical education was scarcely the fault of his teachers. They had not disposed of this information, which permitted Lavoisier to make a science of their lore. Nor had they yet been taught that weight is the chemist's quantity. The chemistry laboratory of the early eighteenth century offered no scene of careful gravimetrics. It boiled over with investigations rather of the processes than the materials of nature.

INEVITABLY Boyle had failed to work the transformations of matter into an atomic chemistry. All ignorant of gases, chemists could never control their evidence, nor make anything determinate of the atomic hypothesis. Thus set back, eighteenth-century chemistry found its salvation, or its purgatory, in the ambiguous, the once-maligned phlogiston theory. Recently, historically-minded chemists, appalled by the complexities which beset their predecessors, have been taking an indulgent view of phlogiston. Eighteenth-century investigators grappled courageously with the phenomena of electricity, heat, and chemistry. Their annals teem with ineffable "principles" introduced by a science which aspired to nothing if not materialism as the physical bearers of otherwise inexplicable effects. Such was phlogiston, the creation of vitalism in G. E. Stahl (1660-1734), and transmuted through long usage into the rationale of Joseph Priestley and the experimentalists, whose fault it was to aspire to objectivity without having mastered quantities. Nevertheless, phlogiston did make sense of chemistry. It was the principle of combustion.

Charcoal, sulphur, phosphorus burn almost to nothing because they are rich in phlogiston. The fumes form acids. Conversely, vitriol (sulphuric acid) plus phlogiston gives sulphur. Similarly, in smelting, phlogiston passes from charcoal to ore. In general, where there is, in fact, a gain or loss of oxygen, Stahl saw a loss or gain in phlogiston, that which deserts the burning mass to leave behind a formless pile of ashes. This, then, was a looking-glass chemistry, a reversed theory which, answering to its purposes in the youth of the science, became as difficult to overcome as any left-handed habit.

The role of such imponderable fluids in eighteenth-century science is a complex problem, insufficiently explored. But clearly distinctions are required between phlogiston, which was a figment, and (say) electricity which, like caloric and aether, has a history leading far. Of those two more later. Even with phlogiston, however, chemistry was well out of the mystic jungle of alchemy. It had become a science, but a qualitative rather than a metrical one. Some such stage occurs with more or less importance in the evolution of every science, during which it explains effects as real qualities occurring out in nature. So pre-Galilean mechanics explained motion by the impetus impressed, and pre-Newtonian optics saw colors as qualities of light really existent. So too, phlogiston, a help so long as it coordinated phenomena in rational fashion, became a hindrance only after 1765, when accommodation of the findings of gas chemistry began to complicate rather than sophisticate the theory.

The French, it often seems, formulate things, and the English do them. In the chemical revolution, at any rate, pneumatic chemistry was the achievement of the English experimental school, while theoretical chemistry expressed the French instinct for formal elegance. In 1727 Stephen Hales, an Anglican clergyman endeavoring to import Newtonian considerations into physiology and chemistry, had demonstrated an "air" to be "fixed" in many organic substances and in certain alkaline earths. In fact this was carbon dioxide. That gas was not, however, specifically identified as a chemical individual, something distinct from "bad air," until researches which Joseph Black reported to the Philosophical (later Royal) Society of Edinburgh in 1755, and published the next year as *Experiments upon Magnesia Alba, Quicklime, and some other Alcaline Substances.*

Joseph Black, drawn from a medical training to chemistry, was of that intellectual circle which started into vitality around the universities in Glasgow and in Edinburgh. James Watt, David Hume, Adam Smith, Henry Home of Kames, Dugald Stewart, Joseph Hutton, who founded igneous geology—they distinguished the northern kingdom, and though remote from the French style in gravity of mien, their radicalism brought Scotland closer in spirit to the Enlightenment than ever was England in her eighteenth-century complacency. Black began his researches as a thesis for a medical doctorate on *magnesia alba* (magnesium carbonate). On heating it (to get the oxide), he found a constant loss of weight, and found, too, that the residue dissolved in acids yielded the same salt as the original *magnesia alba*, though without effervescing. Limeburners had long produced quicklime from chalk by the same technique. Turning to this more common material, Black decided that the fire, far from adding some "principle" of causticity, simply drove off an elastic component of the alkaline earths, something aeriform as air itself. "Fixed air" he called this first found gas, thus perpetuating with Hales's terminology an ambiguity which his exact research might otherwise have dispelled.

It was, indeed, with Black that the balance as the symbol of the chemist's science displaced the still and the retort as symbols of the chemist's craft. His theoretical conclusions are uninteresting—a naïvely Newtonian model of affinities running between corpuscles. But the scrupulosity of his gravimetric methods, his attention to the purity of his reagents, the patient cogency of each inquiry, the sound tactics of his experimental march—these were altogether admirable, and they were his own. For he is to be taken as the founder of quantitative rather than pneumatic chemistry. Though designed to identify carbon dioxide, his experiments read a little like one of those skillful plots in which the main character never appears, but is known by precise description of his effects on others, until ultimately he stands quite revealed. For gases are elusive.

Black took, for example, 120 grains of chalk, dissolved it in 421 grains of hydrochloric acid, and determined the weight of "fixed air" lost by effervescence to be forty percent. Then he took a second 120 grains of chalk, burned it to quicklime at a loss of forty-three percent, and found that to neutralize it—no effervescence this time—required 414 grains of acid. The quantities balanced within his limits of error, and demonstrated that chalk is forty percent fixed air. But Black in his classic *Experiments* seldom collected the fixed air itself, or studied its properties directly. Rather, he became a professor, and "a load of new official duties was laid on me, which divided my attention among a great variety of objects." He retrieved it to formulate in later years the principles of latent and specific heat.

Ten years later, in 1765, the direct, quantitative study of gases was initiated in London by the Honorable Henry Cavendish, a nobleman, cousin to the Duke of Devonshire, and a scientific recluse and bachelor of the most extreme persuasion, and the most precise, the most finicking temper. His technique substituted mercury for water in the pneumatic trough of Stephen Hales, that little tub in which students still make their acquaintance with gas chemistry. He was able thereby to collect carbon dioxide without loss to solution. There, too, he imprisoned and identified the light "inflammable air" (hydrogen) which chemists had long elicited from vitriol (sulphuric acid) by its reaction with chips of iron or tin, and more recently of zinc.

Now the search became a chase. The most ardent in the hunt, and in his innocence one of the most attractive figures in all the history of science, was the Reverend Joseph Priestley of Birmingham. That industrial city, its practical, Puritan tradition softened now into Unitarianism and humanitarianism, made with London and Edinburgh the apex in the triangle of the British technical community. In March, 1772, Priestley presented his first paper on gases before the Royal Society. He announced the discovery of "nitrous air" (nitric oxide) and of "marine acid air" (hydrogen chloride). He had found, too, another "nitrous air" (nitrous oxide or laughing gas) which might be used to test the quality of common air. In modern terms, two volumes of nitrous oxide plus one of oxygen yield two volumes of nitric oxide. Since air is only one-fifth oxygen, the maximum contraction works out (as Priestley found empirically) at two volumes of air plus one of nitrous oxide to give 1.8 volumes of residual gases. Not only was the product smaller in volume than the sum of the reagents; it was twenty percent smaller than the air alone. And the degree to which this result was approached might be taken as a measure of the "goodness" of a sample.

Priestley's distinguishing characteristic was a kind of manual imagination, an experimental unconventionality. Had he known any chemistry, he is reported to have remarked, he never would have made any discoveries. Oxygen was the most famous, of course. He had obtained an "air" by heating mercuric oxide. In it he put a candle—which flared into a torch:

> I cannot, at this distance of time, recollect what it was that I had in view in making this experiment; but I know I had no expectation of the real issue of it. Having acquired a considerable degree of readiness in making experiments of this kind, a very slight and evanescent motive would be sufficient to induce me to do it. If, however, I had not happened, for some other purpose, to have had a lighted candle before me, I should probably never have made the trial; and the whole train of my future experiments relating to this kind of air might have been prevented.

Such was Priestley's engaging frankness about a "crucial experiment." Ingenious and enthusiastic, he cannot, perhaps, be justly called incoherent, but certainly he was the most discursive of scientists. His *Experiments and Observations on Different Kinds of Air* fill six volumes; his theological and metaphysical writings twenty-six. But even his lengthiness cannot irritate. He always redeems our liking by taking us so utterly into his confidence. He will not, for example, draw quantitative conclusions about the respirability of one of his airs: "Not having had the precaution to set the vessel in a warm place, I suspect that the mouse died of cold." But it did not signify; the creature had survived long enough to prove the sample "better" than common air.

❖

It is difficult to imagine a more diametric contrast of intellectual personalities than that between Priestley and Lavoisier, the protagonists of the chemical revolution: the one ingenious, enthusiastic, naïve, and literal; the other skilled, reserved, sophisticated, and critical; the one a Unitarian preacher, generous, incautious, and predictable in the political radicalism of his kind; the other ambitious to serve himself and the state in the official tradition of French *expertise*, planning his career within an ordered framework as carefully as any series of experiments. And however engaging, Priestley's scientific style did betray that want of judgment and elegance which is often the penalty attaching to the worthiest of radical educations. For eighteenth-century England divorced vigor of mind from urbanity of taste by excluding the dissenting community from the universities of Oxford and Cambridge.

Lavoisier, on the other hand, was educated in the most assured of cultures, at the Collège Mazarin in Paris. His mind emerged one of the finest critical instruments ever formed by French secondary education, that remarkable process which, continuing in the great *lycées*, instills in the French intelligentsia something of the Cartesian spirit, something of its imperative toward order and unity of doctrine. Paris was not the city to appreciate Priestley's qualities. He tells in his *Memoirs* of an only visit there, of the skeptical French eyebrows raised on his assuring the gathering in some salon that, though a Unitarian, he counted himself a believing Christian. Much less could his hosts find common ground with Priestley's haphazard, not to say happy-go-lucky, philosophy of discovery, nor accept

> the truth of a remark, which I have more than once made in my philosophical writings, and which can hardly be too often repeated, as it tends greatly to encourage philosophical investigations; viz. that more is owing to what we call chance, that is, philosophically speaking, to the observation of *events arising from unknown causes*, than to any proper *design*, or pre-conceived *theory* in this business. This does not appear in the works of those who write *synthetically* upon these subjects; but would, I doubt not, appear very strikingly in those who are most celebrated for their philosophical acumen, did they write *analytically* and ingenuously.

Lavoisier's milieu never assimilated the analytical to the ingenuous. The difference is notable in the approach to research. For as Lavoisier entered the world of science in the late 1760's, through geology at first but drawn increasingly to chemistry, he did not move among preachers and pharmacists, noblemen and doctors. He moved amid the French Academy of Sciences, choosing his associates in its mathematical sections, and particularly Laplace, Lagrange, and Monge. These men judged ideas with rigor. They felt, and no doubt they expressed, the condescension for the old chemistry that a theoretical physicist might now feel for the weatherman or the veterinarian presuming on some solidarity of science. Physics had long since learned better than to make theories out of vague principles like phlogiston. Thus it was to the unsatisfactory state of theory, and not to capturing new gases, that Lavoisier turned his attention, and particularly to its skeleton in the cupboard. For if phlogiston were the principle of fire in bodies, then its elimination in combustion should diminish the mass. But it was well known that metallic "calxes" outweigh the virgin metal. Practitioners in the pharmacist's shop, or in the teacher's chair, never cared whether loose ends met, so long as the subject made sense. And Lavoisier was the first to undertake an exact, a systematic study of combustion.

He chose the most combustible of ordinary chemicals, and was rapidly rewarded. On 1 November 1772 he confided, as was his right, a sealed note into the hands of the permanent secretary of the Academy:

About a week ago I discovered that sulphur, in burning, far from losing weight, on the contrary gains it; that is to say that from a *livre* of sulphur one can obtain much more than a *livre* of vitriolic acid, making allowance for the humidity of the air; it is the same with phosphorus; this increase of weight arises from a prodigious quantity of air that is fixed during the combustion and combines with the vapours.

This discovery, which I have established by experiments that I regard as decisive, has led me to think that what is observed in the combustion of sulphur and phosphorus may well take place in the case of all substances that gain in weight by combustion and calcination: and I am persuaded that the increase in weight of metallic calxes is due to the same cause. Experiment has completely confirmed my conjectures: I have carried out the reduction of litharge [lead oxide] in closed vessels, with the apparatus of Hales, and I observed that, just as the calx changed into metal, a large quantity of air was liberated and that this air formed a volume a thousand times greater than the quantity of litharge employed. This discovery appearing to me one of the most interesting of those that have been made since the time of Stahl, and since it is difficult to prevent something from slipping out in conversation with friends which might put them on the track of the truth, I have thought it right to make this deposition into the hands of the Secretary of the Academy against the time when I shall publish my experiments.

At that time Lavoisier was twenty-nine years old. He refined his experiments. He studied and repeated those of Black, whose disciple he was in gravimetrics. He perfected his methods. His essential technique then consisted in the combustion of reagents under a glass bell by means of a burning glass. He was fortunate in his instruments, a fine lens belonging to the Academy, "le verre ardent du Palais Royal," and an even more powerful glass belonging to one of the noble houses patronizing science, "le grand verre ardent de la Tour d'Auvergne." With a clarity of mind which cannot be too much admired, Lavoisier saw what had to be done to create a modern science of chemistry out of that chaotic legacy in which were jumbled the Greek doctrine of elements, old Stoic vestiges of fire and flux, Paracelsian principles of salt, sulphur, and mercury, alchemistical distillations and purifications, mineralogical and metallurgical lore, and the proliferating laboratory discoveries of his own time. And to clear his mind—perhaps, too, to fix it—he wrote down in his laboratory register what he meant to do for the rest of his life. Scholars have not resolved an uncertainty about the date of this document, whether it was composed early in 1772 or early in 1773, just before or just after the experiments reported in the sealed note. The earlier date fits the pattern of his thought, the later of his researches. The issue poses in miniature, therefore, that most interesting problem of Lavoisier's career: which came first? For whichever date may be correct, this document remains the most prescient, not to say *a priori*, program of research in all the history of science:

> Before commencing the long series of experiments that I intend to make on the elastic fluid that is set free from substances, either by fermentation, or distillation or in every kind of chemical change, and also on the air absorbed in the combustion of a great many substances, I feel impelled to set down here some considerations in writing, in order to outline for myself the course that I ought to take.
> It is certain that there is liberated from substances, under a great many conditions, an elastic fluid; but there are in existence several doctrines as to its nature. . . .

It is important to be clear about the complexities that Lavoisier confronted. On the one hand, he was convinced that combustion and respiration involve combination with something atmospheric. On the other hand, "fixed air" (carbon dioxide), the one gas firmly identified in experiment, presents properties almost contrary to those required. It is rather the product than the food of respiration and combustion, and very different from common air.

> These differences will be exhibited to their full extent when I shall give the history of all that has been done on the air that is liberated from substances and that combines with them. The importance of the end in view prompted me to undertake all this work, which seemed to me destined

to bring about a revolution in physics and in chemistry. I have felt bound to look upon all that has been done before me merely as suggestive: I have proposed to repeat it all with new safeguards in order to link our knowledge of the air that goes into combination or that is liberated from substances, with other acquired knowledge, and to form a theory. The results of the other authors whom I have named, considered from this point of view, appeared to me like separate pieces of a great chain; these authors have joined only some links of the chain.

What those authors had mainly neglected was the *source* of air found in many substances. He, therefore, would repeat and extend all the experiments in which air is taken up, so that knowing the origin, he could trace what became of it. "The processes by which one can succeed in fixing air are: vegetation, the respiration of animals, combustion, in some conditions calcination, also some chemical changes. It is by these experiments that I feel bound to begin." It is obvious in retrospect that he needed, first, to distinguish between reactions which "fix" carbon dioxide (vegetation, slaking) and those which require oxygen. Secondly, he had to establish the chemical similarity of burning, rusting, and breathing. Finally, he would recognize the atmosphere as a mixture containing oxygen as a distinctive member of the whole population of gases. For Lavoisier, it would prove the most interesting member by far.

Twenty years later Lavoisier had done the experiments he planned, all but those on respiration and chemical physiology, which he was about to undertake when interrupted by the guillotine. He had put the links together. He had created his body of doctrine and formed the theory of his science. He had worked the revolution, in chemistry at least. And perhaps it amounted to a revolution even in physics to recognize chemistry as a science continuous with itself.

What career has ever been planned in such perfection?

It required money, of course, and Lavoisier was not rich. But he secured a place in the corporation of financiers which farmed the taxes, devoted a portion of each day to acquiring wealth, and acquired, too, his patron's fourteen-year-old daughter as his wife. She was an heiress. She was intelligent. She knew English. She became interested in chemistry, and served as amanuensis in the laboratory. Laboratories were private enterprises in the eighteenth century. Lavoisier proposed an improved technique for making gunpowder, was given charge of the commission created in 1777 to practice it, and was installed in an official apartment and laboratory in the Arsenal at the charge of the state. Madame Lavoisier has left an account of his day which has recently come to light. It was written many years after the death he met in company with his colleagues of the tax farm. For at last accounts that proved an unprofitable association.

Each day Lavoisier sacrificed some hours to the new affairs for which he was responsible. But science always claimed a large part of the day. He arose at six o'clock of the morning, and worked at science until eight and again in the evening from seven until ten. One whole day a week was devoted to experiments. It was, Lavoisier used to say, his day of happiness. A few enlightened friends, several young men proud to be admitted to the honor of cooperating in his experiments, would gather in the laboratory in the morning. There they would lunch. There they would hold forth. There they would work. There they performed the experiments which gave birth to that beautiful theory which has immortalized its author. Oh, it was there that one had to see that man to understand him, gifted with so fine a mind, so sure a judgment, so pure a talent, so lofty a genius. It was by his conversation that one could judge the beauty of his character, the nobility of his thoughts, the severity of his principles. If ever any of the persons who were admitted into his intimacy should read these lines, the memory will not, I believe, retrace itself without emotion in their souls.

There, in the laboratory, Lavoisier might, perhaps, be known. For us, it is more difficult. We may visit the instruments preserved in Paris at the *Conservatoire des arts et métiers*. We may admire the beautiful balances, the elegant pumps, the graceful bell-jars, which lent his results a precision to which no chemist had ever dreamed aspiring. Scholars may go to the Archives of the Academy to peruse the registers from his laboratory. They open with the program of 1772 (or 1773). They might still serve as object lesson in any laboratory of instruction. Every quantity is entered, over all the twenty years. His works are gathered in six volumes. His correspondence is just now in course of publication. Step by step, we may follow his career. Nevertheless, these voluminous, these utterly impersonal remains, they conceal, somehow, the most inaccessible of the great scientists. The man himself escapes us. Something is lacking in that too perfect career. It is not ambition. Every act bespeaks ambition. But outside the laboratory, at least, Lavoisier seems a cold man. Not that he avoided polemics, but his writings in this vein are chilly and clear. There is none of the heat of conflict, no welcome self-betrayal, only that "severity of principles." The defense of phlogiston is too forlorn to tempt the historian. Lavoisier's opponents never seem worthy of his steel. He never spared them for that. He controlled, it may be, everything but the power of his own mind. And the only thing he failed to foresee was the vulnerability of qualities like his, those of the expert, those of the intellectual, the man who knows and is right, at a time when the feelings of the common man (Rousseau's common man, "who knows nothing and thinks none the worse of himself") were at their most intense political pitch in all history, when in the metaphor Burke borrowed from Lavoisier's science, the "fixed air had broke loose, . . . the wild gas was abroad."

But that disastrous incompatibility apart, there was a passion in Galileo and Descartes, in Kepler and Newton, and even under the seemly Victorian guise of Darwin, which one misses in Lavoisier, whose superior lucidity does not quite fill the same office. Is this why, alone among the great founders, he was not a great discoverer? Did that clarity of mind exclude some element of imagination, or simple curiosity, some sense of sympathy for the unexpected inwardness of things? The question can only be conjecture. But there was that in the perfection of his experimental method which might be counted a limitation. The luminosity of the great research memoirs of the 1770's and '8o's, in which he executed the program of 1772, has been often and justly praised. In retrospect, to use his own figure, or pre-figure, each seems a link in the chain supporting the theory of combustion and the conception of chemical reactions supposed by that theory. Knowing the role of oxygen, the constitution of the atmosphere, the composition of water, we do not see how to withhold assent, although we know that some more qualified than we to judge, among them Priestley and Cavendish, never did feel compelled by that fine reasoning.

But to us it is persuasive. Does phosphorus emit or does it combine on burning? Weigh it, and the choice must come down with the balance. Does or does not the additional mass come from the atmosphere? Measure, and the diminished volume speaks the answer. Just so every experiment was designed with unprecedented elegance to answer some absolutely relevant question with a yes or a no. And there, perhaps, was the flaw. For Condillac's logic, which Lavoisier applied, is that of a taxonomy, which is to say of a choice. It is a logic of the discovered, which makes no room for the psychology of discovery, for the adventure into the unknown.

Perhaps there is always a danger that it will impoverish inquiry to elevate the logic of existing science into precepts of method. Those magnificent experiments, they were almost laboratory exercises in the syllogism, deductions from known facts in accordance with maxims like conservation, rather than inquiries after the manner of Newton and his prism, or of Priestley and his tub of mercury, its bubbles yielding what next? Not so Lavoisier, whose experiments, how infinitely superior in design, served finally to clarify what was rightly in the premises of 1772. If so, this too would be consonant with his mathematical inspiration, deductive and rigorous rather than adventitious or random.

Chemistry profited, therefore, from the curious, the almost symbiotic relationship between Priestley and Lavoisier, however unwelcome to both. If Priestley's lack of theoretical taste disqualified him from understanding his discoveries, Lavoisier's lucidity disqualified him from making them. By his own program, combustion, calcination, and respiration, all involve fixation of something from the atmosphere. Lavoisier knew where to look, but not what to look for. Thus in this essential instance did all his method prove incompetent as an instrument of discovery. For it was Priestley who told him.

It is an affliction of immaturity in the historiography of science that many of its historians give themselves over to quarrels about priorities with as much, if not more, bitterness than the discoverers themselves, in whom the failing is more forgivable, if no less inelegant. By now the facts about oxygen seem tolerably clear. That gas betrayed its existence by a peculiar property of its oxide with mercury. On moderate heating, mercury will form its red oxide. On stronger heating, but at temperatures still well within the laboratory range, mercuric oxide will decompose. It was the one relatively common "calx" which might easily be reduced in the absence of charcoal. Only this reduction, therefore, would liberate oxygen instead of carbon dioxide.

Given all the activity in pneumatic chemistry and combustion, there must surely be a sense in which the discovery of oxygen was inevitable. In February, 1774, for example, Bayen in Paris roasted mercuric oxide—and reported the gas as fixed air! Nor was he the first. It is now known that at Upsala, Carl Wilhelm Scheele had been experimenting with oxygen, prepared from mercuric and also from silver oxide. "Fire air" he called it, but published nothing until later. Lavoisier had sent a copy of his first book, and Scheele did mention his discovery in the closing sentences of his letter of thanks, dated 30 September 1774. Lavoisier never acknowledged this rather cryptic communication. The phrasing would not command attention. And he must have received it just when Priestley was in Paris, exchanging experiences with his French colleagues, and shocking them by his religious belief.

Priestley first liberated oxygen on 1 August 1774, distinguished it from carbon dioxide by its insolubility and support of combustion, and mistook it for his familiar laughing gas. Shortly after, he tried the same technique with red lead. In October he was in Paris, and "As I never make the least secret of anything that I observe, I mentioned this experiment also, as well as those with the mercurius calcinatus, and the red precipitate, to all my philosophical acquaintance at Paris, and elsewhere, having no idea at that time, to what these remarkable facts would lead." Nor did he suspect until the following March that he had a new and quite extraordinary air: "But in the course of this month, I not only ascertained the nature of this kind of air, though very gradually, but was led by it to the complete constitution of the air we breathe." Trying all the means of testing air, quite at random at first, he found that his new air supported combustion, and respiration, even *after* the test for diminution with laughing gas: "Thinking of this extraordinary fact upon my pillow, the next morning I put another measure of nitrous air to the same mixture, and, to my utter astonishment, found that it was farther diminished to almost one half its original quantity." Starting, then, with a fresh sample, he concluded from the laughing gas test that it was "between four and five times as good" as common air. And this, of course, was a way of saying, with characteristic indirection, that common air is no simple substance, that it contains about twenty percent by volume of "pure" or "dephlogisticated air." So Priestley always called his great find. He was middle-aged and never changed his ideas. His new air supports combustion with such energy because its utter freedom from phlogiston gives it five times the capacity of ordinary air for blotting up that principle from burning bodies.

Lavoisier seized upon the mercuric oxide experiment more quickly than its author. He repeated it in November, 1774, and chose it to communicate the next spring at the Easter meeting of the Academy. This was always a formal occasion, when the public was admitted to the bar of science, and when, therefore, the Academy took care to offer something interesting. Lavoisier was appearing for the first time in a leading role. He had begun, he told his audience, with one ounce of oxide. After calcination under the lens of the Palais Royal for two and one-half hours, pure mercury was recovered to the amount of seven gros, eighteen grains. The difference was fifty-four grains, and the volume of "air" evolved seventy-eight cubic inches, "from which it follows, that supposing all the loss of weight is attributed to the air, each cubic inch must weigh a little less than two-thirds of a grain, which does not differ much from the weight of ordinary air."

For once Lavoisier let eagerness betray him into wishful accuracy. He distinguished those seventy-eight cubic inches of gas from carbon dioxide, grasped at them as the object of his search, and mistook oxygen, not like Priestley for laughing gas, but for air itself: "And it is very likely that all metallic calxes, like that of mercury, would give only common air if they could all be reduced without addition." For all the circumstances convinced him that what he had in his receiver "was not only common air, but that it was more respirable, more combustible, and consequently that it was more pure than even the air in which we live." Lavoisier paid the price for this, almost his only venture into ambiguity. He had made, not perhaps a crucial experiment, but certainly a crucial mistake. He reported of the last of his tests that this air "was diminished like common air by an addition of a third of nitrous air." Priestley had, therefore, one last contribution to make to the founding of the new chemistry. It was to set the founder right.

Priestley felt a certain pique on reading the Easter Memoir. Nowhere did it mention that Priestley had told of the gas evolved from mercuric oxide. Yet his pleasure in correcting Lavoisier's misidentification is expressed with all the moderation of extreme good nature. He had got it right himself only in the meantime.

As a concurrence of unforeseen and undesigned circumstances has favoured me in this inquiry, a like happy concurrence may favour Mr. Lavoisier in another; and as, in this case, truth has been the means of leading him into error, error may, in its turn, lead him into truth. It will have been seen, in the course of my writings, that both these circumstances have frequently happened to myself; and indeed examples of both of them will be found in my first section concerning this very subject of dephlogisticated air.

It is pleasant when we can be equally amused with our own mistakes and those of others. I have voluntarily given others many opportunities of amusing themselves with mine, when it was entirely in my power to have concealed them.

But I was determined to shew how little *mystery* there really is in the business of experimental philosophy, and with how little *sagacity*, or even *design*, discoveries (which some persons are pleased to consider as great and wonderful things) have been made.

Lavoisier was not, perhaps, one to be amused at a comedy of errors, much less if cast for a major role. He never alluded to this rebuke. But he gave over premature attempts at doctrine, and turned to occupying the ground for a new chemistry in a series of memoirs of which the strategy was as formal as that of an eighteenth-century military campaign.

❖

THOSE decisive memoirs develop three lines which Lavoisier made converge on a new chemistry. The first explores the nature of acids. The second leads to the proposition that combustion, calcination, and respiration are all oxidation. The third advances the problem of heat. It is the second line which now seems central, having continued straight through the history of chemistry and grouped the science around its most frequent class of reaction. But if Lavoisier's deviations on the constitution of acids and the "matter of heat" appear oblique, they are at least as interesting. Unlike the gay and spirited mistakes of the tolerant Priestley, they are the defects entrained by the most notable rationality to illuminate science since that of Descartes himself.

On 26 April 1776, just a year after his Easter Memoir had exposed him to the humiliation of correction, Lavoisier read before the Academy a "Memoir on the Existence of Air in Nitric Acid, and on the Means of Decomposing and Reconstituting that Acid." He reminded his colleagues of his early experiments on combustion of sulphur and phosphorus, and of the formation of acids by absorption of the vapors. Reflection had convinced him that air enters into all acids, and that they are differentiated by other principles specific to each. "The which, at first only a plausible conjecture, was soon converted into a certainty when I applied experiment to theory." For Lavoisier's memoirs take their structure and style from the pattern of his thought, and often begin with the conclusion. Nevertheless, he had learned his lesson. He specifies "not simply the air, but rather the purest portion of the air," as that which enters into all acids and which "constitutes their acidity."

The facts and experiments, he carefully recognizes, were all Priestley's. Nevertheless, the conclusions he draws are so different, that "I hope that, if I am reproached for having borrowed my proofs from the·works of this celebrated scientist, at least my property in the consequences will not be contested me." He chose mercury to make his salt, since its reducibility without carbon eliminates the masking of results by carbon dioxide. Having, therefore, dissolved mercury in nitric acid, he evaporated the solution and heated the residue in two stages. First, he decomposed mercuric nitrate, and collected the nitric oxide. Then he roasted the mercuric oxide and collected oxygen. It is a mark of his purposive method that he distinguished the two gas fractions and held them separate. As always, he kept strict account of quantities. But perhaps the most interesting feature of this memoir is the deployment of yet a further element of sophisticated technique, one which became a hallmark in the memoirs to follow. Nitric acid once decomposed, the complement of the demonstration is "to reconstitute it by recombining the same materials, and in this I have succeeded:

> Nitric acid . . . is nothing other than nitrous air, combined with an approximately equal volume of the purest part of common air, and with a considerable quantity of water.

Nitrous air, on the contrary, is nitric acid deprived of air and water. No doubt there will be some who will ask at this point whether the phlogiston from the metal does not play some role in the operation? Without daring to decide a question so fraught with consequences, I shall only reply that, since the mercury emerges from the reactions in precisely the same state in which it entered, it does not appear to have lost or regained phlogiston—unless it be claimed that the phlogiston which has reduced the metal passed right through the walls of the vessels. But that would be to admit a particular kind of phlogiston different from that of Stahl and his disciples. It would be to return to the fire principle, to some fire combined with bodies, and that is a system of thought very much more ancient than Stahl's, and very different in spirit.

And so Lavoisier flicks at phlogiston, lightly for now, hinting on the one hand that a test of the objective existence of a substance for chemistry is the possibility of taking it out of bodies, and then restoring it, *handling* it in short; inferring on the other hand that to evade this criterion is to fall back out of science onto Heraclitus and the Stoics, onto fire as activity and the world as flux and as becoming.

A year later, on 16 April 1777, Lavoisier read a second major paper, "Memoir on the Combustion of Kunckel's Phosphorus." It extends to phosphoric and sulphuric acid the analysis which makes "pure air" the principle of acidity. But the main point is to apply the double movement of analysis and reconstitution to the atmosphere, which "as I have already suggested several times, is composed of about one quarter of dephlogisticated, or eminently respirable air, and of three quarters of noxious, poisonous air, a type of gas of unknown nature."

Lavoisier was busy all the while with a great number of minor experiments, of which record remains in the laboratory registers. He tried guinea pigs in bell jars of oxygen, and performed autopsies when finally they suffocated. He examined the changes in the blood stream on respiration. He made against Priestley a minor but a telling point— air expelled from the lungs has been converted to "fixed air" (carbon dioxide) and not to "phlogisticated air" (hydrogen) as in Priestley's terms it should be. In one day, he deposited at the Academy twenty-six memoirs on acids, salts, and the incarnations of oxygen. His fidelity and exactness were beyond reproach. And they served a purpose higher than the purist's.

By November, 1777, Lavoisier felt ready to advance an intermediate "Memoir on Combustion in General." In the lucid preamble which always orients his reader to his purpose, he makes one of his most explicit and characteristic avowals on the role of theory:

> Dangerous though the spirit of systems is in physical science, it is equally to be feared lest piling up without any order too great a store of experiments may obscure instead of illuminating the science: lest one thereby make access difficult to those who present themselves at the threshold; lest, in a word, there be obtained as the reward of long and painful efforts nothing but disorder and confusion. Facts, observations, experiments, are the materials of a great edifice. But in assembling them, we must not encumber our science. We must, on the contrary, devote ourselves to classifying them, to distinguishing which belong to each order, to each part of the whole to which they pertain.

In the case of modern chemistry, the facts might be ordered into four classes. 1) Combustion always evolves a "matter of fire," apparent as light. 2) Bodies burn only in very few airs, perhaps only in "pure air." 3) In all combustion, "pure air" disappears, while bodies gain weight by just that amount. 4) In all combustion, the product is transformed into an acid by further addition of "the substance which has increased its weight." By the phlogiston hypothesis, combustible bodies are said to burn because they contain a principle of combustion, and to contain that principle because they burn. So combustion accounts for combustion. But "I dare venture the assertion in advance that the hypothesis I propose explains in a very happy, in a very simple manner, the principal phenomena of physics and chemistry."

Now, Lavoisier turned to maneuver his acids into place. In November, 1778, he presented a paper complementary to that on combustion, "General Considerations on Acids." As always, he reviewed the state of the question. Of old, chemists had pushed analysis only to the point of distinguishing oils, salts, earth, and water, conceived less as discrete types of matter than as quality-bearing principles. Eighteenth-century chemistry had immensely improved its description and theory of neutral salts. Henceforth, chemists must do "for the constituent principles of neutral salts, what the chemists our predecessors accomplished for the neutral salts themselves, which is to say to attack the problem of acids and bases and to push back the boundaries of such chemical analysis yet one more stage." He feels himself ready for a first level of generalization on acids. And relying on his recent experiments, he advanced in a more important way than heretofore the hypothesis "that pure air, eminently respirable air, is the principle constituting acidity."

So confident was Lavoisier of this consequence, this "truth which I regard as very solidly established," that he took that step which was decisive for an eighteenth-century science, to be ventured only in the presence of a truth, the step of naming. All chemistry since has immortalized that error. "Henceforth I shall designate dephlogisticated air, or eminently respirable air in the state of combination or fixity, by the name of *acidifying principle*, or if the same signification be preferred in a word from the Greek, by the name of *oxygenic principle*"—that which generates acidity.

Nor had Lavoisier yet arrived at the notion of elementary oxygen. As he broke loose from the chemistry of principles, he carried with him for a little a diminishing residue of those old notions, more verbal than real. It is in chemical combination that oxygen is the acidifying principle. As a gas, the oxygenic principle is otherwise combined, with matter of fire—caloric—to put it in the aeriform state. But this was peripheral to chemistry. What was grave was that once again, in the full flush of theory, Lavoisier ignored a warning from that laboratory where he was the master of them all, more subtle this one than that which might have saved him the Easter Memoir, but a fact no less thorny. Acids, he wrote, are more complex than their neutral salts, and correspondingly troublesome to analyze. They demand great virtuosity of technique: "I hope, however, to be in a position to show in the future that there is no acid, *with the possible exception of that from marine salt*, which may not be decomposed and reconstituted, and from which one may not remove the acidifying principle and restore it at will."

The possible exception, of course, is hydrochloric acid. The italics are not Lavoisier's.

From the configuration of his work, it would appear that Lavoisier meant to press forward along the acid front. He was diverted by an unexpected opportunity. He himself had raised the question of what the product is when hydrogen burns? Finding none, he left the problem in abeyance. Nevertheless, the difficulty was real. A vanishing combustible was more helpful to an emission than to a combination theory of combustion. Then, in 1781 Priestley tried exploding inflammable air with common air by an electric spark. He was seeking to determine the weight of heat, and only mentioned that the inside of the vessels became dewy. In 1783 Cavendish repeated these experiments, on too small a scale to collect the dew, though he did discern that the reaction reduced the volume of the air involved by one fifth. Interested in the dew, he enlarged the scale, arranged to admit the two gases continuously into a receiver where the combustion was sustained, collected the moisture, and found it to be pure water.

The attendant polemics over priorities have been, if anything, even more ferocious than over the discovery of oxygen. In England James Watt, of the steam engine, in France Gaspard Monge, founder of descriptive geometry and then an impecunious instructor at the school of military engineering at Mézières, made the same discovery, independently, almost simultaneously, and in the case of Monge with no less clarity. Once again, therefore, the interesting question is not who made water, but who understood what had happened? For though the composition of water was the last of the great discoveries of the English pneumatic school, Cavendish interpreted his results as the condensation of water, not its synthesis. Indeed, the light weight of hydrogen encouraged certain of the English chemists, militant in their last ditch, to identify it for a time with phlogiston itself.

Lavoisier, on the contrary, instantly saw water, seemingly the simplest of substances and classically the most intuitive of elements, rather as the oxide of that gas which is—to anticipate the rest of nomenclature—hydrogenerative. He seized upon Cavendish's results as upon a piece of tactical fortune. Or rather upon the word of those results, for this reached Paris before Cavendish had published, borne by Charles Blagden, who visited Lavoisier at the Arsenal in June, 1783. On the 24th, Lavoisier assembled a company of academicians, among them Laplace and Meusnier, and tried the experiment. He burned fifty pints of inflammable air with twenty-five of oxygen, and collected 660 grains of water. Next day Lavoisier and La-place repeated the experiment and announced the results to the Academy. Monge communicated a memoir to the same effect in August. And now Lavoisier enveloped the question in his own method. Synthesis having come first, he proposed to demonstrate the reversibility of the results by analysis. Working with Meusnier, he wrapped a musket barrel in a sheet of copper (to prevent oxidation of the surface), and cradled it on an angle through a charcoal brazier. The lower end was connected through a condenser to a gas receiver. Drop by drop, water was admitted at the upper end, and the quantity compared to that condensed below. The difference "according to the theory which guides us" should be equal to the increase in weight of the gun barrel through oxidation of the bore added to the weight of hydrogen in the receiver. Unfortunately, the high temperatures, together with the smallness of the difference compared to the mass of the barrel, precluded real precision. Lavoisier found water to consist of one-third to one-sixth hydrogen by weight.

A memoir to this effect was read by Meusnier on 21 April 1784, just in time to be published (what with the chronic lag in the Academy's schedule of publications) in the volume for 1781. In all his own writings on this subject, Lavoisier made very fleeting mention of Cavendish. One would suppose the findings to have been the continuation of his own abortive trials of hydrogen, and confirmations of the work of Monge. And perhaps one would be right, in principle if not in fact. For this dramatic demonstration that water is an oxide, though not a "crucial experiment" in the Baconian sense, was nevertheless decisive historically in the campaign to exorcise phlogiston in favor of a positive concept of combustion as that chemical reaction in which oxygen combines.

Thus it fell out that Lavoisier's rationalization of chemistry around the combining role of oxygen appears in retrospect as the linear continuation of his early experiments on combustion. This chapter opened with the first paragraph of the *Reflections on Phlogiston* of 1785. In reality, that beautifully argued, that truly magisterial memoir, represents the completion of the case, the summing up and not the opening argument. It came as no surprise. Lavoisier had not, after all, been able to restrain his impatience with the old doctrines, his intolerance for untidy facts. The self-denial imposed after the Easter Memoir weakened as his confidence grew. Expressions of scorn for phlogiston chemistry escaped him in each successive memoir. Now, the assault is frontal:

> Chemists have made of phlogiston a vague principle, which they in no way define rigorously, and which in consequence is adaptable to any explanation they please. Now this principle has weight, and again it is weightless; Now it is free fire, and again it is fire combined with the element earth; Now it penetrates right through the pores of vessels, and again it finds bodies impenetrable. It explains at once causticity and its opposite, translucence and opacity, colors and the absence of colors. It is a veritable Proteus, changing form at every instant.
>
> It is time to bring chemistry to a more rigorous way of reasoning: It is time to strip off everything merely systematic or hypothetical from the rich store of facts to which that science is every day adding: It is time, in short, to exhibit the stage to which chemical knowledge has attained, so that those who follow us may start from that point and proceed with confidence to the advancement of the science.

Wielding Occam's Razor, Lavoisier cut phlogiston right out of the heart of chemistry, right out of its consciousness. And the chemical revolution consisted in the extension of Lavoisier's conception of combustion to all chemical reactions. That act made of chemistry a modern science.

Forgotten was the lingering animistic instinct for cosmic digestion as the archetype of chemical processes. No longer was the allegiance of chemistry ambiguous as between the two great camps of organic and physical science. Its object became the positive combinations and separations of materials rather than qualities, its events reactions rather than fluxes, coctions, and processes. Henceforth, the chemist weighs amounts. He does not distill out principles. For more intimately in chemistry than in any other science, the foundations of objectivity were embedded in a quantitative conscience.

Scientists have sometimes written that Lavoisier formulated the law of conservation of matter. The reality was simpler. He assumed it. It was for him what it had been for the ancient materialists, a precondition but no finding of his science: "We must lay it down as an incontestable axiom," says the *Elements of Chemistry*, "that in all the operations of art and nature, nothing is created; an equal quantity of matter exists both before and after the experiment." From this principle, he writes again in the Memoir on Water, derives "the necessity to perform experiments with more exactness, and on a larger scale."

Social scientists, for their part, sometimes like to see science drawing inspiration from society and politics. It has even been suggested that Lavoisier derived his chemical "philosophy of the balance sheet" from the accounting practices of the corporation which farmed the taxes. The economic interpretation of creativity takes special pleasure in Lavoisier's hobby of agricultural reform. With the Frenchman's feeling for the soil, with the theorist's interest in new techniques, he bought a manor at Fréchines, not far from Blois. There he realized the physiocrat's dream of a model farm. Madame Lavoisier tells how an account was opened for every field, and a record kept (as in the laboratory) of seed, fertilizer and labor, of yield, price and ultimate return. But surely the historian need not go thus far afield to find the origin of Lavoisier's input-output chemical procedure. Lavoisier, a Parisian, had studied Joseph Black before he became a gentleman farmer or a pillar of the tax farm. His was the spirit of the *philosophe*, shining the bright light of scientific method into dark corners of routine.

Philosophe and scientist, he turned from the Academy, before whom he had read *Reflections on Phlogiston*, to the public whom he wished to educate, repudiating in eighteenth-century style the erring past along with error itself, placing his confident hopes for his science rather in that still malleable youth destined to pay the first installment on posterity's debt to the Enlightenment. The closing sentences of that memoir invoke this future:

> I do not expect my ideas to be adopted all at once. The human mind gets creased into a way of seeing things. Those who have envisaged nature according to a certain point of view during much of their career, rise only with difficulty to new ideas. It is the passage of time, therefore, which must confirm or destroy the opinions I have presented. Meanwhile, I observe with great satisfaction that the young people who are beginning to study the science without prejudice, and also the mathematicians and physicists, who come to chemical truths with a fresh mind—all these no longer believe in phlogiston in Stahl's sense. They regard that whole doctrine as a scaffolding more embarrassing than useful for continuing the edifice of chemical science.

The pattern of 1772 still determined Lavoisier's career. First he had discerned the shape of the new science. Then he had spent the fifteen intervening years in the laboratory comprehending "chemical truths." Now those truths must be disseminated and clothed in rational dress. In 1787 the public meeting of the Academy fell on 18 April. Lavoisier chose the occasion to present a "Memoir on the Necessity of Reforming and Perfecting the Nomenclature of Chemistry." Nothing, he remarked in a private note, seemed to him more urgent for the advancement of the sciences.

> That method which it is so important to introduce into the study and teaching of chemistry is closely linked to the reform of its nomenclature. A well-made language, a language which seizes on the natural order in the succession of ideas, will entail a necessary and even a prompt revolution in the manner of teaching. It will not permit professors of chemistry to deviate from the course of nature. Either they will have to reject the nomenclature, or else follow irresistibly the road it marks out. Thus it is that the logic of a science is related essentially to its language.

Consistently with the purpose, at once pedagogical and methodological, Lavoisier stepped back into the company of his colleagues to let the enunciation of chemical linguistics appear as the work of a school, a French school. In no circumstance was the chemical revolution more intimately related to the Enlightenment. Chemists had often wished to specify their materials more precisely than by the picturesque names of old tradition. But the first to propose a sweeping reform was Guyton de Morveau, trained in the categories of the law rather than of science, and experienced in the affairs of the Parlement of Dijon earlier than in the laboratory, to which he came by way of the industrial, literary, and philosophical aspirations of a provincial man of affairs. In the 1770's the publisher Panckoucke put in hand a revision and systematization of Diderot's Encyclopaedia. In this *Encyclopédie methodique* each department of science and affairs was to have a subsidiary collection to itself. Guyton took charge of the chemical volumes. His proposal for a systematic re-ordering of chemical terminology had the object, therefore, of reducing that science to the encyclopedic state.

Guyton came to Paris to consult the chemists of the

Academy. Lavoisier tells how the magnitude of the project was borne in upon them all, and of how Guyton, "destined to be, in a sense, the spokesman for French chemists," generously sacrificed his private intellectual property in the project to the better part of a collaboration. "Only," Lavoisier continues, "after having reviewed all the parts of chemistry several times, after having meditated profoundly on the metaphysics of language, and on the relation of ideas to words, have we ventured to form a plan."

They published their *Method of Chemical Nomenclature*, as a symposium, in 1787. The tenor is far less doctrinaire than its philosophic inspiration might lead the reader to expect. Common names of so-called simple substances—iron, sulphur, ammonia—were retained wherever they did not entrain really false ideas. If they did—"inflammable air" for example—a word was substituted, drawn usually from a Greek stem, so as to express the most characteristic property of the body in question. The principle was both mnemonic and pedagogic. Students were to be accustomed from the outset to admit no word without attaching an idea. Many "simple" substances would, no doubt, yield to analysis. But so long as the science was organized around a method of naming, rather than a dogmatic nomenclature, its future need not be embarrassed. And indeed, the scheme did provide for the possibility of reading, for example, sodium chloride for salt, potassium nitrate for nitre.

As to compound substances, a classification had first to be agreed on. "In the natural order of ideas, the name of class and genus is that which evokes the properties common to a large number of individuals, while the name of species is that which leads to the idea of the properties peculiar to certain individuals. That natural logic pertains to all the sciences. We have tried to apply it to chemistry." Thus, acids are composed of two substances. One constitutes acidity. As a group, acids assume the status of a class or genus. The other is peculiar to each acid. The specific name must differentiate as modifier. But certain acids, those formed by sulfur for example, contain their two principles in varying proportion: they will be distinguished by varying the suffix, sulfurous or sulfuric, according to the proportion of oxygen. So, too, would oxides, salts, and bases find their places, and the very few carbides and sulphides then known to chemical science. Nor is it necessary to insist on the principles which have guided chemical nomenclature ever since.

Nomenclature, conservation, quantity, economy, synthesis-analysis, rationality, enlightenment in truths of nature—such were the materials which in 1789 were assembled in the *Elements of Chemistry*, Lavoisier's first general work, and perforce his testament. And it has seemed best to let the organization of this chapter itself exemplify a Lavoisier analysis: decomposing, distinguishing those elements, and recomposing in the mind of history and—one hopes—in its understanding and even in its sympathy.

❖

THAT sympathy will go out to Lavoisier more naturally, perhaps, in his failures, and in his death, than in his precisions or his triumph.

Reading about combustion, heat, and acidification in Lavoisier's memoirs, the modern chemist will, of course, follow the strand of combustion. It is the warp thread about which his science has woven its fabric. Of the two lines that led astray, the theory of acidification has seemed the more venial a mistake. Most acids do contain oxygen, after all, even if it is in the wrong place, and the notion of acidification as super-oxidation has been indulged as a natural over-extension of theory. But caloric has been handled with some severity. It was Lavoisier's belief that heat is the manifestation of a subtle, elastic fluid—light too, perhaps, though as a chemist Lavoisier was more interested in heat. On the strength, or rather on the weakness, of this view, the most recent of Lavoisier's historians, Maurice Daumas, questions the reality of a distinction between Lavoisier's chemical doctrine and that of "principles" which he supplanted. Caloric, imponderable and permeating, was no more positive than phlogiston. "In spite of his genius," says Daumas in the final sentence of an illuminating book, "Lavoisier could act and think only as a man of the eighteenth century."

But not, one feels, in the retrograde sense which M. Daumas (in this connection only, it is true) implies; and it will, perhaps, be conformable to the tactic of that century to argue toward a reversal of his emphasis from an etymological analysis of Lavoisier's two cardinal principles. For the word "oxygen" honored in the breach the caution which presided over the systematic nomenclature of 1787, ensuring that names should be noncommittal against the chance of future discoveries. Ignoring the inconvenience of hydrochloric acid, Lavoisier fell into mortal sin against fact. He put himself (if the historian may borrow the logician's banality) in the position of a naturalist questing in the Antipodes who should ignore his fleeting glimpse of a black swan in deference to the proposition that swans are white.

The word "caloric" entailed no such fatality. Unlike oxygen, it was coined in a spirit which combined the precepts of Condillac with a Newtonian prudence which knows bodies by their effects. Guyton explained the term in a footnote to the *Nomenclature*. It denotes that which is perceived as heat without implying anything about its nature. Nor will its function appear as a derogation from Lavoisier's chemical wisdom. For caloric does not act as a chemical agent. It is excluded, absolutely excluded from the practice of chemistry, by the most elementary consideration of method. Like light, heat passes right through the walls of laboratory vessels. The bounds of the experiment are no boundaries for heat. It may not be weighed. Its effects are like some action which remains unknown to common law, not because it is nonexistent, but because of legal conventions. Just so a science which found its metric in weighed masses might not take cognizance of caloric.

Surely this consideration must restore all the sharpness of the distinction on which Lavoisier himself insisted. For Priestley as formerly for Stahl, and indeed for the whole school which Lavoisier set his face against, phlogiston constituted the essence of theory. Nor could this concept be outgrown. It had to be turned on its head. No mathematical statement of the quantities embraced by phlogiston could possibly be abstracted from the question of its nature. Not so caloric, which (like Newton's aether) entered permissively rather than constitutively into the structure of theory. Caloric disappeared from science in the nineteenth century. Nevertheless, the theory of oxidation, and by extension the conception of chemical reaction, remain what they were left by Lavoisier, and not by Priestley.

Drawing out the analogy with aether may help. In a note of 1774 Lavoisier had conjectured that under proper conditions of pressure and temperature, any body may assume each of the three states of matter. What transforms a solid to a liquid, and over a higher threshold a liquid to a gas, is permeation by caloric, which counters Newtonian attraction:

Thus, [according to the *Elements of Chemistry*] the particles of all bodies may be considered as subject to the action of two opposite powers, the one repulsive, the other attractive, between which they remain *in equilibrio*. So long as the attractive force remains stronger, the body must continue in a state of solidity; but if, on the contrary, heat has so far

removed these particles from each other as to place them
beyond the sphere of attraction, they lose the adhesion they
before had with each other, and the body ceases to be solid.

If (to render his thought by a terminology which was not
his) the gaseous state argues permeation by heat as anti-
attraction, then caloric its medium was anti-aether. And
Lavoisier introduced a medium for the same reason as
Newton had done: not as a constituent of positive theory,
but as a suggestion designed to situate theory in an intelligible picture of the world. Newton came to that last,
closed the *Principia* with the aether, and sorely puzzled
his readers. Caloric, on the other hand, comes first in the
Elements of Chemistry. That book is nothing if not good
pedagogy, and Lavoisier opened it with the most familiar
of physical phenomena, the expansive effects of heat and
the complete reversibility of those effects.

Caloric, then, belonged rather to Lavoisier's picture of
the universe than to his chemistry. It played the part, not
of a chemical body, but only of a physical or better a
mathematical body—in the same sense in which Newton
left gravity a mathematical and not a mechanistic force.
Throughout his career Lavoisier's predilection was for the
intellectual style of his mathematical colleagues. Studying
heat, he went beyond consultation to collaboration with
the strongest mind among them, with Laplace. On 18
June 1783, Laplace read to the Academy their "Memoir
on a New Method of Measuring Heat." That method employed the ice calorimeter. It took advantage of Joseph
Black's principle of latent heat to measure the specific
heats of various bodies by the weight of ice melted across
a measured range of cooling. Insulation was insufficient
to permit good results except when the external temperature was near the freezing point, and continuation of the
experiments had to await the return of a winter colder
than is usual in France. Nevertheless, they made a num-

ber of determinations, and their memoir was the fountainhead of calorimetry.

And what is very instructive is that those measurements
succeeded, and those calculations were performed, in the
face of a fundamental disagreement between the two col-
laborators on the nature of heat:

> Physicists are divided on the nature of heat. Some regard
> it as a fluid permeating all nature, which penetrates bodies
> to a greater or lesser degree in proportion to their tempera-
> ture and capacity. It may combine with bodies, and in this
> state it ceases to affect the thermometer or to flow freely
> from one body to another. It is only in the free state, which
> permits it to reach equilibrium in bodies, that it forms what
> we call *free heat*.
>
> Other physicists think that heat is only the result of im-
> perceptible motions in the molecules of matter. Everyone
> knows that all bodies, even the densest, are filled with in-
> numerable pores or little cavities, of which the total volume
> may considerably surpass that of the matter they surround.
> This leaves the parts freedom to oscillate in all directions,
> and it is natural to think that these parts are in continual
> agitation, which if it increases to a certain point, may dis-
> integrate and decompose the body. It is this internal motion
> which, following the physicists of whom we speak, consti-
> tutes heat.

They do not say (though from other sources it is known)
that Lavoisier was the partisan of heat as caloric, and Laplace, the Laplace who completed Newton in celestial
mechanics, of heat as motion. Surely, their dilemma is
bound to appeal to our own science, which must live with
its principle of complementarity, even as Newton in his
day had lived with the duality of light, now periodic and
again corpuscular. So, too, Lavoisier and Laplace never
needed to resolve this issue in order to fulfill their object.
Their calorimeter would know no distinction between
conservation of caloric and conservation of kinetic energy
—*vis viva* in the terminology of the time.

We shall not decide between the two preceding hypotheses. Some phenomena seem favorable to the latter. Such, for example, is that of the heat produced by friction between two solid bodies. But there are others which are more simply explicable in the former. Perhaps, they both take place at once. But however that may be, since only these two hypotheses are possible on the nature of heat, then we should admit the principles common to both. Moreover, according to the one or the other, *in simple mixtures of bodies, the quantity of free heat is a constant.*

And in the *Elements* Lavoisier, speaking now only for himself, develops the advantages of this mode of treatment:

Wherefore, we have distinguished the cause of heat, or that exquisitely elastic fluid which produces it, by the term of *caloric*. Besides that this expression fulfills our object in the system which we have adopted, it possesses this further advantage, that it accords with every species of opinion, since, strictly speaking, we are not obliged to suppose this to be a real substance; it being sufficient, as will more clearly appear in the sequel of this work, that it be considered as the repulsive cause, whatever that may be, which separates the particles of matter from each other, so that we are still at liberty to investigate its effects in an abstract and mathematical manner.

Any science of matter must concern itself with two sorts of problems: on the one hand the configurations of particles, and on the other the propagation of phenomena in space. It was to serve the latter necessity that Lavoisier introduced caloric. It is a continuous medium for the streaming of heat. And there, perhaps, we may leave it for the moment, in conceptual equilibrium with aether its complement, there where the nineteenth century was to adopt these notions, aether to be identified with fields of force and vanish into relativity, caloric to merge rather into the continuum of energetics and to be swept away on the great, irreversible stream of thermodynamics. Formulating the Second Law of that enigmatic science in 1822,

Sadi Carnot still handled caloric as flowing from a real reservoir of heat down a continuous gradient.

Strategically, therefore, caloric was no retreat from science. It might, indeed, be argued—this book in fact does argue—that it was the essential instrument by which Lavoisier achieved objectivity for chemistry. That achievement is a precise intellectual act in the history of each science. Criticism should be able to say what constituted it, even when the scientist is himself too pressed into his problems to be so explicit, or so self-conscious. And Lavoisier made between caloric and heat (most signally the heat of combustion) that distinction which Newton drew between aether and light, which Galileo interposed between change and motion, and which Darwin was to make between the origin and preservation of organic variations. Thus he projected old qualities of matter out from chemistry onto caloric, or read them into the perception of its effects. For though no chemical agent, caloric was physically innocent, serving conservation laws rather than some mystical or romantic, some personal or sympathetic nature.

❖

Nor is it caloric which will open in the armor of Lavoisier's rationality some chink through which our sympathy may slip. Only in the theory of acidification does the appealing light of a really interesting error break through the fault in that tightly articulated career. In 1782 appeared a final memoir in the series on acids, "General Considerations on the Dissolution of Metals in Acids." But for the accidental decisiveness of water, it is this memoir (it seems clear) which would have prepared the ground for *Reflections on Phlogiston*. And in that case, *the* Lavoisier theory of oxidation would appear in history as the wrong conception of acidification instead of the right one of combustion.

The opening paragraphs announce the imminence of a definitive attack upon phlogiston. They review his doctrine on combustion and calcination (rusting) as oxidation: "But what is not yet sufficiently known, and what I propose to prove in this memoir, is that a perfectly similar *humid* calcination occurs when metals dissolve in acids, that in all such cases there is decomposition of the acid or the water, and that there is united with the metal a quantity of the oxygenic principle approximately equal to that which the metal is capable of removing from the air in a dry calcination." To exemplify that proof, Lavoisier returned to his early experiments on the "air" in nitric acid, repeated them with all the sophistication now at his command, and argued the results with unwonted urgency. Once again he dissolved mercury in acid. He neutralized the residual acid to determine by the difference what amount had reacted with the metal. He knew the percentage of oxygen contained in nitric acid. He determined the increment of weight upon the metal. And he found the last two quantities to answer. Just so did his precision seal him into an error springing from another source than weights and undetectable by any balance.

But what is most interesting is Lavoisier's mode of representing these results, so correct in quantity, so wrong in principle. In order to see what he was about, one has to follow him into some detail. It is obvious, he writes, that acidification of a metal involves many variables—heat, concentration, chemical affinities, etc.—each of which is a force acting with characteristic energy. Therefrom results a problem complex and difficult of solution:

Better to exhibit the state of the question in this respect, and in order to show at a glance the result of what happens in metallic solutions, I have constructed formulas of a sort, which could at first be taken for algebraic formulas, but which have not the same object, and do not derive from the same principles . . .

Let any metallic substance be **S. M.**

Let any acid be ∿

Let water be ▽

Let the oxygenic principle be ⊕

Let nitric oxide be △Ɨ

Let nitric acid be ⊖Ɨ

Then, we will have, for the general expression of any metallic solution

$$\text{S. M.} (\triangledown \, \sim)$$

This general formula will vary according to the nature of the acid and that of the metal. Thus, for example, to express the solution of iron in nitric acid, we will have

$$(\,\male\,) \, (\triangledown \, \ominus\!\!\text{Ɨ} \,)$$

But, since nitric acid is itself a compound, we must substitute the proper values, and the formula then takes the form:

$$(\,\male\,) \, (\triangledown \oplus \triangle\!\!\text{Ɨ})$$

Let the quantity of iron be a, it is clear that it will require a determinate quantity of acid to dissolve it, that there is a relation between the quantity of acid and that of iron; and that in designating this relation b, I shall have $a b$ as the quantity of acid required for this solution.

Further, it is clear that a quantity $a\,b$ of nitric acid contains a certain proportion of water, which I could call .... $\dfrac{a\,b}{q}$

A certain proportion of oxygenic principle, ... $\dfrac{a\,b}{s}$

A certain proportion of nitric oxide .... $\dfrac{a\,b}{t}$

Finally, to prevent excessive effervescence, a certain amount of water must be added. Thus the formula becomes:

$$(a\,\sigma) + (2\,a\,b\,\nabla + \frac{a\,b}{q}\,\nabla) + (\frac{a\,b}{s}\,\oplus + \frac{a\,b}{t}\,\triangle).$$

This expression represents the solvent and the solute before mixing. But then [to paraphrase for brevity] the metal takes from the acid that amount of oxygen which will saturate it. For each metal, this quantity bears a constant proportion to the quantity of the metal. Since a is the quantity of metal, $\dfrac{a}{p}$ may represent the quantity of oxygen needed to saturate it. Clearly, when solution is complete, this quantity should be added to that of iron in the formula, which then becomes

$$(a\,\sigma + \frac{a}{p}\,\oplus) + (2\,a\,b\,\nabla + \frac{a\,b}{q}\,\nabla)$$

$$+ (\frac{a\,b}{s}\,\oplus - \frac{a}{p}\,\oplus + \frac{a\,b}{t}\,\triangle)$$

This gives Lavoisier all his parameters, and we will not follow the successive steps, in which he subtracts the air evolved as approximately equal to the weight of oxygen absorbed by the iron, supposes for simplicity that the quantity of acid used is unity, and replaces his coefficients with the actual quantities of an experiment, to end with this expression:

$$(0^{\text{livre}}, 2\,\text{♂} + 0^{\text{livre}}, 058\,\text{⊕}) + (2^{\text{livres}}, 5\,\triangledown)$$

$$+ (0^{\text{livre}}, 192\,\text{⊕} + 0^{\text{livre}}, 192\,\triangle\text{⊟}).$$

One step more, and this might seem the first chemical equation. But instead of placing the formulas for reagents and products on either side of an equality sign, Lavoisier wrote them as two sums to demonstrate the equivalence of mass before and after.

Is this not the most tantalizing memoir in the history of chemistry, and in Lavoisier's the most appealing? Here alone, one gets a sense of modesty. He never claims for these expressions the dignity of algebra. "We are still very far from being able to introduce mathematical precision into chemistry, and I beg, therefore, that no one consider these formulas . . . as more than simple annotations, of which the object is to ease the labors of the mind." (But compare this disclaimer with what he says of algebra itself in the *Method of Chemical Nomenclature*: "Algebra is the analytical method *par excellence*: it was invented to facilitate the labors of the mind, to compress into a few lines what would take pages to discuss, and to lead, finally, in a more convenient prompt and certain manner to the solution of very complicated questions.")

Here alone, one gets a sense of true novelty, of a Lavoisier who is groping beyond the program of 1772 toward the mathematicization of chemistry. These embryonic chemical equations, they clothe his experiment in a far more structural formalism than that of nomenclature, a formalism like that in which Laplace and he had just quantified exchanges of heat and differentials of temperature in their joint *Memoir on Heat*. And surely it is a fair likelihood that Lavoisier found in that collaboration the idea of expressing chemical quantity in "the same abstract and mathematical manner." "To know the energy of all these forces," he concludes, "to succeed in giving them a numerical value, to calculate them, that is the goal which chemistry should propose itself. We are moving by slow steps. But it is not impossible that we shall win through."

Here alone Lavoisier touches our fellow feeling by his very mistakes, for here they appear, not as the retribution properly awaiting an overbearing dogmatism, but as risks freely run in an adventure of the mind. Here he plays the innovator rather than the lawgiver. And those mistakes, indeed that failure in this memoir, emerge as inevitable defects of the very great qualities of order and reason which put his science on its objective footing.

For it has not been appreciated how Lavoisier's life ended in an intellectual as well as a personal tragedy, how in both aspects he fell victim to grave faults in that eighteenth-century philosophy of science which he had made his own. That philosophy prepared the reforming spirit in the Revolution which took his life. It is true that the democratic resentment which then raged around the Academy, whose spokesman he had made himself, erupted rather out of popular romanticism, playing on that sentimental vulgarity which lurks in Baconian utilitarianism, never far beneath the surface. Nevertheless, the educational philosophy of science—which was to develop into positivism—had also made its compromise with progress and utility. It had dressed out the method of Newtonian physics in the logic of Baconian natural history, which classifies the forms and species of things. Naming ("Codifying," as Bentham would say, "like any dragon") it had reached in practice for the naturalist's easy instrument of classification, in preference to the theoretical physicist's exacting instrument of abstraction and mathematicization.

Now it should be apparent how much more closely Lavoisier's error over acids was entrained by his philosophy of science than was his hypothesis of caloric. Let us read him in the *Elements* on ordering by nomenclature:

> To those bodies which are formed by the union of several simple substances we gave new names, compounded in such a manner as the nature of the substances directed; but, as the number of double combinations is already very considerable, the only method by which we could avoid confusion was to divide them into classes. In the natural order of ideas, the name of the class or genus is that which expresses a quality common to a great number of individuals: the name of the species, on the contrary, expresses a quality peculiar to certain individuals only.
>
> These distinctions are not, as some may imagine, merely metaphysical, but are established by nature. "A child," says the Abbé de Condillac, "is taught to give the name *tree* to the first one which is pointed out to him. The next one he sees presents the same idea, and he gives it the same name. This he does likewise to a third and a fourth, till at last the word *tree*, which he first applied to an individual, comes to be employed by him as the name of a class or a genus, or an *abstract idea*, which comprehends all trees in general. But, when he learns that all trees serve not the same purpose, that they do not all produce the same kind of fruit, he will soon learn to distinguish them by specific and particular names." This is the logic of all the sciences and is particularly applied to chemistry.

Is it not obvious, from this model of botanical linguistics, that what we finally have to do with, even in Lavoisier, is a chemical taxonomy? Most notably does this appear in the collective *Method of Chemical Nomenclature*. Into the heart of that book is folded a tabulation of chemical substances arranged to exhibit the order of the science on a single sheet. This, too, of course, is the object of the Mendeleev Periodic Table, to which every modern chemistry class may lift its eyes should it grow bored with its professor. But Guyton's principles are altogether different. His table has five columns, each subdivided to juxtapose the old name against the new. Column I divides the "simple substances" into five classes. There are only fifty-five of these individuals in the entire chemical population. In the first class fall four principles, which have in common only their extreme activity: light, caloric, oxygen, and hydrogen. Class II comprises twenty-five acid radicals, Class III sixteen metals, Class IV five earths, and Class V three alkalis. Finally, resisting classification, is a miscellany of seventeen organic composites which seem to act as simple substances.

But it is in the organization of the columns that the modern student may see at a glance how, siren-like, Lavoisier's beautiful principles had led him past the theory of combustion, past the objectification of his science, down a wrong path to a dead end. Column II tabulates the physical action of caloric in putting certain simple substances into the gaseous state: thus oxygen gas is the combination of the oxygenic principle with caloric, and so, too, of hydrogen, nitrogen, and ammonia. Column III assembles the combination of each simple substance with oxygen. Here fall water, all the acids of course, and the old "calxes" now become metallic oxides. To the acids correspond in Column IV the "gaseous oxides"—nitrous gas, carbonic acid gas (the old fixed air), etc. Finally Column V contains the very few composites which do not traverse the acid state: sulphides, carbides, and the metallic alloys.

Thus, the chemical population is tabulated according to two criteria of reference, two "coordinates" if one wished for the moment to grant Lavoisier's claims for this process as a true analysis. (One would be wrong to do so, except in sympathy.) The one, the horizontal arrangement of the table, finds the place of each simple substance, each species, in the natural order. The other, the vertical arrangement by columns (neglecting Column II which has to do with physical state rather than chemical combination) relates every compound to the mode by which some simple substance combines with oxygen. Oxidation becomes, therefore, the critical reaction. One might even call it the privileged reaction, in analogy to the place made by Galileo for circular motion in his not dissimilar rationality, his really quite comparable determination to order a science within the closed scheme of the intelligible. Just so did Lavoisier's science fall victim to the ultimate incompatibility of the algebraic and the taxonomic procedures. That incompatibility vitiated the entire philosophy of science of the Enlightenment. There it lurked under those good intentions, concealed as in Lavoisier's case by the commitment to education and humanitarianism. The preface to the *Elements* contains a remark about his predecessors which might, perhaps, be taken as his own epitaph: "All these chemists were carried along by the influence of the genius of the age in which they lived, which contented itself with assertion without proofs; or, at least, often admitted as proofs the slightest degrees of probability, unsupported by that strictly rigorous analysis required by modern philosophy."

Aspiring to that rigor at the end of his career, Lavoisier turned in the last memoir on acids to the mathematicization of chemistry—and that fine project aborted for having been conceived on the wrong ontology. He grasped at the wrong door. The final terms of his analysis were not such that numbers of general bearing could be assigned them. He took his coefficients from the weights of the reagents in each experiment. Even if he had formally equated reagents to products, instead of simply comparing the sums, his equations still would have had no generality. For what he was numbering were the forms of matter, not its particles. And throughout all history, from Aristotle on, the mathematicization of forms has yet to succeed. Historically speaking, only extension on the one hand,

and parts on the other, atoms and the void, had yet been brought to the test of number, the one by true analysis, never by classification, the other by geometric synthesis, both together by analytical geometry. And this is too bad. For Lavoisier believed in atoms. But

> All that can be said upon the number and nature of elements is, in my opinion, confined to discussions entirely of a metaphysical nature . . . I shall, therefore, only add upon this subject that if by the term *elements* we mean to express those simple and indivisible atoms of which matter is composed, it is extremely probable that we know nothing at all about them; but, if we apply the term *elements*, or *principles of bodies*, to express our idea of the last point which analysis is capable of reaching, we must admit, as elements, all the substances into which we are capable, by any means, to reduce bodies by decomposition.

He believed in atoms. But he never thought to count them, never thought to match or mate them, not even in imagination. It is not only too bad: it is very ironic. Lavoisier treated caloric mathematically. But not matter. Matter he only weighed and classified.

Tragic would be too melodramatic a word for this wrong turning, although Lavoisier, who could little bear to be wrong, might not have thought so. But thus to end in a blind alley was at least an indignity. It was an indignity that Lavoisier left an epilogue to be played out, on another stage, in a different vein. That new chemistry, so proudly and so justly claimed by Lavoisier as a French science, was consummated not in the elegant Gallic style of Condillac, not according to Lavoisier's own standards of theoretical distinction, but in the concrete English manner of Boyle, and under the banner of Newton.

For if Boyle was the founder of atomic physics, John Dalton, not Lavoisier, was the first recognizable physical chemist.

❖

THAT EPILOGUE is quickly told in its essentials. It has to do, of course, with the foundations of chemical atomism. Without that theory the development of modern chemistry is simply unthinkable. But in the over-arching structure of the history of science, the atomic theory goes beyond the chemical. For chemistry is the science which first clothed the ancient atomic assertion with scientific meaning, which made it a theory and not just a philosophical policy. Lavoisier's mind did not run much on these matters. Nevertheless, what decisively vindicated his conception of chemical combination, and his assumption of conservation, was its reducibility to atomism and its consequent accessibility to numbers. For the Mendeleev Table and the concept of valence depend upon atomic number rather than upon chemical typology: oxide forming, acid forming, base, or whatever.

After Lavoisier's death, in the emergence of French science from the Revolutionary torment, discussion turned to the mechanism of reaction, the nature of the chemical bond, and ultimately to the structure of matter. The first important suggestions were advanced by Berthollet. It is extremely unfair to that really notable laboratory chemist that he should appear in this history of scientific ideas as the author of those that were wrong. In 1801 he did, it is true, adumbrate the law of mass action. He did not express it in quite the modern form, but he had it clear from his experiments that the rate and extent of reactions are functions, not just of chemical properties, but of the amounts of the reagents, of the concentrations. From there, he saw the problem of chemistry as one of affinities, (the word is much older). He would seat the science in the relative reactivity of chemical agents. Then on the basis of specific affinity and relative concentration, the chemist would predict the amount of one reagent that will combine with another. For example, the amount of oxygen which will combine with a given quantity of copper depends on the relative affinities, which are fixed, and the concentration of the oxygen. There was, in Berthollet's view, an infinite series of oxides of copper, or indeed of any substance, up to certain saturation points. Nor was this view unsupported by evidence. Not to mention amalgams and glasses, copper and iron do oxidize gradually. There was even a certain warrant in Lavoisier for predicting the extent of a reaction according to the circumstances of the reagent. His tabulation lists alloys under Column V, and elsewhere he alludes to a variable composition of nitric acid up to the saturation point of oxygen.

Nevertheless, Berthollet's view was disputed by Proust, drawing on the same evidence, but disagreeing in principle. For despite occasional inconsistences in Lavoisier's own usage, it was implicit in his chemistry that a given compound always contains its components in a fixed proportion by weight. Proust made the distinction, the first to do so clearly, between mixture and true combination, and explained the apparent variation in metallic oxides as the mixture of two oxides of definite but differing proportions, ferrous and ferric oxide, for example. And he laid down, as an axiom of chemistry rather than as a finding, the principle now known as the Law of Definite Proportions. That law accords well with scientific instincts. It is fair to say that it was displacing Berthollet's conception even before Dalton explained it by the hypothesis that every chemical combination consists in discrete, characteristic atom-to-atom linkages.

Like Priestley's, John Dalton's milieu was the earnest Nonconformity of the English Midlands. Manchester was his home instead of Birmingham, the Society of Friends his religion instead of Unitarianism. The son of a weaver, his circumstances were even humbler. His mind, too, was formed by self-education. It was a strong mind, ingenious rather than distinguished. He expressed it quite without elegance or charm.

Some obscurity hangs about the genesis of Dalton's atomic theory. Nor is it likely to be dispelled, since certain of his private papers were a casualty of the second World War. Nevertheless, it is quite sure that he came to it through physical rather than chemical considerations. The *New System of Chemical Philosophy* of 1808 drew together findings of the previous five or six years. Like Lavoisier's *Elements* it opens with a discussion of heat, not however because of the familiarity of the phenomena, but because thermometry, expansibility, specific heats, and states of matter were what interested Dalton. He reached the atomic hypothesis of combining matter only on page 213, and then rather by way of explaining than introducing it. His first book had dealt with meteorology. Nor did he ever become adept in the laboratory. Even by the standards of the time he tolerated a very generous margin of error. He had never studied chemistry until he began to consider gases, probably in search of models of atmospheric phenomena on a laboratory scale. Neither does one experience in his writings the inwardness of that science, as in reading Priestley or Lavoisier. On the other hand, his scientific style gives a notably graphic instance of thinking in models. And the success of the chemical atomic theory testifies very persuasively to the advantage, on occasion, of trying a fresh model.

The advantage lay in the model and in Dalton's fidelity to the mechanical philosophy in Boyle's sense. For compared to the deep researches of the French school, the problems he resolved with it seem almost trivial. In 1802 and 1803 he was investigating the solubility of different gases in water. A colleague, William Henry, had just announced the law that the weight of gas dissolved is directly proportional to the pressure, and in the case of a mixture, to the pressure of that gas alone. Relying on certain rather inconclusive experiments, and upon the mechanical hypothesis, Dalton supposed that the separation of gas particles one from another in the vapor phase bears the ratio of a small whole number to their interatomic distance in solution. Henry's law follows as a consequence if this ratio is a constant for each gas at a given temperature.

Trying his solubilities further, Dalton found that light gases and elementary gases are least absorbable, and that solubility increases with density and chemical complexity. Yet not knowing Avogadro's hypothesis that equal volumes of gas contain equal numbers of molecules, he could not go directly from density to relative atomic weights. Instead, he turned to chemical combining equivalents to estimate the relative weights of gas atoms. Thus did he import the atomic hypothesis into chemistry, together with his "rule of greatest simplicity." According to this assumption, if two elements combine in only one compound, they form a binary compound AB; if in several, the first is binary and the next two ternary (AB_2, A_2B). Taking hydrogen as 1, his information gave him 5.66 for oxygen, 4 for nitrogen, 6.66 for water, etc. These results were much improved in later years, until he arrived at 8 for oxygen.

Apparently Dalton lived with this picture for some years before he saw it as applying primarily to chemical combination. The immediate use he made of it extended it rather from the solubility to the mixture of gases. In Proposition 23 of Book II of the *Principia* Newton had derived Boyle's law from the definition that an elastic fluid is composed of small particles repelling each other by a force inversely as the distance. But a difficulty had arisen. All unknown to Newton, there are at least three gases in the atmosphere. Why is oxygen not a bottom layer, water vapor in the middle, and nitrogen bringing up the top? Dalton worried this problem for several years.

He tried combining his atoms into composites, but they would never come out even. He tried making each atom a little world with an atmosphere of caloric around it, evening things up to bring all the atoms to the same specific gravity by making the affinity for heat proportional to weight. But this conflicted with the tendency of caloric to seek its own level. Therefore—this is a real step, of course—he gave up heat as a repulsive mechanism. Then it occurred to him that perhaps atoms repel only their own kind. Kinetics would then explain diffusion. From this he supposed that it is because of differences in size and weight that atoms elect only their own species to repel, "the particles of one kind being from their size unable to apply properly to the other at their common surfaces of contact." This is obscure enough, but again it led Dalton back to chemistry. Once again that science could answer his questions. Not as to size, but chemistry does deal with weights:

> This led the way to the combinations of gases, and to the *number* of atoms entering into such combinations, the particulars of which will be detailed more at large in the sequel. Other bodies besides elastic fluids, namely liquids and solids, were subject to investigation, in consequence of their combining with elastic fluids. Thus a train of investigation was laid for determining the *number* and *weight* of all chemical elementary principles which enter into any sort of combination one with another.

And a plate in the *Chemical Philosophy* shows how that must be (See facing pages 256 and 257).

In this *New System of Chemical Philosophy*, Dalton simply drops the unanswerable question of particle size, and concentrates on weights and numbers.

> Now, it is one great object of this work, to show the importance and advantage of ascertaining the relative weights of the ultimate particles, both of simple and com-

ELEMENTS

Simple

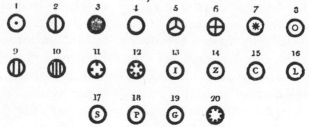

Binary

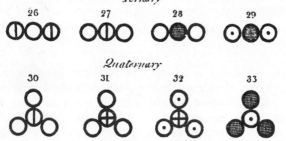

Ternary

Quaternary

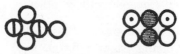

Quinquenary & Sextenary

Septenary

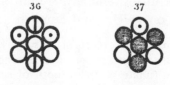

This plate contains the arbitrary marks or signs chosen to represent the several chemical elements or ultimate particles.

Fig.						Fig.					
1	Hydrog. its rel. weight				1	11	Strontites	-	-	-	46
2	Azote,	-	-	-	5	12	Barytes	-	-	-	68
3	Carbone or charcoal,		-		5	13	Iron	-	-	-	38
4	Oxygen,	-	-	-	7	14	Zinc	-	-	-	56
5	Phosphorus,	-	-		9	15	Copper	-	-	-	56
6	Sulphur,	-	-	-	13	16	Lead	-	-	-	95
7	Magnesia,	-	-	-	20	17	Silver	-	-	-	100
8	Lime,	-	-	-	23	18	Platina	-	-	-	100
9	Soda,	-	-	-	28	19	Gold	-	-	-	140
10	Potash,	-	-	-	42	20	Mercury	-	-	-	167

21. An atom of water or steam, composed of 1 of oxygen and 1 of hydrogen, retained in physical contact by a strong affinity, and supposed to be surrounded by a common atmosphere of heat; its relative weight $=$ - - - - - 8

22. An atom of ammonia, composed of 1 of azote and 1 of hydrogen - - - - - - - - - 6

23. An atom of nitrous gas, composed of 1 of azote and 1 of oxygen - - - - - - - - 12

24. An atom of olefiant gas, composed of 1 of carbone and 1 of hydrogen - - - - - - - - 6

25 An atom of carbonic oxide composed of 1 of carbone and 1 of oxygen - - - - - - - 12

26. An atom of nitrous oxide, 2 azote $+$ 1 oxygen - 17

27. An atom of nitric acid, 1 azote $+$ 2 oxygen - - 19

28. An atom of carbonic acid, 1 carbone $+$ 2 oxygen 19

29. An atom of carburetted hydrogen, 1 carbone $+$ 2 hydrogen - - - - - - - - - - - 7

30. An atom of oxynitric acid, 1 azote $+$ 3 oxygen 26

31. An atom of sulphuric acid, 1 sulphur $+$ 3 oxygen 34

32. An atom of sulphuretted hydrogen, 1 sulphur $+$ 3 hydrogen - - - - - - - - - - 16

33. An atom of alcohol, 3 carbone $+$ 1 hydrogen - 16

34. An atom of nitrous acid, 1 nitric acid $+$ 1 nitrous gas - - - - - - - - - - - - - 31

35. An atom of acetous acid, 2 carbone $+$ 2 water - 26

36. An atom of nitrate of ammonia, 1 nitric acid $+$ 1 ammonia $+$ 1 water - - - - - - - 33

37. An atom of sugar, 1 alcohol $+$ 1 carbonic acid - 35

pound bodies, the number of simple elementary particles which constitute one compound particle, and the number of less compound particles which enter into the formation of one more compound particle.

Thus Dalton completed the chemical revolution. The principle of simplicity is wrong, of course. For Dalton, water was HO, and oxygen had the atomic weight of eight. Many of his gravimetric determinations were inaccurate. But though often wrong in fact, Dalton was right in principle. And we may stop here, therefore, short of Gay-Lussac's law of combining volumes, which by the rule of simplicity Dalton could never accept, short too of Avogadro's Hypothesis which reconciled the differences, and far short, finally, of the valence theory which gave Mendeleev the numbers he needed for that tabulation of chemical elements which is truly an analysis of their properties.

All that is interesting, and of gathering technical significance. But chemistry moved into its strategic place in the progress of science with Dalton. He it was who brought Newton and Lavoisier face to face, eliminating the mediation of Condillac. We are done now with the educationists in science. "The novice," says Dalton, brushing him aside in a footnote rather than writing for him, "will all along understand that several chemical subjects are necessarily introduced before their general history and character can be discussed." All the terrible difficulties that Lavoisier had overcome, in the theory of combustion and reaction, in the quantitative technique, in the rationalization of the language—all these Dalton simply took for granted, along with the chemical information that accompanied them. That was his right, of course, as of any scientist. And to them he brought old wisdom straight from his reading of seventeenth-century corpuscular philosophy:

It has been imagined by some philosophers that all matter, however unlike, is probably the same thing, and that the great variety of its appearances arises from certain powers communicated to it and from the variety of combinations and arrangements of which it is susceptible. . . .

This does not appear to have been his [Newton's] idea. Neither is it mine. I should apprehend there are a considerable number of what may properly be called *elementary* particles, which can never be *metamorphosed* one into another.

For in Dalton it is for the first time obvious why it is better to think of streams of particles than streaming fluids. They can be numbered.

CHAPTER VII

THE HISTORY OF NATURE

BOTH CHEMISTRY AND BIOLOGY are eighteenth-century sciences in inspiration. Chemistry, the science of matter, has remained true to the rationalism which gave it form. Biology, on the other hand, etymologically the science of life, has redirected (or betrayed) the impulse it took from romantic idealism. It offers a more general illustration than oxygen of the treachery of facts to names. The word biology was coined by Lamarck in 1802 to lend cosmic unity to natural history, that descriptive study of living nature which classified detail until the mind reeled in boredom along ordered rows of trivia. But the theory of natural selection, by which Charles Darwin explained the evidence for organic evolution, assimilated the development of life to its circumstances in objective nature. The theological issue obscured this deeper question at the publication in 1859 of *On the Origin of Species*. Now, however, that the implications of that deceptively unassuming book have been understood in their full generality, biology is no longer to be partitioned off as the science of life. It is a science of nature, and the boundary between life and nature becomes one of narrowing ignorance rather than of principle. Quite generally, indeed, the historical movement of modern science has transferred the arena where unity reigns from nature into science itself, until in positivism the ancient assertion that there are no boundaries or jumps applies rather to science than to the nature it objectifies, or alienates, or even (as romantics would say) annihilates.

Because the theory of natural selection is what turned the study of all living nature into an objective science, the historian of ideas interested in the strategy and structure of progress will, when he turns to biology, fix attention upon the background and consequences of evolutionary thought. He will place a different emphasis from that of the historian of technique or medicine. And if his compass is brief, he is likely to slight the deep nineteenth-century researches, the beautiful laboratory investigations, the admirable toughening of experimental rigor, to which are owing much knowledge and health. Perhaps it is unjust that the founders of evolutionary thought, Lamarck and Darwin, Cuvier and Lyell, Huxley and Mendel, should be so much more familiar to the general reader of history than those who in a technical view were (it may be) greater scientists: Xavier Bichat and François Magendie, Johannes Müller and Karl Ernst von Baer, Claude Bernard and Louis Pasteur, Theodor Schwann and Matthias Jakob Schleiden, Rudolf Virchow and Robert Koch. Thanks to their detailed work, the great specialties of modern biology build, confident of their foundations: histology and physiology, cytology and embryology, bacteriology and pathology. Yet no one of these fine subjects afforded biology a vantage point from which to reorient its whole posture. They invited, and indeed demanded, penetrating inquiry. They were less suited to stimulating broad ideas.

Neither will a focus for interpretation be found in the notorious, the singularly inconclusive debate between vitalism and mechanism, nor yet in the doubly quixotic conflict in which scientists and theologians charged one another's windmills. Indeed, these issues have been confused too often, which is easier than to keep them distinct since both were very vague. As will appear, Lamarck's theory of evolution was no refuge for theology. It was an application to taxonomy of Diderot's organismic and metamorphic philosophy of nature. But even if the theological issues be kept distinct (as they seldom are), discussions of vitalism still contribute a measure of misunderstanding to the history of biology—a larger measure than to biology itself, which has got past it. To take the same example, Lamarck is often set over against mechanism as a vitalist. But the real question in criticizing Lamarck, and indeed the whole tradition of biological romanticism, is not whether an animal is more than a classical machine. Obviously it is. So too, since relativity, is the solar system, and this is a circumstance which makes physics neither more nor less humane, neither more nor less accessible to theology.

The serious question is what model of order biology will contemplate, and what instruments it will employ. Is it to suppose an order of things different from the order embraced by physics? Is it to seek out different laws of nature? And from this point of view, there is little to choose between mechanism and vitalism. Both propose the organism (whether or not it be only a machine) as the ultimate object of inquiry. Both address biology to a special kind or level of organization superior to what physicists find, and vaguer. And this organismic conception of order was that on which the great continental biologists in practice acted, grooved as they all were into specialties, constantly confronted by organisms in the laboratory, their attention fixed on cells, embryos, diseases, or whatever, rather than on the course of nature.

Only for limited investigations like Harvey's of the circulation, had life scientists ever assumed a physical order, or quantified their information. The failure was critical. It had confined natural history and medicine to dependence on classification and dissection. Then once biology was named, it found its field of phenomena being fought over by champions of two quite incompatible conceptions of order. They were incompatible but equally unphysical, since one derived ultimately from the Stoic and the other from the Aristotelian traditions. On the one hand, the romantic nature philosophy of the Enlightenment had revived the ancient, the pagan sense of cosmic organism. Such was Goethe's innate, indwelling order, the bodily expression of identity and personality. Individual animals participate in life process, in this sense, as organs do in the life of the single body. But (as will appear in the example of Lamarck) there is no correlation between this and a theological view of nature. On the contrary, the organismic is a self-sufficient order. It may be a moral order, but such morality will be naturalistic, never theistic. Diderot was an atheist, Goethe was no Christian.

In Christian culture, on the other hand, the little finger of Aristotle was still (despite the Enlightenment) thicker than the Stoic loins. The providential conception of divinely created order was far more deeply ingrained than romantic naturalism. Against this, indeed, whoever held the Christian view of nature was bound to set his face, not because of its physical arguments, but because of its moral pretensions. In Christianity, God is creator, teacher, and judge, not nature itself. And in the Platonic-Aristotelian sense of order, Christianized in the Middle Ages, which transmuted our culture, all nature is a divine artifact. Every effect in nature is a device for carrying out the purposes of the supreme artist, which may surpass our understanding as they certainly surpass our skill. Archdeacon William Paley, the dean of English natural theology, would often point out the superiority of divine to human craftsmanship. "How difficult it is," he wrote, "to get a wig made even. Yet how seldom is the *face* awry!" But we worship the artist and not the work of art. Natural science, therefore, is only one of the servants of theology, the queen of the sciences. In serving, it becomes natural theology, which demonstrates the existence of the designer from the evidence for design, of the watchmaker from the watch. "Natural theology," wrote Bacon, "is rightly also called Divine Philosophy. It is defined as that spark of knowledge of God which may be had by the light of nature and the consideration of created things; and thus can be fairly held to be *divine* in respect to its object, and *natural* in respect to its source of information."

The universe, then, is one vast design, contrived of an infinite complex of subordinate expedients, all intended to work just as they do work with the ultimate purpose of promoting the welfare of created beings, and fulfilling the destiny of mankind. All this seems simple-minded enough nowadays in the twentieth century—now that worship of God has softened into devotion to man, so that religion in the mass becomes less and less distinguishable from social service, while the masses contrarily prefer to serve themselves with the fruits of science and technology. (For there, it may be, in the material plenty which science provides, and not in some intellectual defeat of theology, lies the real victory.) Perhaps it was simple-minded. But it was also Newton's view of nature. It was, indeed, the view assumed by most scientists until well into the nineteenth century. At least this was so in Protestant countries, and most notably in Britain. There scientists sprang from the industrious middle classes, and there the injunction to individual interpretation of experience marched comfortably with Bacon's supposition that to investigate the working of nature is to study the works of God. Nor did their findings yet embarrass a religion close to the Bible.

The fault in this structure of scientific support for religion lay deeper than the Biblical creation story. For God did not create the world in order to ignore it or (which amounts to the same thing) to withdraw behind the laws of nature. He governs it, and His government takes the form of Providence. "We know Him," wrote Newton (to recall a passage already quoted in part),

> only by His most wise and excellent contrivances of things, and final causes; we admire Him for His perfections; but reverence and adore Him on account of His dominion: for we adore Him as His servants; and a god without dominion, providence, and final causes, is nothing else but Fate and Nature. Blind metaphysical necessity, which is certainly the same always and everywhere, could produce no variety of things. All the diversity of natural things which we find suited to different times and places could arise from nothing but the ideas and will of a Being necessarily existing.

And in practice eighteenth-century natural theology tended to illustrate the governance of God from interruptions in the course of nature, from cataclysms like the punishment meted out in the Biblical Flood, for example. Newton himself supposed in passing that God must occasionally intervene in the motions of the planets to set right certain anomalies which he thought cumulative. Thus, theology put itself at a quite unnecessary disadvantage as compared to science, so that (as Whitehead once noticed) whenever scientists are forced by evidence to modify their theories, it seems a triumph for science, whereas when theologians find themselves under the same necessity, it is taken as a defeat for religion.

This continuing humiliation is a heavy price to exact for the fault of the first generations who had to live with modern science. With Newton, they fell into the unfortunate habit of arguing the existence of God from what science had discovered, and His governance from what it had not. Inevitably, therefore, any new territory embraced by science was withdrawn from the domain of Providence. And it was the descriptive sciences, geology and biology, which had to incur the strain and bear the odium. They were just then emerging from natural history to become historical in both the ancient descriptive sense and the modern temporal sense. Newtonian physics has nothing to say about the development of nature. For all of Newton, the universe might be the same today as on the day of its creation. Geology for its part was the first of the sciences to become historical in compass, and to touch in passing on the question of the Creator's control of events. And living nature had always been the favorite repository whence illustrations might be drawn of "all the diversity of natural things which we find suited to different times and places."

What argument could be more persuasive than that God's infinite attention to detail should be exemplified in the adaptation of the forms of life to the lives they have to lead: that the butterfly should have been created to resemble the leaf, the tiger given its sabre tooth to seize the doe, and the giraffe endowed with its reach to browse off tree tops? How otherwise than by purposeful creation might the phenomena of adaptation be explained? Nothing is more striking to the student of organic nature. As a young man, Darwin accepted Paley's reasoning: "The logic of . . . his *Natural Theology* gave me as much delight as did Euclid . . . I did not at that time trouble myself about Paley's premises; and taking these on trust I was charmed and convinced by the long line of argumentation." It was not, therefore, just a question of religious commitment, though that embittered the discussion and saddened many. Rather, the theological explanations of adaptation, employing the instrument of classification by form or species, were the last manifestations of Aristotelian science, that magnificent structure of thought which after two thousand years was still serving the sciences of organic nature in the nineteenth century—and still does serve, if only as a habit, in breaking which biologists grow strong and wise.

❖

THE *Natural History of Invertebrates* in seven volumes by Jean Baptiste de Lamarck was published in Paris between 1815 and 1822. *The Animal Kingdom Arranged in Conformity with its Organization* in four volumes by Georges Cuvier was published in 1817, and was followed by a second systematized edition of Cuvier's *Researches on Fossil Bones* in seven volumes from 1821 to 1824. To turn over the pages of these immense and lucid compilations is to traverse the painstaking detail in which natural history passed over from Aristotelian classifications to those of modern zoology. This new science links old nature with living in paleontology. It establishes relationships in space and time by the rigorous techniques of comparative anatomy instead of in the more obvious distinctions between four-leggedness or two, infant-bearing or egg-laying, slithering, swimming, or flying, and the like, which had typified animals since Aristotle. Botany had already made that transition in the work of Linnaeus. It is, indeed, possible to consider the work of Cuvier and Lamarck as an extension to zoology of Linnaean taxonomy, with its four levels of class, order, genus, and species. As in botany, refinements have been introduced into the ordering, along with immense accretions of detail. Contemporary species and very old ones in their thousands and their millions have fitted into place, or moved from here to there. But the categories that modern biology treats as the classes and orders of living nature are evident in principle, and often in definition, in Linnaeus, Cuvier, and Lamarck. The difference is that evolutionary biology has superimposed arrangements into phyla. It seeks pedigrees. It looks for lines of descent amongst the species which Linnaeus and Cuvier saw as discrete populations, fixed in form and capable only of creation or extinction, and which Lamarck saw as eddies in the one stream of life.

Cuvier and Lamarck were colleagues at the great

Muséum d'Histoire Naturelle, Buffon's old *Jardin du Roi* nationalized, rationalized, and splendidly supported by the Revolution in its enthusiasm for those sciences it took to be humane and democratic. Their works, indeed, were essentially publications of the arrangement they established among its splendid collections, some inherited from the old regime and others appropriated for the Republic by the revolutionary armies from the foremost natural history cabinets of Europe. Nor did they work alone. There were twelve chairs in this, the most lavishly endowed scientific institution in Europe. Geoffrey Saint-Hilaire and Brongniart addressed themselves to particular genera of mammals, Latreille to insects, and Lacépède (though superficially) to fish. There were demonstrators, laboratory assistants, and students. There were constant visitors from Italy, from Germany, and from America. Alexander von Humboldt nearly became a Parisian. Nevertheless, the contribution of the *Jardin des Plantes* (to give the Museum its popular name) to modern zoology and paleontology was polarized around the uncomfortable relationship of Cuvier and Lamarck. Their work was complementary. They could agree on details of actual taxonomy. The technique they employed Lamarck learned from Cuvier. But they could never agree on the structure of nature. They were like some pair of sixteenth-century astronomers, one Ptolemaic and the other Copernican, who differed not about practice, but only about fundamentals. They exemplify, indeed, the two great pre-Darwinian conceptions of biological order, Lamarck the romantic and metamorphic, and Cuvier the Aristotelian and providential.

Cuvier was Lamarck's senior in authority but his junior in years. In 1827, it fell to Cuvier as Secretary of the Academy to compose the *éloge* in which that body pauses for a moment in making the history of science to reflect on the accomplishments of each deceased member. Cuvier was not generous in his tribute to Lamarck: "It is his observations on shells and polyps, . . . the sagacity with which he has circumscribed and characterized their genera, . . . the perseverance with which he compared and distinguished the species, fixed the synonymy, and gave detailed and clear descriptions," by which Lamarck "had in the end raised to himself a monument as lasting as the objects on which it rests." This was not the credit Lamarck wanted. He often exhorted his students to aspire beyond the detail and pedantry of taxonomy, way beyond it to a philosophy of nature. He offered himself to his students and his colleagues, indeed he thrust himself upon them, as natural philosopher to his generation.

Cuvier's judgment was cruel (his colleagues required him to omit certain strictures from the printed tribute) and scientifically correct. It was Lamarck's contributions to paleontology which entered into the developing structure of biology and helped pave the way for Darwin, exhibiting a concrete succession of organic forms. Lamarck's evolutionary theory, on the other hand, became famous only after the *Origin of Species*, when certain of Darwin's critics fell back upon it, either to discredit Darwin's originality, or to substitute a humane alternative to natural selection, or for both these reasons mixed together.

Lamarck became a zoologist late in life, upon his appointment in 1793 to fill a chair in the reorganized *Jardin des Plantes*. Until then, he had been known to science only as an anti-Linnaean botanist. He first advanced his theory of organic evolution in 1800, when he was fifty-six years old. Few men develop a fundamentally new outlook at that age, and Lamarck's evolutionary theory was, in fact, simply the transfer to his new concern with the animal kingdom of speculations with which, in succession to Buffon and Diderot, he had long been preoccupied in writings on chemistry, geology, and meteorology. And so embarrassing were Lamarck's theoretical ventures, not as he resentfully imagined to the security of some vested orthodoxy, but simply for the light in which they placed their author, that he was never refuted. He was not even dignified by unjust treatment. The only conspiracy against him was one of silence. During all his lifetime, no scientific judgment was ever published on these writings. Officially they did not exist. "I know full well," he once observed bitterly, "that very few will be interested in what I am going to propose, and that among those who do read this essay, the greater part will pretend to find in it only vague opinions, in no way founded in exact knowledge. They will say that: but they will not write it." And one can hardly fail to admire his fidelity and sympathize with him as he toiled through the last twenty years of his life under the shadow of encroaching blindness, on what was expected of him, on what he did superlatively well, and on what he considered to be work of inferior dignity.

Nevertheless, it is as an evolutionist that Lamarck is famous—for who cares about an invertebrate paleontologist? Lamarck's post-Darwinian notoriety may even seem to have prevailed over Cuvier's judgment. But history has the last word over science only at the risk of getting it wrong. It will be well, therefore, to reconstitute Lamarck's evolutionary philosophy in order to retrieve it from his reputation as an unappreciated precursor of Darwin, one who was right in principle but wrong about the inheritance of acquired characteristics. For Lamarck meant his work to establish, not simply the subordinate fact of transmutation, but a view of the world. His theory of evolution was the last serious attempt to make science out of the old instinct that the world is flux and process, and that science is to study neither the configurations of matter nor the categories of form, but the manifestations of that activity which is ontologically fundamental, as bodies in motion and species of being are not. And to put his thought in his own perspective, it will help, perhaps, to move from what is familiar in Lamarck to what is less so, and thus to trace the formation of his theory of evolution from its final application to taxonomy back to its origin in the general pattern of romantic resistance to physical science, and in his specific case, to the conception of chemistry as a physical science which Lavoisier shared with Priestley and all its founders.

Lamarck's *Zoological Philosophy* of 1809 contains the full development of the evolutionary theory which he later exemplified in the *Natural History of Invertebrates*. Concise statement was never his own way. Nevertheless, it is possible to abstract a summary from the *Zoological Philosophy*. The book has three divisions. Part I treats of natural history, Part II of physiology, and Part III of psychology. In living nature, according to this philosophy, inheres a plastic force—indeed living nature *is* a plastic force—forever producing all varieties of animals from the most rudimentary to the most advanced by the progressive differentiation and perfection of their organization. If this action of organic nature were omnipotent, the sequence would be altogether regular, a perfect continuum of organic forms from protozoa to man. But the innate tendency to complication is not the only factor at work. Over against it, constraining it into certain channels of necessity which we mistakenly take for natural species, works the influence of the physical environment. The dead hand of inorganic nature causes discontinuities in what the organic drive toward perfection would alone achieve. These appear as gaps between the forms of life. Changes in the environment lead to changes in needs; changes in needs produce changes in behavior; changes in behavior become new habits which may lead to alterations in particular organs and ultimately in general organization. But the environment cannot be said to act directly on life. On the contrary, in Lamarck only life can act, for life and activity are ultimately one. Rather, the environment is a shifting set of circumstances and opportunities to which the organism responds creatively, not precisely as the expression of its will (although Lamarck's admirers interpreted him in that fashion), but as an expression of its whole nature as a living thing. And it was rather as a consequence than as a statement of his view of nature that Lamarck laid down two corollaries which he described as laws: that of the development or decay of organs through use or disuse, and that of the inheritance of the characteristics acquired by organisms in reacting to the environment.

Seven years before, in 1802, Lamarck published *Researches on the Organization of Living Bodies*, a treatise based on the course he had been giving at the *Jardin des Plantes*. Here, too, the reader will find the main evolutionary principles. But the emphasis is different. The position is simply that species do not exist, and what interests Lamarck is rather the whole tableau of the animal series. We are to see it, not as the chain or ladder, but as the escalator of being. For nature is constantly creating life at the bottom. And the life fluids are ever at work differentiating organs and complicating and perfecting structures. And there is a perpetual circulation of organic matter up the moving staircase of existence, and of its lifeless residue spilling as chemical husks back down the other side, the inorganic side. The series in nature is indeed regular. But it resides, not in unreal species, but in "masses," a conception of great attraction for Lamarck, which he defined as the principle *systems* of organization.

It was in 1800, however, in his opening lecture for the year, that Lamarck first asserted the mutability of animals.

Here the emphasis takes still another form. The theses of 1802 appear, as do most of the evolutionary principles of 1809, but they do so in very summary statement and are introduced simply as subsidiary propositions illustrating the main contention. This is that natural history must begin with the fundamental distinction between living and non-living bodies, between organic and physical nature.

Now, not only is this the argument of Lamarck's debut as a zoologist. It is also the argument of the final diatribe in the campaign he had been waging for over twenty years against the new chemistry—ever since the early papers of Priestley and Lavoisier. In *Memoirs of Physics and Natural History* of 1797, Lamarck refers in passing to the immutability of animal species. But here the central dynamical proposition is that all inorganic composites are residues of life processes, perpetually repairing the decay and disintegration which are the basic tendencies of physical nature. Returning for a moment to *Zoological Philosophy*, this is perfectly consistent with the ultimate theory of evolution, which describes the departures from regularity in the animal scale as consequences of the conflict between organic nature and the brute nature of the environment. This relationship between organic nature as order and physical nature as disorder, a situation both of opposition and dependence, is fundamental to Lamarck's thought, which in this respect is almost dialectical.

Nor is the inconsistency on species other than trivial. In a short essay of 1802, he tells us how he came to alter his views on what he then saw as a detail. All he did between 1797 and 1800 was to assimilate the question of animal species—or rather their nonexistence—to that of species in general. For in Lamarck the word has not lost its broader connotations. He had long been impressed with the perpetual decay of the earth's surface and had long shared the opinion that there are no permanent species among minerals. The only entities in inorganic nature are the "integral molecules" and the masses which form in the play of circumstance and attraction.

There is a striking parallel with the view which Lamarck came to hold of the living world. In both organic and inorganic nature nothing but process links the individual—the particular animal, the particular molecule—and the system of organization—mammalian quadruped, granitic structure—into which it is temporarily cast. This explains Lamarck's pleasure in the concept of masses as the links in the double chain of systems along which materials move from mollusc to man, from limestone to granite. It was natural to think of the principle of mammal in the same fashion. The old intuition that minerals are molded by some plastic force, that they are bred in the earth, was still widespread.

In Paris, Lamarck complains, the chemists teach that the integral molecule of every compound is invariant, and consequently that it is as old as nature. It follows that species are constant among minerals. As for himself, he continues, he is convinced that the integral molecule of every compound can change in its nature, i.e. in the number and proportion of the principles which constitute it. To deny this is to deny the reality of the phenomena of chemistry, the fermentations, the dissolutions, the combustions, which leave the molecules in some different condition, as to form or density or other characteristics.

Moreover, the other two aspects of the *Zoological Philosophy*, the physiology and the psychology, these too will now appear as derivative from an archaic chemistry, both in manner and substance. This was a contemplative chemistry conducted far from laboratories. Lamarck lumped Priestley and Lavoisier together in the "pneumatic" school. His was rather the sympathetic chemistry of Venel and

Diderot. Its cardinal principle was that only life can synthesize. Conversely the physiology of growth consists in retention during youth of what is needed from the materials which the organism passes through its system. Aging and death follow on the progressive hardening of the pliant organs by this life-long digestion of the environment. Later, Lamarck adapted his principle of an equilibrium balancing life against mass to provide evolution with a mechanism. It was analogous to erosion. (Lamarck hit upon the idea of evolution at the time he was writing his uniformitarian treatise on geology.) The property of life fluids is to wear away new channels, new reservoirs, new organs in the soft tissues, and thereby to differentiate structures and specialize functions. The individual organism silts up and dies, but leaves more highly complicated descendants.

Lamarck composed his first chemical treatise in 1776. It was his earliest scientific essay, and it contains an interesting note. In order to explain the origin and mechanism of the universe, he writes, we need to know three things: the cause of matter, of life, and of that *activity* everywhere manifest. In all his chemistry, Lamarck attached primary importance to the element of fire, and regarded oxygen as a perfectly gratuitous postulate. Not only has it never been seen, but combustion is explicable as the action of fire, which *can* be seen in the act of burning or shimmering over a tile roof in the sun of a summer day. For fire is the principle of activity in nature. It exists in many states, of which Lamarck undertook the taxonomy. Conflagration is fire in a state of violent expansion, penetrating a body and ripping it to shreds. Evaporation occurs when fire in a state of moderate expansion surrounds molecules of water, and bears them upwards, so many molecular balloons, to rejoin the clouds. Lamarck also aspired to found meteorology. Finally there is a natural state, to which fire strives to return, and this striving explains the phenomena of light and heat, of the atmosphere and the sun.

Nor did Lamarck ever abandon his commitment to fire. It provided him with a physical basis of feeling and of life itself, and this will make clear the mistake of those who have taken him for a vitalist. His dichotomy of organic and inorganic nature provides no escape into transcendentalism, and that has always been the door through which vitalists have slipped from science into mystery. Life is a purely physical phenomenon in Lamarck, and it is only because science has (quite rightly) left behind his conception of the physical that he has been systematically misunderstood, and assimilated to a theistic or vitalistic tradition which in fact he held in abhorrence. In his view spontaneous generation was no continuing miracle. Life was activated by the stirring of fluids. Lamarck hinted that this process is quickened by fire, and on the mechanism of sentience he was explicit. Its physical basis is the nervous fluid, the same substance as the electrical fluid, which itself is only a special state of fire. The pyrotic theory, therefore, embraces matter, life, and activity, and in that theory lay the common origin of the three aspects of the *Zoological Philosophy*—its psychology, its physiology, and its evolutionary view of species.

In no serious sense, therefore, is Lamarck's theory of evolution to be taken as the scientific prelude to Darwin's. Rather, as epilogue to an attempt to save the science of chemistry for the world of the organic continuum, it was one of the most explicit examples of the counter-offensive of romantic biology against the doom of physics. He escapes the consequences of the particulate views everywhere accepted by denying the molecule, the ultimate particle, that permanence required by the doctrine of chemical fixity. For Lamarck's attack upon Lavoisier was of a piece with Goethe's *Farbenlehre*, extending even to their mutual resentment of the claims of mathematics to speak as the language of science. His theory of evolution simply sucks the animal kingdom into the vortex of universal flux. Lamarck's, therefore, is bound to remain an unenviable position in the history of science. He is a truly outstanding figure. But this ambiguity inevitably hangs about the merit of his achievements, and his career seems to oscillate between poles of futility and pathos. The futility is that of any victim of a plot, however fortunate the outcome. And the tacit agreement to ignore his theories did have the effect of turning him into a distinguished taxonomist, not perhaps against his will, but certainly against his inclination. The pathos is of one who achieved recognition for what he held in small esteem and never for what he prized.

No pathos spoiled the success of Cuvier, who went from strength to strength. It is characteristic that Lamarck should have ended among the humbler forms of life, classifying the "white-blooded" creatures, all vaguely lumped together by Linnaeus in the catch-all of "Vermes—worms"; while Cuvier, though he cut his scientific teeth on the molluscs, should have moved up the scale to appear as impresario of the dramatic and outré: the mastodons and pterodactyls, the fossil pachyderms and sabre-tooths. Cuvier was the better showman. But he was also the better scientist.

He was never altogether French, and the reorientation which he brought to his subject was only partly technique. Partly, too, it was perspective. Cuvier was born in Montbéliard in the border region of the Jura, which is culturally and ethnically ambiguous as between France, Switzerland, and Germany. His language was French. His religion was Calvinist, full of the old Huguenot sense of personality and responsibility, still untouched in that country region by Unitarian inroads. His sovereign was the Duke of Württemberg, and he studied as a royal pensioner in the Caroline Academy of Stuttgart. There he fell under the influence of Kielmeyer, and of German *Naturphilosophie,* which encouraged his passion for nature study without taking possession of his scientific soul. Nature study was only his joy, however. He trained for the civil service of that small German state. While awaiting an appointment, he accepted a post as tutor to the son of a Protestant nobleman of Normandy, the Comte d'Héricy. Cuvier arrived in Caen in 1788 at the age of nineteen. Events soon rendered the administration of Rhineland duchies an irrelevant career, and he passed the Revolution in peaceful Norman seclusion, going only to the local natural history society. For he spent his time collecting and contrasting the animals of the beach, comparing them to fossil shells from certain quarries, dissecting cuttle fish, doing anatomies on the molluscs, and for his own satisfaction undertaking a systematization by anatomical criteria. "It was thus," writes a friend and early biographer, an American lady whose style conveys the spell he cast, "from an obscure corner of Normandy, that that voice was first heard, which, in a comparatively short space of time, filled the whole of the civilized world with admiration,—which was to lay before mankind so many of the hidden wonders of creation,—which was to discover to us the relics of former ages, to change the entire face of natural history, to regulate and amass the treasures already acquired, and those made known during his life; and then to leave science on the threshold of a new epocha."

His genius was discovered by a Parisian naturalist, the abbé Tessier, rusticating during the anticlericalism of the Terror. And in 1795, at the moment when the great French scientific institutions were being formed, Cuvier on the strength of two or three small publications, together with the impression made by his correspondence, was summoned to his post in Paris. His success was immediate. Confined hitherto to birds and molluscs, his interest expanded among the collections of the Museum. His first paper in Paris was on the *Skeleton of an Immense Species of Quadruped, Unknown until the Present, and found in Paraguay.* This was the megatherium, a giant sloth (which Thomas Jefferson had hopefully taken for a giant lion, in his determination to expose Buffon's slander about the degenerative effect of the New World on the forms of life). Cuvier's second paper was a *Memoir on Living and Fossil Elephants.* From the moment of his arrival, he addressed himself to the most interesting class, the mammals, and within that class to the largest members, not precisely assigning, but gradually relegating "lower" orders and classes to his colleagues (his elders, many of whom thus became his juniors), and to Lamarck the lowest. No year went by without its three or four major memoirs. He had no time to draw his course together into a systematic treatise. One of his students published it for him, working from his lecture notes, and *Lessons of Comparative Anatomy* appeared in 1800. The same year Cuvier was named to a chair at the *Collège de France.*

There was that in Cuvier's talent which harmonized with the new dispensation in Napoleonic France. The Emperor, ordering the world after his own fashion, turned to the scientific community for counsel and assistance, and found there the sympathy denied him by political philosophers and men of letters. So it was Napoleon who profited from the training Cuvier had received to serve the Duke of Württemberg. He was appointed an inspector-general of education. He put in hand the creation of *lycées*, or higher schools, in the chief cities of the South of France. He became permanent secretary of the scientific division of the Institute, the reincarnation of the Academy. In 1809 and 1810 he organized the educational system of the Italian provinces. In 1811 Napoleon made him councillor to the Imperial University, the holding company, so to say, for all the faculties of France. Nor did these dignities disappear with the Restoration. A Protestant, he ruled for a time as chancellor over the Sorbonne. He was a member, too, of the Council of State, that grey eminence in committee which keeps to the shadows to preserve the profound continuity of French administration through all the superficial changes of ministry and regime. In 1819 Louis XVIII created him a baron; in 1832 Louis Philippe raised him to the dignity of peer of France; in 1833 he died, as the document which would have made him President of the Council of State awaited the king's signature.

All this is a little less pleasing than the prospect of great talents recognized and skillfully employed ought to be. To say precisely why it jars is difficult. It is not that Cuvier fell victim to the temptation which seduces many scholars who have tasted power, dined on fame, and are never able to go from service back to work. Thus has many a creative scientist been turned, not unwillingly, into a statesman of science, a public figure, loudly lamenting errands that he secretly embraces as a shield from the fear of confronting his political self alone, there in the laboratory where he made his reputation. But it is not that. Cuvier never abandoned science for importance. Perhaps it is that Samuel Smiles might have found in him too willing an example of self-help. In Cuvier, the historian of science has to come to terms abruptly with the nineteenth century, where nothing edifies like success. The Protestant ethic pays off quickly now. The pleasure of the country boy in the seats of the mighty is ill concealed.

There is something overtly magisterial about the authority he won, and more than magisterial, prophetic. It carried over into his scientific style. The imperative mood seems to govern passages in which the genus *Equus* (horses, zebras, etc.) is assigned a whole class to itself in the order pachyderm, while the Indian elephant is divided from the African species, and both are related not, as commonly supposed, to the extinct mammoth, but (of all things) to living rodents, though distantly. One wonders why these strange facts seem familiar. One has the sense of having been here before. And then one recognizes why. The style is what echoes. This is how we were first told things about the world. It is as if the injunction "Let!" prefixed these pronouncements about the earth and its creatures. It is exposition by fiat. There is no room here for high-flown, man-made systems of theory. Cuvier is very severe about the spirit of system and the ungoverned imagination. As historian of nature, his relation to his subject is that of Moses to the people of Israel, or of Michelet to the French nation. He enters in, and keeps transgressors out. Its triumphs are his, its disasters epic. He turns the Sahara green and causes tropical forests to flourish in Siberia. He populates the earth with strange beings, pterodactyls, paleotheriums, anoplotheriums, and megatheriums; sweeps them aside in a great cataclysm (there is something here, too, of the Titan gods); summons to replace them mastodons, sabre-tooths, and great hyenas; overwhelms their penultimate successors in a deluge; and finally allows the present species to crawl from obscure corners and radiate across the continents. Not that Cuvier was so naïve as to reconcile science and the Bible. It is rather his manner than his detail which is Biblical, his scientific personality rather than his method or results which was formed by that book, according to which nature is to be taken as given, as provided in the very fullest sense. He tells of his state of mind confronted by the fossil finds unexpectedly turned up in the gypsum quarries of the Paris basin:

I was in the position of one who has been presented pell-mell with the incomplete and mutilated debris of hundreds of skeletons belonging to twenty kinds of animals. It was required that every bone should go find the beast to which it belonged. A miniature resurrection was called for, and I did not hold at my disposal the omnipotent trumpet. But the immutable laws prescribed for living beings filled that office, and at the voice of comparative anatomy, every bone and every fragment leaped to take its place.

That voice of comparative anatomy was his. He worked in a fuller awareness of his own method than many scientists do, and it will, therefore, be best to let him explain it himself:

Fortunately, comparative anatomy possessed a principle which, fully developed, proved capable of making all the difficulties vanish. It is that of the correlation of forms in organic beings, by means of which every sort of being may be rigorously recognized by any fragment of any of its parts.

Every organic being forms an ensemble, a unique closed system, of which all the parts mutually correspond, and co-operate in any definitive act through reciprocal reactions. No one of these parts may change without the others changing also, and consequently each of them, taken separately, indicates and gives all the others.

Thus, . . . if the intestines of an animal are organized to digest only meat, and fresh meat at that, its jaws must be constructed to devour its prey; its claws to seize and tear; its teeth to divide and cut; the entire system of its organs of locomotion to pursue and catch; and its sense organs to perceive from afar; Nature must even have implanted in its brain the instinct for knowing how to hide and set traps for its victims. Such will be the general conditions of the carnivorous diet. Every animal destined to that diet must infallibly exhibit these characters, for its race could not have subsisted without them. But particular conditions exist subsidiary to the general ones, having to do with the size, species, or habitat of the prey, to which the animal is affected. And from each of these particular conditions result modifications of detail in the general forms. Thus, not only class, but order, genus, and even species are found expressed in the form of every part.

It is a little difficult, perhaps, to appreciate the power of the principle of correlation of parts if one insists on comparing it to so-called principles of physics like inertia, for example, or conservation of energy. These are rather axioms about nature, and taken as an axiom about the design of the animal economy, Cuvier's principle is tendentious if not misleading. But if it be accepted rather as the account of a method than a statement about the laws of nature, then clearly it is positive and scientific. It permits prediction of the whole animal from a part (if it is the right part, that is, for the claims of being able to reconstruct an eagle from a pin-feather were a little enthusiastic).

In Cuvier's positive work, the correlation of parts served him rather as a regulative principle than as a theory embracing some given set of phenomena. One is not to be misled by the emphasis on the wholeness of the animal. This works out innocuously enough in practice, for it is only the starting point of the analysis, and not its terminus. It is a complex fact rather than a question-begging explanation. The analysis itself fixes on each system of organs and ultimately on each particular organ. Dissection compares the structure of these parts, their variation, and (what is most significant) their relative importance to the situation in other animals. And once these subordinate facts are determined, they are related, not to the unity of life, much less of nature, but to objective circumstance: to an animal's environment, its diet, its natural enemies, its destined victims. And if one may anticipate for a moment the Darwinian theory of natural selection, it will begin to appear why it was Cuvier and providentialism, rather than Lamarck and organic emanation, who set the stage for evolutionary biology. For already in Cuvier the creature's organization is a function of its way of life. In a sense Darwin will only have to switch places between the dependent and independent variables in order to reduce the equation of life and circumstance to solvable terms.

In practice Cuvier usually found the most revealing indications in the structure of teeth and feet, parts which might be thought to have less to do with each other than most others. Their interdependence yielded him some of his most unexpected correlations:

> Just as the equation of a curve entails all its properties, so the form of the tooth entails the form of the knuckle, and the shape of the shoulder-blade that of the claws. And reciprocally, just as any element of a curve may be taken as the basis of the general equation, so too the claw, the shoulder-blade, the knuckle, the femur, and all the other bones taken separately, give the tooth as well as each other— so that whoever has a rational command over the laws of the organic economy may reconstruct the whole animal from any one of them. . . .
>
> We see clearly, for example, that animals with hooves must be herbivorous, since they have no way of seizing a prey. We see further how, having no other use for forefeet than to support the body, they do not need so powerfully developed a shoulder. . . .
>
> Their herbivorous diet requires teeth with a flat crown for grinding grains and grasses. The crown must be rough, and to that end the enamelled segments alternate with bony. Since this type of crown requires horizontal motions for champing and macerating, the jawbone cannot be so tightly hinged as in carnivora. The joint must be flattened and correspond to a flattened facet of the temple bone. The chamber of the temple, having only a small muscle to house, will be neither wide, nor deep, etc., etc.

Biology must always be preoccupied with the adaptation of form to function, and what redeems the apparent teleology of Cuvier's approach from the organismic fallacy is that, though he studies function, he orders by form. He accepts the animal as a given phenomenon. He approaches it as an engineering student might approach some machine to be analyzed. He teaches about it as a master-sergeant teaches some recruit the functioning and nomenclature of the rifle. The trainer disassembles the piece. He demonstrates the place and purpose of every part. He shows how each fits with all the others and conduces to the operation of the whole. He may compare one model to another, the manual bolt-operated to the semi-automatic gas-fired. But a good sergeant does not distract his charges by relating his lesson to the evolution of firearms, much less to the nature of war. Nor is the analysis turned into theology or philosophy or mysticism just because the rifle is provided by high authority, in whose name the teacher speaks.

Sometimes one hears biologists admit a little sheepishly that they still find themselves employing teleological considerations. No doubt this is true in part. No doubt it is in a limited sense necessary. Nor would they need feel guilty about it if they would distinguish between teleology as a philosophy of all nature, and the structural-functional analysis of particular organisms. The latter simply shows how the hoof serves the horse. This is their business to practise, as it was Cuvier's. His teleology, on the other hand, which was essentially that of Aristotle, has indeed become obsolete as a philosophy of nature. In his own time, however, it may well have been helpful insofar as it led him to the study of each animal as an objective contrivance, and to the practice of intensive, detailed dissection. For this was Cuvier's great positive strength. He was a deep-cutting anatomist rather than a natural philosopher. He was the first zoologist to do his taxonomy from fundamentals in the economy of each species. He never stopped with observation of external characters. Lamarck's philosophy was not all that weakened him in the strife of their partnership. Neither did Lamarck practise dissection. He used the collections but not the laboratories of the Museum. No doubt it is impossible to experience vicariously what Cuvier gained from life-long habits of minute attention, from the prolonged, the self-imposed exercises which put him in absolute command of the facts of anatomy. He truly knew that of which he made comparison. And it is only just to recognize how he earned the authority he wielded.

Nevertheless, the authoritarian strain in Cuvier implied a certain scientific conservatism. He seldom used the term biology. Natural history remained his subject, and *The Animal Kingdom* opens with a discussion of the sciences. *Science naturelle*, or *physique*, has two branches, *physique générale* and *physique particulière*. The former, consisting of mechanics, dynamics, and chemistry, admits quantification and employs experiment. The latter, *physique particulière*, is synonymous with natural history. It studies the particular object and may not aspire to abstract theory. "In the former, phenomena are studied under controlled conditions, so that analysis may result in general laws. In the latter, phenomena occur under conditions which escape control, and the scientist must, therefore, strive only to distinguish amidst their complexity the effects of general laws already known." This is because experiment is impossible in natural history. The naturalist has to accept his problem as a whole, and analyze it only in his mind. To experiment upon a living organism is to alter or destroy it, and in natural history comparison must take the place of experiments:

> The differences among bodies are, so to say, experiments already set up by nature, which adds to or removes from each different parts, as we should wish to do in our laboratories, and shows us herself the results of these additions or removals.

Thus are established those laws which govern these relations, and which are employed like those which have been determined by the general sciences.

Now, this must not be interpreted as an assertion of vitalism, or an escape from science into mystery. It is the methods which differ according to the complexity of phenomena, and not the laws of nature according to whether one studies life or matter. What Cuvier is really expressing (for once) is a certain modesty, or even timidity, about natural history, which patiently discerns the operation in its realm of laws which the experimental and theoretical physicist has discovered in his. And *Researches on Fossil Bones* explains how one is to understand the uniformity of the laws of nature. Anatomical relations are constant, and

they must necessarily have a sufficient cause, but since we do not know what it is, observation must supply the default of theory. By this means we establish empirical laws, which become almost as certain as rational laws when they rest on observations often enough repeated. So it is today that whoever notices simply the footprint of a cloven hoof may conclude that the animal which passed that way chews its cud, and this conclusion is absolutely as certain as any in physics or moral philosophy. The spoor alone tells him who observes it the form of the teeth, the form of the jaws, the form of the vertebrae, and the form of all the bones of legs, thighs, shoulders, and pelvis of the animal which went by. It is surer than all the marks of Zadig.

In yet another respect, Cuvier's conservatism had the virtue of its defects. His concentration on the particular, his despair of any instrument sharper than the taxonomic eye (or lens, for he was no Goethe to abjure the microscope), would have put a paralyzing limitation upon biological science, for all the surety of his anatomical comparisons. His was an attitude older than biological romanticism and idealism. Indeed, it was as old as Aristotle. Yet it was a fortunate survival. For it preserved against the idealists the proposition that life is not one but many. Cuvier could never discern that unity of plan which Goethe's archetypal biology asserted, and which Lamarck's emanationist evolutionary process strained to realize. Not that Cuvier adduced evidence for successive acts of creation. His science never aspired so high. But he could not in practice find points of anatomical comparison between the four main divisions among which he distributed the populations of zoology, the *vertebrata*, the *mollusca*, the *articulata*, and the *radiata*. Each seemed utterly independent from the others in point of organization, with nothing more in common than any of them had with plants. Nor was there any question of genealogical relationship among the classes, orders, genera, and species into which these divisions fell.

From the moment of his arrival in Paris, Cuvier had been fascinated by the fossil bones which had found their way into the collections of the Museum, spectacular relics picked up here and there from Paraguay to Siberia, which bespoke giants in the earth. As he perfected his techniques of comparative anatomy, he tended to find here their most interesting application. The science of paleontology, indeed, was created as the comparative anatomy of extinct species, building out from the kernel of some pelvis those vast, skeletal, plaster-of-Paris dinosaurs which still rear up so uneconomically in the two-story halls of natural history museums. Gradually the distribution of Cuvier's interest shifted their way. By the time of his last work comparative anatomy had become the means rather than the end of his researches. As with other great books of the nineteenth century, the full title gives away the plot: *Researches on Fossil Bones, Reëstablishing the Character of many Animals, of which the Species have been destroyed in the Revolutions of the Earth*. And a couplet from Delille on the title page sets the mood of these seven volumes. One is to imagine it declaimed, perhaps, in those tones of sepulchral soulfulness which the French reserve for historical or funereal recitative, to be heard nowadays in the sound-and-light spectacles illuminating monuments of the glory that was France:

> Triomphante des eaux, du trépas et du temps;
> La terre a cru revoir ses premiers habitans.

In this, his last appearance on the scientific stage, Cuvier was "a new kind of Antiquarian," compiling the "Catalogue of Lost Beings." Or possibly he was the hero of a cosmic detective story, with a tooth as the all revealing clue. As in a good detective story, the evidence was right under foot, requiring only to be looked at correctly. He need not go to Paraguay. He need not go beyond the Paris basin, where the gypsum quarries were yielding a treasure hitherto ignored of fossil carnivora, pachyderms, and reptiles, belonging to genera whose nearest living relatives prowl the wilds of Africa. To a Frenchman it was a profoundly moving thought that the Ile de France, the cradle of his civilization, should have been the habitat of these savage beasts. What was more, it was a profound convenience that he could tour nature back through time without ever leaving home. For it is thus that a Frenchman prefers to travel.

With Brongniart, Cuvier investigated the geological structure of the Paris basin. They published their *Essay on the Mineralogical Geography of the Environs of Paris* in 1811. It is one of the earliest systematic treatises of stratigraphy, in which successive formations of chalk, limestone, clay, and sand are distinguished lying layer over layer. Cuvier perfected and published the *Essay* as part of the *Research on Fossil Bones*. There he generalized the argument. Fossils seem to exhibit an order in their occurrence—the older the strata, the higher the proportion of extinct forms. But in Cuvier this is no order of morphological progression. Indeed, what interested him was less the succession in the animals he had studied all his life than the story of what had undone them, the panorama of cataclysm and catastrophe, and especially floods, which had suddenly overtaken the world, there in Siberia encasing a mastodon in a block of ice, here in Paris burying a crocodile in gypsum.

> If there is any circumstance thoroughly established in geology, it is, that the crust of our globe has been subjected to a great and sudden revolution, the epoch of which cannot be dated much further back than five or six thousand years ago; that this revolution had buried all the countries which were before inhabited by men and by the other animals now best known . . . that the small number of individuals of men and other animals that escaped from the effects of that great revolution, have since propagated and spread over the lands then newly laid dry; and consequently, that the human race has only resumed a progressive state of improvement since that epoch, by forming established societies, raising monuments, collecting natural facts, and constructing systems of science and learning.

And Cuvier wrote for the *Research on Fossil Bones* a lengthy preface, a treatise in itself, frequently republished separately as *Discourse on the Revolutions of the Surface of the Globe*. For it was in studying these revolutions that science joins on to history, and the history of nature becomes the history of man. The traditions of all early peoples confirm quite independently the geological evidence for a natural disaster standing at the beginning of recorded experience. "These ideas have pursued, I might almost say tormented, me all the while I worked at my researches on fossil bones." And here, thought Cuvier, in the physical history of man, lay the frontier of a new science, awaiting still its Newton, who should situate our knowledge of nature in time, as Newton had done in space.

❖

IN THE CLOSING PASSAGE of his *Discourse on the Revolutions of the Surface of the Globe*, Cuvier reproached geologists with paying too little attention to these more recent events:

> But how fine it would be to possess the chronological order of the organic productions of nature, even as we know that of the principal mineral substances. Organic science itself would profit. The developments of life, the succession of its forms, the precise determination of which had first appeared, the simultaneous birth of certain species, their gradual destruction—all this might probably teach us as much about the essence of the organism as all the experiments that we could try on living species. And mankind, to whom has been allotted only an instant on the earth, would have the glory of recreating the history of the thousands of centuries which preceded his existence, and the thousands of beings which have not been his contemporaries.

Geology is the subject which introduced a historical dimension into science: its own history, indeed, scarcely goes back beyond the nineteenth century. Even then, doubts were sometimes expressed about whether it could properly be a science at all. The geologist, like the historian, had to rely largely on interpreting relics of change. He could neither experiment nor quantify. Nor in the early stages might he even test the predictive value of his generalizations, as Cuvier, for example, could do. His conclusions were bound to remain matters of opinion in a way that those of the physicist, or even the anatomist, did not. And indeed, the science of geology represents a coming together of lore from the ancient practice of mineralogy with speculations about the origin of the earth, seventeenth-

and eighteenth-century cosmogonies which have in them
more of science fiction than of science. No doubt the historical bent taken by scholarship in the early nineteenth
century stimulated a more sober approach to the question
of how the world has come to be the place it is. But the
essential technique for ordering the information about
rock structures and topography into a science came from
paleontology.

For lack of that, or any determinate technique, the con-
troversies which attended the birth of the science were
violent, immature, and inconclusive. The prevailing school
at the turn of the century followed Abraham Gottlob
Werner, professor of mineralogy in the ancient and famous
mining school at Freiberg-im-Sachsen. He was one of those
superb teachers whose magnetism turns students into disciples, and who is led by the necessity to lecture down the
primrose path of systematization and organization by head-
ings. It was the belief of the Wernerian or "Neptunist"
school that all the rocks had been precipitated from pri-
meval seas which had covered all the earth: chemically at
first, when the slates and granites crystallized out; then
mechanically, as the waters lowered and chalks and limestones settled; finally by silting as great torrents came and
went and laid bare the mountains and the continents.
Volcanic action occurred late and incidentally, activated
by ignition of coal deposits, twisting the strata here and
there out of the horizontal. In general, however, the earth
was girdled in layers of rock as uniform as the leaves of
an artichoke.

The picture was possible only at a time when geological
information was mineralogical rather than structural. Even
then it was barely possible, though its appeal is obvious.
This "Neptunist" interpretation marched comfortably
with a providential view of nature. It required no excessive
run of time. There was enough water for any number of

floods. Living species appear only after the primary rocks, and then in the order of Genesis: fish, mammals, man. These vast fluctuations must have wiped out hordes of individuals, and even whole species and genera, so that modern forms might well represent distinct creations.

The opposing school, the Vulcanists, offered no such advantage to tradition. "We find," wrote James Hutton, "no vestige of a beginning—no prospect of an end." Hutton was a Scottish rationalist, a colleague and friend of Joseph Black and James Watt in the Edinburgh circle. He worked among the insoluble granites of Scotland, and was perhaps the first student of the earth who may properly be called a geologist. He published his *Theory of the Earth* in 1795. It enjoins upon the historian of the earth the full self-denial of science. Past events can be described only by inductive analogy to processes which we observe working in the present, and by the evidence of the rocks. There are two sorts in the crust of the earth, one of igneous and the other of aqueous origin. The primary igneous rocks (granite, porphyry, basalt, etc.) usually occur under the aqueous, except where formations are overthrust, or where molten dikes and plugs have intruded through the limestones. Weathering and erosion are constantly carrying a fine silt of sandstones, clays, and topsoils down the rivers and out to settle in the ocean bottoms. Some agency must have transformed, and must still transform, these loose deposits into the rocks around us. It cannot have been water. They are utterly insoluble. That agency must, therefore, have been heat. The intense heat of the earth's heart, acting under enormous pressure, consolidates the rocks, and its expansive force uplifts the continents thus born in the bed of the sea.

This hypothesis accounts, as the Neptunist could not, for bent and tilted strata, and for the evidence of widespread volcanic activity. Particularly damaging to the

Neptunists were the *puys*, the extinct volcanoes, of central France. But more valuable than Hutton's igneous hypothesis was his assumption of the secular uniformity of nature. Nature does not grow old or lose its powers. His process was no past occurrence. Even now rocks are being consolidated at vast pressures under the bottom of the seas. Even now some lands are rising, and others being worn away. No perceptible change had taken place in all recorded history. But instead of summoning abnormal forces from the vasty deep, Hutton simply assumed the existence of all the time he needed. Nature has been swinging through cycles in a time as inconceivably old as the space is large through which the planets run their courses. Nor is geology anymore concerned with "questions as to the origin of things," than physics with the reason for the law of gravity.

It is obvious that the Vulcanist position was the expression of the spirit of science and rationalism addressed to geology, while the Neptunist was the personal and sectarian spirit of a particular school, drawn not from the Enlightenment but from the crabbed tradition of a German mining school. Nevertheless, without some means of establishing a sequence among the rocks, the discussion need never have been fruitful. In one respect, the Neptunists helped. Their doctrine emphasized stratigraphy, succession rather than cyclic change. For it is not possible to order rocks chronologically according to their mineral or chemical composition. Fossils gave the key to classification and hence to system. It was picked up (such was the nascent state of the science) not by some one of the disputing geologists, but by an obscure English drainage engineer, William Smith, a consultant in the construction of canals, reclamation of fens, and the search for minerals. As early as 1791, he noticed that particular species of fossils occur in certain groups of strata, and in no others.

He seized upon this fact as offering a means of identification of the main rock systems. He could, he wrote in 1799, easily trace the outcropping of each English stratum from the chalk of Dover in the South right across the island to the East Anglia Coast and down to the coal measures of Wales. Smith was not a writer or a scientist. Not till 1815 could his friends prevail on him to reduce his information to the surface of a map. *The Delineation of the Strata of England and Wales* (1815) shares honors with the Brongniart-Cuvier memoir on the Paris basin as the first stratigraphical analysis to use paleontological indices. But it appears that Smith knew nothing of Parisian paleontology, and indeed approached the subject from the opposite end. They moved from the study of fossils to that of geological structure, and he the other way about.

Like Werner, Smith was nearly incapable of literary composition, and the first connected account of his researches reached the public in 1813 in a book by a friend, the Reverend Joseph Townsend, entitled *The Character of Moses Established for Veracity as an Historian, Recording Events from the Creation to the Deluge*. For it fell out curiously. The merger of paleontology and stratigraphy, the former serving the latter in geology, cast the new science under the spell of Cuvier's catastrophism rather than Hutton's uniformitarianism. Personal circumstances reinforced this early association of theology and geology. Many of the first generation of structural geologists were Anglican clergymen, who as dons had to be in orders, or as parsons combined the bluff, outdoor spirit of the gentry with the Church's vague sponsorship of genteel learning.

In this embryonic stage of geological science, the main categories of classification of rocks were primary, secondary or transition, and tertiary. Within these groups, it early became the practice to name the systems and periods either according to a traditional descriptive appellation (carboniferous, cretaceous), or in the case of the formations isolated in this, the "heroic age" of geology, after the region of the world where the system first displayed its development—for example, Devonian, Jurassic, Pennsylvanian, Permian. Thus as befits a science which developed out of history, the nomenclature of geology preserves a record of its own history, in contrast to chemistry which took its form from rationalism. The leadership of English investigators is commemorated in the number of formations which bear regional or traditional names from the British Isles. Smith himself, a man of practice but no education, simply adopted the local words for the characteristic rocks as he moved across England ticketing the more recent strata: Lias, Forest-Marble, Cornbrash, Coralrag, Portland Rock, London Clay, Purbeck marble. These, it turned out, came between the systems which Werner had described as Muschelkalk and Cretaceous, and the same series were soon found outcropping in the same order in other parts of Europe.

In the 1820's, the Reverend William Daniel Conybeare, working with William Phillips, carried the English tertiary down from the Cretaceous to the Carboniferous. They distinguished between the Upper and Lower Cretaceous systems, the upper, middle and lower oolite, and established the divisions of each as well as of the carboniferous, which they brought to a close with the Old Red Sandstone. Below this all was tangle and confusion. The older formations were overturned and interspersed in far greater complexity. Many more accidents had happened to the original sequence. Their fossil population was much more sparse, much less studied, and ever less analogous to living species. By chance, moreover, they appear in regions which seem historically as well as geologically remote—in Wales and Cornwall, where Celtic vestiges survive from a different order of things. (In France, too, or rather in

Brittany, the Arthurian mist clings about the geologically antique.) Ordering these systems, down through the era since known as the Paleozoic, was the major work of the 1820's and 1830's. The most important chapters were carried out by the Reverend Adam Sedgwick, professor of geology at Cambridge, and Sir Roderick Murchison, a Scottish gentleman turned geologist. At first they worked together, going off every summer with map, hammer, and specimen bag to tramp and clamber about the mountains of Wales or highlands of Scotland and the isles. Murchison eventually unravelled the upper systems in South Wales, which he called the Silurian after the tribes whom Caesar had found inhabiting that region. The work was if anything more difficult in North Wales, where Sedgwick was assisted one summer by a student, Charles Darwin. These systems, which he could establish with less clarity, he called the Cambrian. Unfortunately, Sedgwick and Murchison quarreled over the boundary between the two and closed their days in extreme old age after one of those interminable Victorian estrangements.

No doubt the amateur status of this generation of geologists helps account for the naïveté of the interpretation they put upon their work. They were forever rescuing geology from the imputation of being inimical to religion, but were never very explicit about who was doing the imputing. Perhaps it was their own Victorian consciences, torn between their desire to explain the history of the earth naturally and their fear that they would succeed. The Reverend William Buckland of Oxford was the most famous teacher and writer of the 1820's. His first book was called *Vindiciae Geologicae; or, the Connexion of Geology with Religion Explained* (1820); and his second, which made his reputation, *Reliquiae Diluvianae; or, Observations on the Organic Remains Contained in Caves, Fissures, and Diluvial Gravel, and on Other Geological*

Phenomena, Attesting the Action of an Universal Deluge, an idyll of the caves which appeared in 1823. It is essentially an application of Cuvier's catastrophism to English fossil vertebrates, and there is good geology in it. Nor did this assimilation of the Biblical flood to geological catastrophism express simple fundamentalism. Accumulating evidence forced natural theology to interpret the Biblical events in a more and more allegorical fashion: the six days of creation became indefinite periods, and only the order of creation was retained; the six-thousand-year span of earth history lengthened out as far as needed in the pre-diluvial epoch, and it was the flood with which recent and recorded history began. In the 1830's even the universal flood had to be resolved into (at best) an epoch of local catastrophes. Still, as Sedgwick often said, truth could not be inconsistent with itself. However fine and long drawn out the central thread of interpretation became, it must ultimately connect God's word with His works. And finally, the notion of a divinity who must continually interfere with his creation to show himself a governing force depended upon the special creation and adaptation of forms of life, and most notably man, the reason for the whole creation. Geology, wrote Sedgwick in 1835,

> like every other science when well interpreted, lends its aid to natural religion. It tells us, out of its own records, that man has been but a few years a dweller on the earth; for the traces of himself and of his works are confined to the last monuments of its history. Independently of every written testimony, we therefore believe that man, with all his powers and appetencies, his marvelous structure and his fitness for the world around him, was called into being within a few thousand years of the days in which we live— not by a transmutation of species, (a theory no better than a phrensied dream), but by a provident contriving power. And thus we at once remove a stumbling block, thrown in

our way by those who would rid themselves of a prescient first cause, by trying to resolve all phenomena into a succession of constant material actions, ascending into an eternity of past time.

By laying hold on fossils, geology had come into control of its materials by the 1830's. What it needed to become a science was to retrieve its soul from the grasp of theology, and to resume Hutton's assumption of the historical as well as the physical uniformity of nature. That was the work of Sir Charles Lyell, whose *Principles of Geology* was perhaps the most famous and certainly the most influential book in the history of the science. In the quietest possible way, its purpose was polemical, "to sink the diluvialists," Lyell wrote privately. It is a very large book, three volumes in the first edition, which appeared in 1831, 1832, and 1833. There is very little original geology in it. The main novelty was systematic—the distinctions of Pliocene, Miocene, and Eocene in the Tertiary are his, to which he later added Pleistocene for the most recent period of glaciation and the appearance of man. But its power and influence derive from the measured and civilized account it gives of the whole state and content of a science. It is a model of what literate exposition may accomplish. The great, the insuperable obstacle which had impeded the development of geology into a true science was the unphilosophical supposition that a different order of things had formed the earth on which we live from that which now pertains.

Time was of the essence. The three skillful and lucid volumes of the *Principles* marshal the evidence that existing forces, given time enough, account for the observable state of man's habitat. Since there must be no exceptions, the book came close to being a *Summa Geologica*. Lyell did not, of course, deny the reality of change, but he insisted that all change has been uniform, proceeding in cycles in time rather like the orbits in space through which the planets swing. Climate, for example, has varied here or there according to the shifting proportions of land and sea. Volume I describes geological dynamics. Familiar examples are adduced of the mode in which various agents behave—weathering, volcanic eruptions, earthquakes, the influence of living organisms on the environment, and above all, the sculpturing by running water—the valley carved by the trickle. After each illustration, historical or contemporary, one is reminded how the cumulative effects of such forces (given time, always given time) had produced the phenomena which Cuvier and Buckland referred to cataclysms (to miracles, that is), and how geology, if it is to be science rather than theology, must reconcile the changes in the surface of the earth, not with the Bible, but "with the existing order of nature."

In retrospect, it has often been observed that uniformitarianism in geology seems to cry out for evolutionism in biology. And certainly geology repaid its intellectual debt to biology almost immediately. That is to say, the seating of the chronology of earth history in paleontological indices, in the succession of fossil forms, gave biologists by way of return the sequence of species in geological time. It was for lack of this information that Lamarck (even if his theory had been free of other faults) had had to establish his order in a scale of increasing morphological complexity, instead of in a secular order of historic events. We know, moreover, that Darwin studied the *Principles of Geology* very closely. It did more to form his scientific outlook than any other book. "He who can read Sir Charles Lyell's grand work," Darwin wrote in the *Origin of Species*, ". . . which the future historian will recognize as having produced a revolution in natural science, and yet does not admit how vast have been the past periods of time, may at once close this volume." As another successor once said of Lyell's influence, "We find the data, and Lyell teaches us to comprehend the meaning of them."

Nevertheless, Lyell set his face against the transmutation of species just as firmly as did Sedgwick, if in a better temper. Volume II of the *Principles* discusses the animal economy. The question was, "Whether species have a real and permanent existence in nature; or whether they are capable, as some naturalists pretend, of being indefinitely modified in the course of a long series of generations?" And Lyell gave his readers the first comprehensive précis of Lamarck's views to appear in English, or for that matter in any language. It should be obvious why he was bound to come down on the side of fixity. Nor is his later influence on Darwin turned into a paradox thereby, for (as will appear) Lamarckism has nothing in common conceptually with Darwinism. Indeed, it would have been contrary to uniformitarian precepts to allow an alteration in species when none had ever been observed. We must rather suppose that life had originated in a number of "foci of creation," spread here and there about the earth, and activated now and again in its history. For it was certain that some species had become extinct, and the globe repopulated from time to time, although Lyell admits to some surprise "that so astonishing a phenomenon can escape the observation of naturalists."

Lyell, indeed, could scarcely feel altogether comfortable with the evidence for a progression of things in organic nature, and sensing his discomfort, his critics among the catastrophists pressed what they rashly took to be their advantage. Their objections make ironic reading. Here, for example, is a paragraph from the review by the Reverend William Whewell, a philosopher of science, in *The British Critic, Quarterly Theological Review and Ecclesiastical Record*, which brought that respectable journal

right to the brink of the ultimate catastrophe which en-
gulfed providential science:

> It is clear . . . that to give even a theoretical consistency
> to his system, it will be requisite that Mr. Lyell should
> supply us with some mode by which we may pass from a
> world filled with one kind of animal forms, to another,
> in which they are equally abundant, without perhaps one
> species in common. He must find some means of conducting
> us from the plesiosaurs and pterodactyls of the age of the
> lias, to the creatures which mark the oolites or the ironsand.
> He must show us how we may proceed from these, to the
> forms of those later times which geologists love to call by
> the sounding names of the paleotherian and mastodontean
> periods. To frame even a hypothesis which will, with any
> plausibility, supply this defect in his speculations, is a
> harder task than that which Mr. Lyell has now executed.
> We conceive it undeniable (and Mr. Lyell would probably
> agree with us,) that we see in the transition from an earth
> peopled by one set of animals, to the same earth swarming
> with entirely new forms of organic life, a distinct manifesta-
> tion of creative power, transcending the known laws of
> nature: and, it appears to us, that geology has thus lighted
> a new lamp along the path of natural theology.

That was in 1831. One thinks of the warning of Sir
Thomas Browne, some two centuries before, about those
who "have too rashly charged the troops of error, and remain as trophies unto the enemies of truth." Thomas
Henry Huxley first read Darwin's hypothesis of evolution
by natural selection in 1858. The force of the concept
simply leaped out at him, like the pattern from the pieces
of this puzzle of biological adaptation. All he could say
to himself was, "How extremely stupid not to have thought
of that." The right answer, it came in that combination of
unexpectedness and irresistibility which has often been
the hallmark of a truly new concept in scientific history.

CHAPTER VIII

BIOLOGY COMES OF AGE

IT IS A FAMOUS TITLE, and one should read it through with a stiff upper lip: *On the Origin of Species by Means of Natural Selection; or, The Survival of the Fittest in the Struggle for Life.* But is there any "great book" about which one secretly feels so guilty? Is there any fundamental scientific generalization which came into the world in so unassuming a guise? So ordinary is the language that it almost seems as if we could be in the midst of reading a lay sermon on self-help in nature. All the proverbs on profit and loss are there, from pulpit and from counting-house—On many a mickle making a muckle: "Natural selection acts only by the preservation and accumulation of small inherited modifications, each profitable to the preserved being"; On the race being to the swift: "The less fleet ones would be rigidly destroyed"; On progress through competition: "Rejecting those that are bad, preserving and adding up all that are good; silently and insensibly working, *whenever and wherever opportunity offers*, at the improvement of each organic being"; On success: "But success will often depend on the males having special weapons, or means of defence, or charms; and a slight advantage will lead to victory"; On handsome is as handsome does: "Nature . . . cares nothing for appearances, except insofar as they are useful to any being"; On saving nine: "I could give many facts showing how anxious bees are to save time;" On reflecting that in the midst of life we are in death: "We behold the face of nature bright with gladness, we often see superabundance of food; we do not see or we forget, that the birds which are idly singing round us mostly live on insects or seeds, and are thus constantly destroying life; or we forget how largely these songsters, or their eggs, or their nestlings, are destroyed by birds and beasts of prey; we do not always bear in mind, that, though food may now be super-abundant, it is not so at all seasons of each recurring year;" On the compensation that all is, nevertheless, for the best: "When we reflect on this struggle, we may console ourselves with the full belief, that the war of nature is not incessant, that no fear is felt, that death is generally prompt, and that the vigorous, the healthy, and the happy survive and multiply."

None but a Victorian Englishman could have written those words. It is, as a German critic said in a remark meant to be scathing, classical political economy applied to biology. Or as Darwin said himself, "This is the doctrine of Malthus applied to the whole animal and vegetable kingdom." So familiar has the argument become, that its magnitude is not on the face of it apparent. For in this unlikely guise there was clothed nothing less than a new natural philosophy, as new in its domain as Galileo's in physics. Darwin, in effect, abolished the distinction which had divided first natural history and then biology from physics at least since Newton, and which rested on the supposition (or defense) that the biologist must character-istically study the nature and the wisdom of the whole rather than the structure and arrangement of the parts. Nevertheless, so ordinary are the terms that it requires interpretation to bring out the originality and force of the argument, more so than if Darwin were a Newton or an Einstein abstracting far beyond everyman's power to follow and to understand. A hero of science should be less accessible.

Darwin never gives his interpreter the benefit of self-dramatization. His style suffers from the damp respectability which spots and kills so many of the lesser flowers of Victorian prose. Nor does his vision of a new science arrest as Galileo's does. His genius has not the daemonic quality of Newton's. There hangs over him the shadow of no tragic destiny like Lavoisier's. There looms, indeed, nothing more ominous than the shadow of Samuel Wilberforce, Bishop of Oxford, and the debate about men and monkeys.

Like T. H. Huxley, his champion and defender, Darwin would face the unknown four-square, taking only science for his guide. And one would find his courage stirring, except for the feeling that the Victorian unknown contained nothing very dreadful. Late in life he wrote a short intellectual autobiography to be circulated among his children. There is a numbing candor in this account of his own mind. He claims for himself not power of abstract thought, but only the worthiest and dullest of intellectual virtues—patience, accuracy, devotion:

> I have no great quickness of apprehension or wit which is so remarkable in clever men, for instance Huxley. I am therefore a poor critic: a paper or book, when first read, generally excites my admiration, and it is only after considerable reflection that I perceive the weak points. My power to follow a long and purely abstract train of thought is very limited; I should, moreover, never have succeeded with metaphysics or mathematics. My memory is extensive yet hazy.

He deplores the "curious and lamentable loss" of his youthful taste for music, poetry, and the fine arts:

> My mind seems to have become a kind of machine for grinding general laws out of large collections of facts, but why this should have caused the atrophy of that part of the brain alone, on which the higher tastes depend, I cannot conceive. A man with a mind more highly organised or better constituted than mine, would not I suppose have

thus suffered; and if I had to live my life again I would have made a rule to read some poetry and listen to some music at least once every week.

And only once does he betray that drive, lacking which no one achieves what he did, and then in what seemly phrases:

My industry has been nearly as great as it could have been in the observation and collection of facts. What is far more important, my love of natural science has been steady and ardent. This pure love has, however, been much aided by the ambition to be esteemed by my fellow naturalists.

Charles Darwin was born in 1809, a younger son in a family whose marriages have drawn it into one of the great ramifications which lend the Victorian intelligentsia an almost hereditary air. By now the Darwins are related to the Wedgwoods, Trevelyans, Macaulays, Huxleys, Arnolds, and Galtons, and number among their connections the names of Keynes and Cornford. Darwin's grandfathers were Josiah Wedgwood, founder of the pottery firm at Etruria, and Erasmus Darwin, both of whom had been members with Joseph Priestley and James Watt of the Birmingham scientific circle in the 1770's and '80's. Erasmus Darwin, a physician and versifier, had composed one of the numerous evolutionary flights of fancy of the eighteenth century. His *Zoonomia* reads not unlike Lamarck. Critics of Darwin have sometimes taken him to task for concealing that he owed an intellectual as well as a genetic debt to his grandfather. As far as he could tell, he said, *Zoonomia* made no impression on his mind at all, any more than did the theories of Lamarck when he later heard tell of them at college. And like most of what Darwin wrote, this was the simple truth.

Nor is it an unimportant truth, for there were two aspects to Darwin's work, the empirical and the theoretical. The *Origin of Species* did definitively establish the mutability of animals in their descent out of the past. Never-theless, it would be difficult—in the face of Erasmus Dar-win, Lamarck, Goethe, and Diderot, to name no others—to claim the fact of variability as a Darwinian discovery. What was truly novel was the theory, the concept of natu-ral selection which explained the facts lying all ready to Darwin's hand in the sciences of paleontology, compara-tive anatomy, geology, and geography. It may bring out the point to anticipate the two kinds of opposition which Darwinism was to encounter. Religious fundamentalists might deny the fact of evolution. But this reaction was intellectually trivial. Where philosophical offense was taken, it was rather the view of the world implicit in the theory of natural selection which wounded humane sensi-bilities more deeply, and which was repudiated as inad-missible or meaningless or both, for like the Cartesian and the Leibnizian objections to the Newtonian theory of gravity, the two complaints come down to the same thing, and turn on the eternal question of what a scientific ex-planation really says.

Darwin's father, too, was a doctor, a crashing Victorian father, who weighed over three hundred pounds and in-spired in his son the kind of unwholesome respect later explained by Freud. Of his mother, he remembered only her deathbed. It is characteristic that he should have mar-ried into the same family as his father. His wife, too, was a Wedgwood and his own cousin. Dr. Darwin, indeed, left his son with a neurosis and with adequate means to support it. In later life his chronic indisposition drove him into the shelter of his family, to long days at the baths, long hours upon the sofa and under the shawl. He, too, was intended for medicine as a youth, and at the age of sixteen was entered for that training in the University of Edinburgh. His formal education was a failure. His stomach literally could not support dissection or the horrors of the operating theater. "Dr. Duncan's lectures on Materia Medica at 8 o'clock on a winter's morning are something fearful to remember. Dr. Munro made his lectures on human anatomy as dull, as he was himself, and the subject disgusted me." And he withdrew, to enter Cambridge with a view to becoming a clergyman.

His father had no notion of allowing him to indulge his love of nature as an idle sportsman. "But no pursuit at Cambridge was followed with nearly so much eagerness or gave me so much pleasure as collecting beetles." Even at Edinburgh he had frequented the naturalists of the Pliny Society, and had read a little paper showing that the so-called ova of Flustra were, in fact, larvae. It was his first discovery. "I gloried in the progress of geology," he wrote, and was drawn at Cambridge into Sedgwick's orbit. In 1831 Sedgwick invited Darwin to accompany him on his annual summer expedition into the Cambrian rocks of North Wales. It was the beginning of a scientific life. The next year the vessel *Beagle*, whose Captain Fitzroy was the nephew of Castlereagh, was outfitting for a voyage of exploration around the world. Darwin was offered the (unpaid) post of naturalist. He accepted, and a kindly uncle, Josiah Wedgwood II, father of his future bride, used his influence to bring his father around.

The voyage of the *Beagle* was the great event of Darwin's life, the making of him as a scientist, and the end of his prospects as a clergyman. It was, indeed, almost the only event, and the account of it which he published on his return is certainly his most charming book. When Darwin embarked, he was twenty-two years old, and far from being as conversant with the background of evolutionary thought as are the readers of this chapter. In his sea-chest he packed a selection from Milton and Volume I of Lyell's *Principles of Geology*. Across the South Atlantic from the Azores to Brazil they went, passed through the Straits of Magellan where they lingered for a bit, and coasted up the Pacific shore to the Equator. The young Darwin was shocked at his brush with slavery in Brazil. In Montevideo he received Volume II of Lyell. In Tierra del Fuego he was horrified at the bestiality of the primitive tribes they encountered, one of which supplemented its diet with old women in severe winters. All these new lands he saw through Lyell's eyes. In Chile and Peru he made geological forays into the High Andes. And gradually he found his attention more drawn by the flora and fauna, particularly the fauna, than by the structure of these regions. "During the voyage of the *Beagle*," he wrote in his *Autobiography*,

> I had been deeply impressed by discovering in the Pampean formation great fossil animals covered with armour like that on the existing armadillos; secondly, by the manner in which closely allied animals replace one another in proceeding southwards over the Continent; and thirdly, by the South American character of most of the productions of the Galápagos Archipelago.

The Galápagos, indeed, were decisive in Darwin's own development. For the islands were young geologically, differing only slightly in climate from the mainland. No doubt the dramatic character of those far fragments of land was peculiarly impressive. Cones of black lava, they thrust out of the Pacific right on the Equator. It was, he noted, "what we might imagine the cultivated parts of the Infernal regions to be." Though related to those of South America, most of the species were peculiar to the islands. Reptiles dominated. Everywhere great tortoises lumbered, "so heavy, I could scarcely lift them off the ground. Surrounded by the Black lava, the leafless shrubs, and large cacti, they appeared most old-fashioned antediluvian animals, or rather inhabitants of some other planet." What was very puzzling to him, the natives could tell by minute variations from which of the islands any particular tortoise had been brought. And from island to island, other species exhibited similar slight variations. The most notable case, which has become a classic of evolutionary studies, is that of the finches. There were several species. The distinctive element seemed to be the beak, some small, some massive, some pointed, some curved. In effect, therefore, these islands were a laboratory ready-made in nature, which isolated the problem of species behind 600 miles of ocean. "It was evident," he wrote later (after he had made it so)

> that such facts as these, as well as many others, could only be explained on the supposition that species gradually became modified; and the subject haunted me. But it was equally evident that neither the action of the surrounding conditions [for these did not differ materially from island to island], nor the will of the organisms (especially in the case of plants), could account for the innumerable cases in which organisms of every kind are beautifully adapted to their habitats of life—for instance, a woodpecker or a tree-frog to climb trees, or a seed for dispersal by hooks or plumes. I had always been much struck by such adaptations, and until these could be explained it seemed to me almost useless to endeavour to prove by indirect evidence that species have been modified.

The subject haunted him. He was back in England in October, 1836, having found on his five-year voyage the education that Edinburgh and Cambridge had denied him. He set to work. He wrote, "on true Baconian principles, and without any theory collected facts on a wholesale scale." He did not long wander undirected in this limbo. In October, 1838, fifteen months after opening his first notebook, he happened to read Malthus on Population—"for amusement," he says, improbably enough. The argument of that dismal essay is surely known to all. Malthus had first advanced it in 1795 in order to define the limits within which material progress is possible. There is a limiting ratio between a population and its subsistence. Population, he laid down, tends to increase at an exponential rate, while food supply can at best be augmented at a linear rate. Under these circumstances, competition is the law of life. Nor need we pause to consider the validity of these assertions. Their historical importance is independent of that question. For Malthus transformed the subject of political economy from Adam Smith's mood of optimistic naturalism to the deterministic naturalism of the early industrial apologia for laissez-faire. Eighteenth-century assumptions of harmony gave way before the nineteenth-century conflict of interests.

Like classical physics (and encouraged by its example), classical political economy assumed an atomistic ontology in its universe of discourse, and became in consequence a sort of social kinetics. More subtly, and more deeply in evangelical England, the Malthusian subsistence doctrine became transmuted into the moral foundation of liberal individualism. It gave that tone to the *Zeitgeist* which identifies an epoch in social history. For the inadequacy of the means of subsistence was the rod of discipline by which Providence, cruel in order to be kind, establishes incentives to thrift, virtue, industry, and continence, and deterrents to their contraries. Only within the rules was progress possible. These would have been the connotations called to Darwin's mind by the essay he read for amusement, "and being well prepared to appreciate the struggle for existence which everywhere goes on from long-continued observation of the habits of animals and plants, it at once struck me that under these circumstances favourable variations would tend to be preserved, and unfavourable ones to be destroyed. The result of this would be the formation of new species. Here, then, I had at last got a theory by which to work."

Work he did, for twenty years, collecting facts less on Baconian principles, perhaps, than on Darwinian principles, from whatever might bear on the origin (which is now to say the variation) of species. He studied the accomplishments of stock-breeders and of pigeon-fanciers. ("Few would readily believe in the natural capacity and years of practice requisite to become even a skilful pigeon-fancier.") He read widely in the literature of geology and paleontology, informing himself in ever more detail on the geographical distribution and the historical succession of animals and plants. He experimented, for Darwin was no mere onlooker, and especially on phenomena of cross-pollination and the hybridization of plants. Here the exchange of old species for new might be demonstrated as explicitly as in Lavoisier's combinatorial analysis-synthesis applied to chemical species (or reagents). Darwin had a special feeling for botany. "It has always pleased me to exalt plants in the scale of organised beings," and his last book shows how the power of movement in plants—the root growing toward water, the leaf turning toward the light—accords with the theory of evolution. Botanical experiments were, moreover, peculiarly suited to his retirement in 1842 to a country-house existence.

His state of mind was curious during all this gathering and winnowing of facts. On the one hand, he was urged and driven forward by his theory. It possessed his scientific soul. On the other hand, he was held back from publication, and even from giving himself joyfully to his conclusions, by a fear of seeming premature. This went beyond scientific caution in Darwin. It is, perhaps, a disease of modern scholarship to hold back the great work until it can be counted on to overwhelm by sheer factual mass. Thus has too many a scholar oscillated through a lifetime between creativity and timidity only to succumb to the tension, his ideas secure from criticism and the great work unwritten. Even so did Darwin shiver on the brink of publication—giving occasional hostages to fortune: a travel book on the voyage of the *Beagle*, a treatise on coral reefs, an immense work on the taxonomy and physiology of Cirripedia (barnacles). In 1842 he ventured to write out a little thirty-five-page abstract of his theory, to try it only on himself. Two years later he expanded this to the dimensions of a small treatise, and had a fair copy made. Only then did he solve a problem which had worried him—that of divergence, which he came to see as a kind of evolutionary momentum pressing the most diverse forms of life into every possible vantage point, and carrying each further from its origins.

Still the notebooks multiplied. Darwin's two closest scientific friends—J. D. Hooker was to him in botany what Lyell was in geology—began urging him to publish. Not that they were convinced, but they feared lest Darwin never try his wings. Nor was he widely known. In becoming his champion, Huxley was to make of Darwin a symbol of science as evocative in the nineteenth century as Newton had been in the eighteenth. But in 1851, commenting on the eminence of two now forgotten naturalists, Richard Owen and Edward Forbes, Huxley barely alluded to Darwin as one who "might be anything if he had good health." In 1856 Darwin yielded to Lyell's exhortations and began writing out his views in detail. The scale on which he worked would have filled many volumes. Then fortune took a hand, or misfortune (as Darwin was bound to feel). In 1858 he received a letter from Malaya, from a little known naturalist called Alfred Russel Wallace, who was working in those far jungles. He enclosed an essay, "On the Tendency of Varieties to Depart Indefinitely from the Original Type." If Darwin thought well of it, he asked, would he be kind enough to forward it on to Lyell for his opinion? Darwin thought agonizingly well of it. He might have written it himself, and he did send it on to Lyell, under a covering letter:

> It seems to me well worth reading. Your words have come true with a vengeance—that I should be forestalled. You said this, when I explained to you here very briefly my views of Natural Selection depending on the struggle for existence. I never saw a more striking coincidence; if Wallace had my manuscript sketch written out in 1842, he could not have made a better short abstract! Even his terms now stand as heads of my chapters. Please return me the manuscript, which he does not say he wishes me to publish, but I shall of course at once write and offer to send to any journal.
>
> So all my originality, whatever it may amount to, will be smashed, though my book, if it will ever have any value, will not be deteriorated; as all the labour consists in the application of theory.

Thus occurred one of the famous cases of independent discovery—Wallace, too, had been put onto the theory of natural selection by recalling a passage from Malthus as he wrestled mentally with the puzzle of species during a bout of malaria. One can only sympathize with Darwin, whose priority was incontestable, caught between the code of an English gentleman and a scientist's feeling about his intellectual property in discovery. It is a commentary on the excessively Baconian cast of Victorian notions of science, reinforced by the gospel of work, that Darwin should always have felt obliged to think that the real merit of his work lay in empirically piling Pelions of fact on Ossas of evidence. For all the while his own feelings as an innovator belied this labor theory of value in science. What hurt was to lose priority in his theory. Nor was there consolation in his pile of notebooks. Yet these were "trumpery feelings," and he must clearly now refrain from publication, lest he seem to be forestalling Wallace. "It seems hard on me that I should be compelled to lose my priority of many years standing, but I cannot feel at all sure that this alters the justice of the case."

Fortunately for Darwin, Lyell and Hooker took over the management of his interests, and those of science. Hooker asked for all the manuscripts and letters. The Wallace memoir was communicated to the Linnaean Society to be published in 1858, side by side with extracts from Darwin's sketch of 1844. Lyell and Hooker explained the circumstances as friends of the court. This was England, and everyone behaved extremely well. Ever after, Darwin and Wallace were meticulous and truthful in their appreciation of the other's qualities. Posterity may feel that Darwin got the better of all this equity. But after all, seniority (and all those facts) did count, and should have counted. It is not likely that Wallace alone would have prevailed, or even won attention. The two papers were little noticed in 1858. Nor is it likely that Darwin without Wallace would have written *On the Origin of Species*. For he undertook as part of the settlement to prepare an "abstract" of the great work of compilation. He began in September 1858, and "though often interrupted by ill-health, and short visits to Dr. Lane's delightful hydropathic establishment," he completed the *Origin of Species* in thirteen months, and it was published in November, 1859.

It does not read like a profound book. "Some of my critics have said," wrote Darwin in his *Autobiography*, " 'Oh, he is a good observer, but he has no power of reasoning.' " This he did think a little hard: "For the *Origin of Species* is one long argument from the beginning to the end, and it has convinced not a few able men. No one could have written it without some power of reasoning." The argument is easier, perhaps, to summarize than to appreciate in its true scope and depth. Chapter I passes in review the familiar varieties that occur in domesticated species. It points out the ambiguity of species. And it adduces many illustrations to show how man has adapted species to his uses, differentiating by selective breeding the dray horse and the racehorse, the whippet and the dachshund. "The key is man's power of accumulative selection: nature gives successive variations; man adds them up in certain directions useful to him. . . . Breeders habitually speak of an animal's organisation as something plastic, which they can model almost as they please." Chapter II transfers the scene to nature, and diminishes the distinction between species and varieties from the status of a boundary to that of a convenience for the naturalist. Chapter III comes to the crux of the argument, which assigns to the struggle for existence the role in nature that the stockbreeder plays in the barnyard. Variations arise by chance in particular animals. "Owing to this struggle, variations, however slight and from whatever cause proceeding, if they be in any degree profitable to the individuals of a species, in their infinitely complex relations to other organic beings and to their physical conditions of life, will tend to the preservation of such individuals, and will generally be inherited by the offspring."

In later editions Darwin was forced by criticism to develop what he understood by the concept of natural selection. And it is in this more than in any other passage that the real assurance of his scientific grasp appears:

Several writers have misapprehended or objected to the term Natural Selection. Some have even imagined that natural selection induces variability, whereas it implies only the preservation of such variations as arise and are beneficial to the being under its conditions of life. . . . Others have objected that the term selection implies conscious choice in the animals which become modified; and it has even been urged that, as plants have no volition, natural selection is not applicable to them! In the literal sense of the word, no doubt, natural selection is a false term; but who ever objected to chemists speaking of the elective affinities of the various elements?—and yet an acid cannot strictly be said to elect the base with which it in preference combines. It has been said that I speak of natural selection as an active power or Deity; but who objects to an author speaking of the attraction of gravity as ruling the movements of the planets? Every one knows what is meant and is implied by such metaphorical expressions; and they are almost necessary for brevity. So again it is difficult to avoid personifying the word Nature; but I mean by Nature, only the aggregate action and product of many natural laws, and by laws the sequence of events as ascertained by us. With a little familiarity such superficial objections will be forgotten.

The authentic voice of science speaks out of the Victorian smog in that last sentence, stiff with hauteur. No biologist had yet thought or dared to write like this, not just about the structure and functioning of some organism or set of organs in his laboratory, but about the whole course of nature. And now it begins to be clear, the sense in which Darwin settled the crucial problem of adaptation. Crucial it was, because the case in favor of purpose, the conception of biology as the science of the goal-directed, rested precisely there, on the ancient and reasonable observation that animals seem to be made in order to fit their circumstances and in order to lead the lives they do lead, with the right equipment, the right instincts, and the right habits. Darwin did better than solve the problem of adaptation. He abolished it. He turned it from a cause, in the sense of final cause or evidence of a designing purpose, into an effect, in the Newtonian or physical sense of effect, which is to say that adaptation became a fact or phenomenon to be analyzed, rather than a mystery to be plumbed.

He himself suggests the comparison to "an author speaking of the attraction of gravity as ruling the movements of the planets." And the turn of the discussion bears out the analogy. For the next chapter deals with the laws of variation. It is full of acute observations on use and disuse, climate, the greater variability of specific than generic characters, and the like. Most of this is either obvious or obsolete in the light of modern genetics, and Darwin himself recognized his limitations here. For the burden of the discussion is that the causal mechanism of variation is precisely what we do not know. But also (to continue the comparison with Newton and the cause of gravity), it is what we do not need to know. All we need is the evidence (which he and geology and comparative anatomy and all of foregoing natural history provided) that variations do occur, and that they are inherited. The theory of natural selection never depended on Darwin's trying to specify the cause of the variations that are selected by circumstance. Indeed, its success hinged precisely on dropping that question. Of Darwin, too, it might be said, "Hypotheses non fingo," and in the same sense in which it is true of Newton —not as a sterile assertion of empiricism, but as a statement that theories (speculations are another matter) must just embrace the evidence.

The final chapters of the *Origin of Species* anticipate and meet the difficulties of the argument. "Some of them are so serious that to this day I can hardly reflect on them without being in some degree staggered." But for all his professed Baconianism, Darwin never failed to maintain his composure as a theorist. He deals with instinct, and with extreme specialization. He handles the (normal) sterility of hybrids, as against the fertility of varieties within a species. He grapples with the apparent absence of transitional forms between species. He discusses the extravagant beauty of the male in certain birds, so difficult to reconcile with the humdrum operation of natural selection. Some accounts have interpreted his emphasis on "sexual selection" in such cases as a partial abandonment of his claims. His language here is a little vague, but it does seem more consonant with his single-mindedness to interpret sexual selection as a special case of natural selection rather than an auxiliary. He made great play with the imperfection of the geological record, emphasized the incompleteness of our knowledge, and reviewed from his experience of the *Beagle* the geographical distribution of variations, and particularly in the isolated populations of the great oceanic islands. Without summarizing all this material, let it simply be said that contemporary evolutionists bear him out on all essential points: they have vindicated him as Laplace did Newton. All these apparent difficulties are like the planetary anomalies observed in the eighteenth century. Once they are explained as instances rather than violations of natural selection or of gravity, they sophisticate the theory, instead of serving as escape hatches. Nor does contemporary science offer any handle to dissent from the concluding paragraph of what he knew to be "the chief work of my life." In it Darwin rose on rudimentary wings of prose as high as ever he could flutter toward lyricism

It is interesting to contemplate a tangled bank, clothed with many plants of many kinds, with birds singing on the bushes, with various insects flitting about, and with worms crawling through the damp earth, and to reflect that these elaborately constructed forms, so different from each other, and dependent upon each other in so complex a manner, have all been produced by laws acting around us. These laws, taken in the largest sense, being Growth with Reproduction; Inheritance which is almost implied by reproduction; Variability from the indirect and direct action of the conditions of life, and from use and disuse: a Ratio of Increase so high as to lead to a Struggle for Life, and as a consequence to Natural Selection, entailing Divergence of Character and the Extinction of less improved forms. Thus, from the war of nature, from famine and death, the most exalted object which we are capable of conceiving, namely, the production of the higher animals, directly follows. There is grandeur in this view of life, with its several powers, having been originally breathed by the Creator into a few forms or into one; and that, whilst this planet has gone cycling on according to the fixed law of gravity, from so simple a beginning endless forms most beautiful and most wonderful have been, and are being evolved.

✧

EIGHTEENTH-CENTURY PHYSICS had required about fifty years from the publication of the *Principia* to assume with ease the Newtonian posture. Biology has had to traverse even wider confusions before orienting itself cleanly around the Darwinian theory of evolution. Indeed, it has clarified its outlook through the instrumentality of modern genetics only since the 1930's. In 1929 Erik Nordenskiold's *History of Biology*, still the foremost text, closed with a reference to "the dissolution of Darwinism." That natural selection "does not operate in the form imagined by Darwin must certainly be taken as proved," wrote Nordenskiold. Exactly the contrary is now thought to be true. Since such shifts of scientific opinion were still to occur in the twentieth century, it is not surprising that the import of Darwin's theory for the whole science was not mastered at the time. Claude Bernard, perhaps the greatest of experimentalists in his skill and sobriety, saw the future of biology as lying in its reduction in physiology to laws of chemistry and physics. There was nothing for him in Darwin, whose work he did not distinguish from *Naturphilosophie*. "We must doubtless admire," he writes in his fine manifesto on *Experimental Medicine*, "those great horizons dimly seen by the genius of a Goethe, an Oken, a Carus, a Geoffroy Saint-Hilaire, a Darwin, in which a general conception shows us all living beings as the expression of types ceaselessly transformed in the evolution of organisms and species,—types in which every living being individually disappears like a reflection of the whole to which it belongs." But he did not admire them. He never thought this science. In the critical tradition of French learning, Darwin's mind and language seemed simply slack.

At the other extreme from Gallic indifference burgeoned the enthusiasm which made a religion of science, mistook nature for God, and adopted Darwin as the prophet. The German for this "ism," *Darwinismus,* best conveys its spirit, always most at home in Germany. There Ernst Haeckel and his like, deploying all the rich capacity of their language for blurring distinctions, worked a syncretism between the Goethean sense of unity in nature and the Darwinian proof of organic evolution. Haeckel's were the voice of Jacob and the hands of Esau, the historicist spirit of romantic idealism and the hairy philosophy of monistic materialism. All biology would be made over into evolution. Thus (to take only one example) embryology was henceforth to be ruled by the doctrine of recapitulation, by which every individual traverses in embryo the evolution of the race, from the single-celled through the invertebrate stage, the gill-breathing, the reptilean, and so on. Everyone, to adapt a saying of Huxley's, climbs his own family tree out of the womb. But all truth, too, was evolution. In Haeckel's words, "Darwin's theory of the natural origin of species at once gives us the solution of the mystic 'problem of creation,' the great 'question of all questions,'—the problem of the true character and origin of man himself." Two chapters in Haeckel's *Riddle of the Universe* handle the "embryology" and the "phylogeny" of the soul, making use of recapitulation. And though enthusiasm is the most forgivable of excesses, in this lavishness evolutionary theory explained too much. Haeckel gave little thought to what precise problems might connect evolutionary theory to the technical reaches of biology, or what positive route through the whole science its reduction to evolutionary terms might take. His fault in judgment was the reverse of Bernard's, and of parochial specialization.

In retrospect it seems obvious that in pushing analysis beyond the point where Darwin left it, the critical question was the mechanism of inheritance. Darwin here confused the clarity of his original admission of ignorance by vague and contradictory speculations about pangenesis in certain later writings. To meet the criticism that Darwinian evolution does not explain inheritance, certain other biologists leafed back the pages to Lamarck. From among all his principles, they chose the inheritance of acquired characters, and in their own extremity made it into the main point of Lamarck's philosophy and the essential supplement to Darwin's. German biological romanticism gave neo-Lamarckism a hospitable climate. Nor was it unreasonable. In the trivial sense, the inheritance of acquired characters is not so much wrong as tautological. Morphological changes in species must be cumulative, and therefore in some sense heritable, or there would be no evolution. The serious question is rather how these characters are inherited, and the relationship of acquisition to inheritance. Indeed, a general failure to distinguish these questions was what kept biological thought moving in widening circles for the rest of the nineteenth century.

Nor were mechanists notably more successful than vitalists in breaking out of these circles. Once again it appears that the real problem was to achieve biological objectivity rather than to choose between vitalism and mechanism, idealism and realism. If one takes the standpoint of modern genetics and compares the work of the foremost idealist, Karl von Nägeli, with that of the foremost mechanist, August Weismann, the structure of their theories seems not dissimilar, and their differences come down to matters of temperament. Nägeli was a botanist and cytologist who had studied philosophy under Hegel in Berlin, and who taught in Munich. His researches on pollen grains and uni-cellular algae, and upon sexuality in cryptogams, were fundamental. The conception of natural selection held for him no meaning whatever. And his own speculation (or as he thought theory) on evolution illustrates how romanti-cism in biology carried on a kind of Cartesian search for an ideal mechanism to serve in explanations.

Nägeli's idioplasm ramifies through time like some webbed vortex. It bears all the characteristics of the im-aginary mechanism—a texture too fine and plastic to be detectable by instruments, films through which influences pass, tensions in the Protean ineffable. Idioplasm was born of the tactical necessity for making a physical dis-tinction between ontogeny and phylogeny, the history of the individual and the history of the race. There are in every being two kinds of substance, tropoplasm and idio-plasm. Tropoplasm builds up the gross structure of parts and organs, and all tropocells die with the organism. Idio-plasm is immortal. It is itself cellular in structure, a fila-ment of "micelles" which crystallize out of primitive albuminoid matter in the ooze where life begins. They arrange themselves in parallel. They are like fibers twisted and tensed into thread, whereas tropocells flocculate in depth like layers of felted serge.

Or perhaps idioplasm might better be understood as an evolutionary counterpart of the Stoic *pneuma*. On the one hand, it is the principle of unity in evolution. But on the other hand, it differentiates the stream of living substance into phyla, species, individuals, organs, and cells. For idioplasm is the bearer of determinants. Every filament fixes some quality upon the organ it penetrates: its color, chemical composition, physical state, physiological function, or whatever. Nothing is left out. Idioplasm branches all through the organism like an undetectable nervous system. All parts of the body are traversed by the whole of the idioplasm. The reproductive organs are no exception, and generation consists in the mingling in the fertilized ovum of idioplasm from both parents. That strain of idioplasm, therefore, which runs through (say) the species horse is nothing less than identical with the nature of horse. Differences between individual horses are caused, not by any change in the substance of idioplasm, but by differential tensions which may develop here and there between particular micelles, or perhaps by accidents which occur in the lifetime of the animal.

Now this is obviously the purest latter-day Lamarckism, and it would be an ungrateful task to follow Nägeli into his hundreds of pages of imaginary detail. What is significant historically is the way in which the romantic mind had still to explain evolution in the decades after Darwin. We are here concerned with the strand of life: "I shall proceed from the primitive, unorganized condition of matter and endeavor to show how organized micellar substance has arisen in it, and how, from this micellar substance, organisms with their manifold properties have arisen." Evolution is the expression of the *"automatic perfecting process* or progression of the idioplasm, and entropy of organic matter." Any adaptive variations caused by external conditions are of the most trivial importance. The environment only provides the organism with food and matter for its life processes. It causes no permanent variation and has only ontogenetic significance. Reproduction represents no break in continuity. Parents continue in the offspring as the "stem continues its specific life in the branch." Finally, what he, the biologist, must study is the nature of the organism as displayed in evolution, the fundamental biological process of the world. It is not a problem which can be broken up into its parts, any more than may the organism itself:

If heredity and variation are defined according to the true nature of organisms, they are only apparent opposites. Since idioplasm alone is transmitted from one ontogeny to the next following, the phylogenetic development consists solely in the continual progress of the idioplasm; and the whole genealogical tree from the primordial drop of plasma up to the organism of the present day (plant or animal) is, strictly speaking, nothing else than an individual consisting of idioplasm, which at each ontogeny forms a new individual body, corresponding to its advance.

Weismann, on the other hand, is usually introduced by historians of biology as the one who imported Darwinian natural selection into the study of heredity. His postulation of the germ plasm (*Keimplasm*—these things lose their magic outside the language of their birth) still figures in the textbooks. There it is taken to be the historical foundation of the case against evolution by inheritance of acquired characters. Weismann coined his germ plasm in 1882, two years before Nägeli's idioplasm. Both were expressions of the tactic, quite generally adopted in theoretical biology of the 1880's, to contemplate the hereditary patrimony apart from the incidents of ontogeny. To the latter, anatomists, physiologists, and pathologists might address themselves, and to the former, evolutionists. Thus, Weismann distinguished between germ-cells and soma-cells, the germ-cells immortal in the sense in which unicellular organisms are that multiply by division. The distinction no longer obtains in all sharpness, but it still serves to dramatize the important difference between gametes and differentiated body cells, and to emphasize that the latter have nothing to do with heredity. That is to say, the body cells of an organism derive from its germ cells, but not vice versa. The line of reproductive cells is indeed direct from generation to generation. But in any one generation they lead a sequestered, one might say a monastic life, if the figure did not seem anti-clerical.

Nägeli was a profound idealist and a Lamarckian; Weismann a profound mechanist and a Darwinian, and his construction of the germ-plasm presented conceptual advantages over Nägeli's idioplasm. He imagined it to be the locus of natural selection, not of an indwelling drive toward progress, man, and beyond. He imagined it, further, to be a particulate plasma continuing from one organism to the next, not an undetectable network ramifying everywhere in life and time. Weismann's temper was more sober, his mind open as well as ingenious. He proved very agile in expedients for adapting his germ plasm to the discoveries that occurred in his lifetime. Of these the most relevant was Roux's identification by staining of the chromosomes, and the extraordinary and beautiful behavior of those structures in cell division or mitosis (to use the later term). Weismann correctly predicted how meiosis would involve first a halving and then a mingling of the hereditary substance. And since Weismann enthusiastically adopted the chromosomes as the containers of germ-plasm, it is not surprising that he is usually credited with having announced the program of contemporary genetics.

If this be altogether correct, however, it is difficult to see why that science should have had to await the twentieth century to learn its business. For a critical scrutiny will leave Weismann closer, perhaps, to Nägeli than to the modern synthesis. Weismann could have saved mechanism simply enough, by making one particle in the germ-plasm the physical basis of every variable quality expressed in the cell. But this would have been to fall into infinite regression. It was, therefore, by an intellectual necessity that the germ-plasm consisted of bundles of organic molecules or "biophores"—"They must exist, for the phenomena of life must be bound to an entity of matter." Each type of biophore governs some part of the cell, and the biophores are, therefore, "the bearers of the characters or qualities of cells." But cells are differentiated by histological function, too, and this aspect of organization is governed by the second order of units within the germ-plasm, the "determinants" into which biophores cluster. Finally, the determinant "knows" to what cell it must migrate in the course of ontogeny because of its position in the third order of plasmic units, the "ids," out of which is fashioned the "inherited and perfectly definite architecture" of the germ-plasm. In this word Weismann acknowledged his debt to idioplasm. His "ids" are to be thought of, perhaps, as functioning like the harness of a draw loom, each of the lashes (or ids) raising a characteristic combination of warp threads (or determinants) to make the design. And all this weaves on below the level of sense inside the rodlike chromosomes.

But what of natural selection? And only now does the full measure of Weismann's ingenuity first appear, and then betray itself. For it is Darwinism he would seat in all these intragerminal entities. Since this is where the stream of living water flows, why it is there that the struggle for existence must obtain, there among the biophores and ids. Each is the root of variations or complexes of variation which appear in ontogeny, but are selected in phylogeny. It is in the germ-plasm that one id wins out over another, as a consequence of its suiting the organism wherein it prevails to changes in climate, habitat, diet, or other external conditions. In such case, the selected germ-plasm will run in greater volume in the widening channel provided by the increased numbers of favored individuals and races, gradually becoming new species.

The notion of germinal selection is what reveals the measure of the gap between Weismann's thinking and that of modern genetics. For the germ-plasm remains a qualitative hypostasis, after all, an imaginary physical agent (even like the idioplasm) which bears properties through historic time. Germ-plasm is nothing but a word for a stream of rudiments within the chromosomes. It never became, it was logically incapable of becoming, that programmer of biological events, that item of organizational information, which is the gene. In effect, therefore, Weismann's isolation of the germ-plasm turned evolution into a process indwelling in the chromosomes. Instead of objectifying evolution in the relation of structure to circumstance, he internalized it in the organism. For Weismann, too, we are beads strung on a rope. Even though it is a rope of sand, evolutionists would still study the rope. Germ-plasm, therefore, was another of those proto-scientific constructions which failed by explaining too much. Weismann would solve both heredity and development by germ-plasm, "on the changes of which development depends, while heredity rests on its continuity."

❖

AND ALL THE WHILE the right answer had blossomed unregarded in the form of Mendel's garden peas. Mendel published his now famous paper, "Experiments in Plant Hybridization," in 1866. But because biologists were busy asking all these wrong questions, it was ignored until 1900. For what is often said is not true, that Mendel's results remained unknown because of the obscurity of the *Brünn Society for the Study of Natural Science*, whose *Proceedings* printed them. One of the supreme ironies of the history of science is the correspondence between Mendel and Nägeli—the modest Sudeten monk in his kitchen garden, the Geheimrat Professor Doktor in the University of Munich. Mendel described his findings in an early letter. Were they not, replied Nägeli (a good Hegel student), "empirical rather than rational?"—and passed on, intending no discussion, to put Mendel onto doing experiments for him on hawkweed, thus treating him like an unpaid laboratory assistant. And it is indeed curious that the two concepts on which biological objectivity rests historically, natural selection and the discreteness of hereditary characters, should have been the work of the last two naturalists, not to say amateurs, to figure importantly in the history of science—Darwin, an English country gentleman, and Gregor Mendel, an Austrian monk.

No more in Mendel's case than in Darwin's, however, did science have to do with a lucky strike. To read "Experiments in Plant Hybridization" is to experience the company of a fine biological intelligence. There is that instinct for the detail of a situation which distinguishes the biologist from the physicist. There is a quality of patience, which the physicist is not called upon to exercise, with the conditions imposed by the organism, and a willingness to set up the experiment and wait and watch while the organism performs it. Mendel's experiment lasted through generation after generation for eight years. There is displayed throughout a gentleness of handling which argues a manual temperament different from the mechanically precise, but no less delicate. He worked with statistically significant numbers of plants, and had to pollinate each one himself before nature in the form of some busy bee should forestall him with uncontrolled pollen: "Artificial fertilization is certainly a somewhat elaborate process, but nearly always succeeds. For this purpose the bud is opened before it is perfectly developed, the keel is removed, and each stamen carefully extracted by means of forceps, after which the stigma can at once be dusted over with the foreign pollen." Then he would tie a little bag over the plant's head, to protect what was left of its chastity from its own kind. There is evident, finally, in Mendel's mind the clearest perception of what precise problem would have what general significance, and of how to pose it.

For the critical mystery of biology was the relationship of the persistence of organic form to the variation of species. In retrospect we can see how neatly Mendel took up this problem from the pole opposite to Darwin's, to whose solution his was the necessary complement. That is to say, Darwin would study variation in species apparently fixed. Mendel would study fixity amid the apparent variation of hybrids, where the problem of persistence and change appears with peculiar urgency. He had at hand the great body of botanical experience in producing novel flowering plants by artificial insemination. What struck his attention was the regularity with which the same hybrid forms recur, and he proposed to find "a generally applicable law governing the formation and development of hybrids." No experiment had ever "been carried out to such an extent and in such a way as to make it possible to determine the different forms under which the offspring of hybrids appear, or to arrange these forms with certainty according to their separate generations, or definitely to ascertain their statistical relations." These were the right questions. To their cogency Mendel owed his success. But neither was he (as is sometimes implied) lost to the large issues: "It requires, indeed, some courage to undertake a labour of such far-reaching extent; this appears, however, to be the only right way by which we can finally reach the solution of a question the importance of which cannot be overestimated in connection with the history of the evolution of organic forms." (Mendel's library contained a heavily annotated copy of *On the Origin of Species*).

Nor was it luck he hit on peas for his experiments. His conception of the problem required plants which should show constant differentiating features, and among which the differences should be marked. The hybrid offspring of two varieties must be either tall or short, not taller or shorter. After many trials, Mendel selected thirty-four varieties of the genus *Pisum*, and bred them to observe the inheritance of seven pairs of contrasting characters: smooth versus wrinkled seeds; yellowish versus green cotyledon; white versus colored seed-coat and flower; plump versus emaciated pods; green versus yellow pods in the unripe state; axial versus terminal arrangement of flowers along the stem; tall plants (six to seven feet) versus short ones (about one foot).

The results are known to the student in the most elementary course in general science, who easily bandies the words dominant and recessive once he is taught by those famous genealogical diagrams:

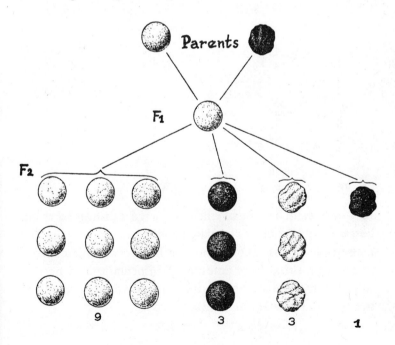

How simple it looks, and how pat those ratios! Like rigid models of atoms, such pictures support the student's memory and destroy his appreciation of the thought the information once required. Mendel had had first to conceive of dominant and recessive characters amid the medley of tall and short plants, or smooth and wrinkled seeds, which succeeded each other generation after generation. Then, in an age when probability was still an occasional instrument of scientific demonstration, he had to resign himself to the search for a statistical order. For the progeny of any one set of parents will not usually exhibit Mendelian ratios.

As an example, he gives the results for offspring of the first ten plants (F-1) in his initial two experiments:

	Experiment 1 *Form of Seed*		*Experiment 2* *Color of Cotyledon*	
PLANTS	ROUND	ANGULAR	YELLOW	GREEN
1	45	12	25	11
2	27	8	32	7
3	24	7	14	5
4	19	10	70	27
5	32	11	24	13
6	26	6	20	6
7	88	24	32	13
8	22	10	44	9
9	28	6	50	14
10	25	7	44	18

In Experiment 1 the extreme cases were 43 round, 2 angular; and 14 round, 15 angular. It will, perhaps, suggest the magnitude of his labors, and the firmness of his control, to tabulate simply the amount of information he needed in order to establish the dominant-recessive ratio in each of the seven characters he studied:

Character	Number of plants (F-1)	Dominant-Recessive Ratio
Smooth vs. Wrinkled seed	7,324	2.96
Yellow vs. Green cotyledon	8,023	3.01
White vs. Colored flower	929	3.15
Plump vs. Emaciated pod	1,181	2.95
Green vs. Yellow pod	580	2.82
Axial vs. Terminal flower	858	3.14
Tall vs. Short plant	1,064	2.84
		2.98
		or 3 to 1

And this was only the F-1 generation. In later generations he had to keep his mind firmly fixed on the *succession* of ratios to make the second order distinction between the true breeders (one-third) and the two-thirds which breed hybrid, once again in the three-to-one proportion among themselves.

This arithmetical maze did not, perhaps, invite mathematical thought at a high level of abstraction. Nevertheless, the assertion may be ventured that Mendel's was the first significant application of mathematics to biology. And notwithstanding his other great merits, this may well be taken as Mendel's cardinal contribution to his science: that its quantification goes back to his experiments. With even greater admiration must one follow the strong thread of his reasoning through the labyrinth of inheritance when several variable characters were united by crossing—tall-ness or shortness, for example, combined with white or colored flowers. Without following into detail, it will illustrate the point to give the expression by which he represents his results:

$$AB+Ab+aB+ab+2ABb+2aBb$$
$$+2AaB+2aab+4AaBb \ldots,$$

where each term represents a class for every possible combination of the four characters, and the coefficients the relative occurrence of that class in the offspring.

The identity of particular characters persists in discrete form from generation to generation, and this was Mendel's basic discovery: that heredity comes, not in blendings, but in packets. It comes in jumps like those which the physicists, when they got down to comparable fundamentals, would call quanta, which too would require statistical ordering. His own interpretation of his results is atomistic in its structure, though he no more offered an hypothesis of wherein the units of heredity consist than did Newton of the cause of gravity:

> Since the various constant forms are produced in *one* plant, or even in *one* flower of a plant, the conclusion appears logical that in the ovaries of the hybrids there are formed as many sorts of egg cells, and in the anthers as many sorts of pollen cells, as there are possible constant combination forms, and that these egg and pollen cells agree in their internal composition with those of the separate forms.

Precisely because Mendel did distinguish the problem of heredity from that of variation, these "cells" were logically capable of reduction to the genes of contemporary genetics. Such was not the case with the blending inheritance of Nägeli's and Weismann's theories, nor with the constructions of the other eminent and puzzled biologists of the late nineteenth century, who had to grapple with Darwin's legacy. Their idioplasms and germ-plasms put them in a position like that in which Fontenelle and the Cartesians had once confronted Newton, trying to explain the cause and describe the operation of gravity by a single concept. The historian of science may, therefore, be pardoned for wondering what might have been the influence on biology had these scientists known the history of science, and whether they might then have noticed the interest of Mendel's work? Suppose they had thought to compare the simple whole numbers of his ratios to Dalton's, by which the chemical revolution was reduced to numerical terms. Suppose they had known of the relationship of the corpuscular philosophy of the seventeenth century to the Newtonian synthesis. Might they not have saved themselves much unprofitable reasoning, and advanced the progress of their science by several decades?

This was not to be, however, and Mendel's paper was stillborn. It lay in a state of suspended animation until 1900, when it was quickened to life, less by the logic of science than the illogic of its history. Several biologists at the end of the century were addressing themselves to an objection frequently advanced against Darwinism: that it was difficult to see how a variation too slight to be detected from one generation to the next could confer enough advantage to be selected and perpetuated. It is now known that this is a false problem, since the significant mutations occur in the gene. Nevertheless, it directed attention to the possibility of gross mutation rather than gradualism. Perhaps evolution was a staircase rather than a slope. The most notable to explore this avenue was Hugo de Vries of Amsterdam, who observed what he took to be brusque transformations into new species in certain offshoots of the evening primrose. As a botanist, moreover, he found unsatisfactory the implication of Weismann's germ-plasm, that the color (say) of the flower is determined by a genetic

substructure governing just that organ. Rather, the whiteness of a flower involves other manifestations in leaf or
stem, and must be governed by what De Vries called
"pangens," discrete but immanent constituents of the entire organism: "The whole character of a plant is built
up out of definite units. . . . To each character there corresponds a particular form of material carrier"—not (be it
emphasized) to each organ or each soma cell, but to each
character. "There are no transitions between these elements, any more than there are between the molecules of
the chemist. . . . We have to recognize that the general
image of the species passes into the background, and that
the idea of the composition of the species out of independent factors comes to the front. . . . The units of the
characters of a species are . . . to be regarded as sharply
separated magnitudes."

When De Vries wrote that, he had already discovered
Mendel's paper. So almost simultaneously had two colleagues, Carl Correns in Tübingen and Erich Tschermak
in Vienna, who were working quite independently on
problems of discontinuous heredity. The triple coincidence arrested attention, the more so as it involved the
pathos of a wrong righted too late to benefit the injured
party. De Vries's unsettled primroses were not in fact true
mutations. Nevertheless, they served to impress upon the
scientific world the proposition that heredity might be
discontinuous in its basis. Only the prefix had to be
dropped from De Vries's "pangens," and biology would
be left with the word and the Mendelian structure of interpretation waiting to accommodate the right evidence.
That appeared in the year 1906 in the laboratory of Columbia University, where T. H. Morgan, breeding generation after generation of pink-eyed fruit flies, found one
whose eyes were white. A disproportionate share of the explosive new science of genetics is born on the wings of that

tiny insect, which is to the geneticist what the mouse was to Boyle and Priestley. And with this the historian must hand over to the biologist the threads he has gathered up. Suffice it to point out how Mendel-cum-Morgan occupy the place in the historic strategy of the biological revolution that Dalton did in chemistry. The geneticist will separate out or combine characters from generation to generation—short-wings, grey bodies, white eyes—as wilfully as the chemist with his reagents. His test of the objective significance of the material units of heredity is the same as Lavoisier's: analysis followed by synthesis. Nor is the point clouded by gathering doubt whether the gene is a real particle or a configuration in the giant molecules of the hereditary patrimony—either hypothesis (as Lavoisier observed in a similar situation) will permit expression of the quantities involved in an abstract and mathematical manner, and perhaps they are complementary truths.

❖

MUTATIONS occur in the gene, either at random or as a consequence of radiation or some other positive injury. But they are selected for preservation according as they confer on the animal advantages in objective circumstance. And such is the movement of evolution against time. Darwin himself closed *The Origin of Species* with a passage which compared the generality of evolution in respect to time with that of gravity in respect to space. Certainly the nineteenth-century awareness of science was conditioned by the theory of evolution as that of the eighteenth century had been by the Newtonian theory. And now the vindication of Darwin's theory of natural selection in the science of genetics puts scholarship in a position to specify what were the elements in its success, to push further a structural comparison between the Newtonian and Darwinian theories, and thus to exhibit the justice of the judgment which attributes to Darwin the importance for biology which Newton has had in physics. For the concept of natural selection, quantified in genetics, has put an end to the opposition between mechanism and organism through which the humane view of nature, ultimately the Greek view, had found refuge from Newton in biology. Lamarck's theory, on the other hand, had originated as the transfer to biology of that old view of flux and process, since become romanticism, for which Lavoisier had made chemistry, the science of matter, uninhabitable. It is for this reason that Darwin was the orderer of biological science, as Newton had been of physics and Galileo of kinematics. Perhaps it will be *apropos* to push the earlier term of the comparison back to Galileo. For the objectification and quantification of motion were the germs of metrical science. And in the over-arching structure of the history of science, Darwinian evolutionary theory stands in the same relation to Lamarckian or neo-Lamarckian as does Galilean kinematics to that of the impetus school.

In mechanics, Galileo achieved objectivity by accepting motion as natural, and considering its quantity as something to be measured independently of the moving body. This he accomplished by treating time as a dimension, after which translational motion is no longer taken as metaphysical change. In Darwin—to draw out the parallel— natural selection treats in the same way that sort of change which expresses itself in organic variation (the motion of species, so to say, through historic time). Instead of explaining variation (or motion), he begins with it as a fundamental fact of nature. Variations are assumed to occur—in the gene they do occur—at random, requiring from Darwin no further explanation and presupposing in his theory no causative agent for science to seek out. This is what opened the breach through which biology could follow physics into objectivity. It introduced the distinction, which Darwin was the first to make, between the origin of variations and their preservation. Variations arise by chance. But they are preserved according as they work more or less effectively in objective circumstance.

Darwinian evolutionary change is analytically analogous to Galilean motion in another and even more impressive respect. There is direction in it, whereas in Lamarck's formulation life simply circles endlessly through nature. H. F. Blum has recently considered evolution as a problem in entropy, and advanced the interesting argument that time as it enters into thermodynamic processes may be considered as a coordinate of evolution. This amounts to saying, on the one hand, that evolution is capable of vectorial description, and on the other, that biological time is a dimensional component of a physical situation and ceases thereby to be a refuge of becoming or a locus of flux. Quite generally, in fact, Darwin's work, though not numerical in expression, was, nevertheless, quantitative in method and matter of thought. As to this, the relationship of Mendel to Darwin repeats that of Dalton to Lavoisier— the one attaching numbers to the other's quantities. Thus, that Darwin began with the Malthusian ratio was of far more significance for his success than was the question of its validity. It was, indeed, of utmost significance. What selection does in Darwin's theory is to determine the quantity of living beings which can survive in any given set of objective circumstances. This aspect of the approach is more evident, perhaps, in Wallace's essay than in Darwin's more diffuse account. For example:

> Wild cats are prolific and have few enemies: why then are they never as abundant as rabbits?
> The only intelligible answer is, that their supply of food is more precarious. It appears evident, therefore, that so long as a country remains physically unchanged, the numbers of its animal population cannot materially increase.

If one species does so, some others requiring the same kind of food must diminish in proportion. . . . It is, as we commenced by remarking, "a struggle for existence," in which the weakest and least perfectly organized must always succumb.

And even more striking is Wallace's passage on natural selection as accounting

. . . for that balance so often observed in nature—a deficiency in one set of organs always being compensated by an increased development of some others—powerful wings accompanying weak feet, or great velocity making up for the absence of defensive weapons; for it has been shown that all varieties in which an unbalanced deficiency occurred could not long continue their existence. The action of this principle is exactly like that of the centrifugal governor of the steam engine, which checks and corrects any irregularities almost before they become evident; and in like manner no unbalanced deficiency in the animal kingdom can ever reach any conspicuous magnitude, because it would make itself felt at the very first step, by rendering existence difficult and extinction almost sure to follow.

No longer need it appear as a paradox, therefore, that it should have been Darwin, Wallace, and Mendel, the old-fashioned naturalists, and not the embryologists and physiologists of the continental laboratories, who brought the revolution in biology. The reason is clear. Nor does it lie only in the nature of their empirical contributions. It lies in the nature of their reasoning, which was concerned with quantity and circumstance. This is why it was they who liberated biology from its limiting dependence on classification and dissection, with the gulf between bridged insubstantially by that metaphor of goal-directed organism which the evidence never could control. For nothing is more arresting in the comparison of the biological to the Newtonian revolution than the reduction of the concept of natural selection to material atomism in the hybrid science of genetics, produced by the crossing of Darwinism and Mendelism. Just as the discontinuity of matter in atoms-and-the-void liberates motion from subjectivity, or indwelling qualities, so biological objectivity was firmly seated in the discontinuity of the hereditary patrimony, where inheritance might be comprised in number. In this perspective, it may appear as a kind of wisdom in Darwin, rather than as a failing, that his theoretical work began as an application to biology of the individualistic assumptions of classical political economy. He had, after all, no other basis for atomism. And the outcome is a conception of biological order no different from the order assumed by contemporary atomic physics—an order of chance to be analyzed by the techniques appropriate to mathematical probability.

Darwin and the Lamarckians speak their parts, therefore, in that endless debate between atoms and the continuum, the multiplicity of events and the unity of nature, which is what has given the history of science its dialectic since its start in Greece. For the continuum of life as the program of nature goes back to that aspect of classical philosophy which was a prolongation of cosmogony, back through the Stoics and Heraclitus to fire and the world as flux and process. But it is cosmology, the opposite of this, from which science derives, rather from the contemplation of being in the light of reason than of becoming in the light of process. And this resolves, perhaps, a final apparent paradox, which is that providentialism and belief in fixity and divine design have in effect been more conducive to positive scientific work—in Newton, for example, in Linnaeus, or in Cuvier—than has belief in process—in Diderot, in Goethe, or in Nägeli. As a constituent of theory, providentialism ends in self-defeat—but only after establishing structure for science to find. It posits the existence of specific entities which may serve as the term of analysis. But in becoming, everything blends into everything, and nothing may ever be defined. And though Darwin certainly did invest science with historical depth, his is the nominalist history of the true historian, and not the immanent process of the Marxian or Hegelian historicist philosopher. It is a mistake to say, as often is said, as even the great Cassirer said, that Darwin brought this Hegelian sense of becoming within the pale of science. What he did was to treat that whole range of nature which had been relegated to becoming rather as a problem of being, an infinite set of objective situations reaching back through time. He treated scientifically the historical evidence for evolution, which had been marshalled often enough before him, but more as a travesty than an extension of science. Rightly understood, therefore, the Darwinian theory of evolution by natural selection turned the problem of becoming into a problem of being, and permitted the eventual mathematicization of that vast area of living nature which until Darwin had been protected from logos in the wrappings of process.

❖

IF THIS BE SO, a further word is needed on the significance of evolutionary thought. Science and religion, evolution and ethics, social Darwinism—these topics form a considerable portion of the nineteenth-century stock-in-trade of the historian of ideas. Will all this, too, appear as a variation on the theme of the Enlightenment and science, transposing the essential misunderstanding of the one for the other into the keys of history and time? And the answer, on the whole, is yes. No doubt Darwinism was more directly translatable into political and social terms than Newtonianism. Rugged individualists, Marxists, militarists, racists, they all found in the cruelty of nature ideological comfort comparable to that which their eighteenth-century forebears had discovered in its harmony. And if the influence of Darwinism was not deeper, its lessons seemed more plausible, however contradictory. For Darwin took the form of his ideas, not like Newton from classical geometry, but from the constructs of English liberalism. We are concerned not with a translation after all, but only with a retranslation.

This, indeed, is the only point that needs to be made about the validity of social Darwinism (as apart from the real and different question of its influence). Consider words in the title alone: "Origin"—it is a theological problem; "Selection"—it implies an act of choice; "Survival"— who can be indifferent? "Favoured Races"—the phrase which Darwin substituted for "Survival of the Fittest" in later editions evokes shadows of Gobineau and Hitler falling before; "Fittest"—but who is to judge of the fit? Is nature harmony or is it conflict? Is the law of the jungle or the law of love the rule of life? Must dog eat dog with relish? And the trouble was, not that Darwin meant to pronounce on all this, but that in the relatively inexact state of biology, he borrowed from common language terms which have human connotations quite different from the meaning they assume in the scientific theory of evolution. Just so does a word like force mean one thing in physics and quite other things in life. Nor could Darwin stop at every paragraph to explain himself. When he did, it is clear that he was capable of thinking clearly even in loose language. The measure of "success," for example, he defines as the amount of progeny, and this is obviously the particular animal's contribution to evolution. But that is not what Andrew Carnegie meant by success, much less Malthus. And we for our part may define social Darwinism as the re-exportation into social science of a language quite speciously fortified with the deterministic vigor of natural science—opinions converted into truths through having traversed science.

In a sense the critics of Darwin saw the point more correctly than did the epigones who associated evolution with some social gospel. For the romantics criticized the theory of natural selection precisely because it denied the history of nature any meaning for man as a social or moral being. The concept of fitness in the organism was simply a tautology, said the romantic biologists, of whom Driesch was the most important. (Just so had Fontenelle objected to Newtonian geometric proofs that they come out exactly even.) What is causation in Darwin, and what in Newton?—only a formless sequence of results extending backward or outward endlessly into a metaphysical limbo. For the romantics of the nineteenth century, like the Cartesians of the eighteenth (in this way, too, their successors), wanted a science which would account at once for the behavior and the cause of phenomena, which would see nature steadily and see it whole. They wanted a science which would seize on the unity of nature, instead of fragmenting it into discrete events connected only by chance and circumstance, and never by reason or purpose or will or the good of man. In the case of Driesch, the heart of the position is that Darwin simply impoverishes biology, that he gives no rational insight into events, that he is simply a recorder posing as a philosopher, a chronicler rather than a historian of nature—and all this because he will not see that the laws of life are different from those of physics, and that in the organism purpose is everything. And this, of course, is why, once Darwin had established the facts of evolution, the Nägelis and Driesches and Bergsons repudiated the very theory of natural selection which carried those facts home to the scientific consciousness of the age, substituted for it an idioplasm or an entelechy or an *élan vital*, and reached back to Lamarck—in the case of George Bernard Shaw back to Methuselah—as to a humane alternative.

Biological romanticism never made much impression in the world of English letters, where Samuel Butler and Shaw have been the most widely read of Darwin's critics. In their case, too, a comparison with certain themes of the eighteenth-century Enlightenment is instructive, for it makes clear that the question is no biological discussion, but simply the continuing expression of a moral resentment which wants more out of nature than science finds there. To read Diderot and Butler together is a curious experience, itself almost a vindication of Butler's *Unconscious Memory*. For one has the impression that this and Butler's other writings upon nature were products of his own rather painful and labored reflection, and yet how unoriginal they are! These, for example, are the four principles of Butler's *Life and Habit*: "The oneness of personality between parents and offspring; memory on the part of offspring of certain actions which it did when in the persons of its forefathers; the latency of that memory until it is rekindled by a recurrence of the associated ideas; and the unconsciousness with which habitual actions come to be performed." Butler follows Diderot's route out of atomistic materialism: "It is more coherent with our other ideas, and therefore more acceptable, to start with every molecule as a living thing, and then deduce death as the breaking up of an association or corporation, than to start with inanimate molecules and smuggle life into them; and, . . . therefore, what we call the inorganic world must be regarded as up to a certain point living, and instinct, within certain limits, with consciousness, volition, and power of concerted action."

But from the point of view of one who admires the intellectual achievements of science, it is Shaw rather than Butler who, by contrast to his pretensions, seems drastically diminished in stature by his ventures into scientific criticism. The famous preface to *Back to Methuselah* presents clichés with the air of lordly malice that Shaw knew how to assume as the right of a superior intelligence which did not mind pointing to its own perversity. But it was an intelligence which, far from transcending science, had never given itself the trouble to understand the force or limitations of scientific demonstrations, and in the perspective of history, Shaw on Darwin will surely find a place side by side with Bellarmine and the papal jury setting the astronomers right about natural philosophy. It does not appear that Shaw ever thought to ask the biologists whether natural selection was true. It was simply "a blasphemy, possible to many for whom Nature is nothing but a casual aggregation of inert and dead matter, but eternally impossible to the spirits and souls of the righteous." Darwin is forbidden to banish mind from the universe: "For 'Natural Selection' has no moral significance: it deals with that part of evolution which has no purpose, no intelligence, and might more appropriately be called accidental selection, or better still, Unnatural Selection, since nothing is more unnatural than an accident. If it could be proved that the whole universe had been produced by such Selection, only fools and rascals could bear to live." And the Shavian word on evolution, therefore, is in fact only a diatribe, another expression of the antivivisectionism—and in a certain sense the vegetarianism—of a personality whose Rousseauist attitude to nature involved more of sentimental hostility to intellect (as to any aristocracy) than is generally appreciated.

With this background in mind, one might have predicted that the latest thrust back to Lamarckism would have occurred in a Marxist context. And this episode should stand as a warning that ideas have consequences, and that to succumb to the very natural and often well-intentioned temptation to bend science to the socializing or the moralizing of nature is to invite its subjection to social authority, which is to say to politics. For the moralist knows what kind of nature he wants science to give him, and if it does not, he will either, like Shaw, repudiate it; or if, like Lysenko, he has the power, he will change it. Once again, as in Diderot, as in Goethe, as in Lamarck, resentment of mathematics (which expresses quantity and not the good) reveals the moralist beneath the natural philosopher —the Michurin school rejects in principle the mathematicization of biology in favor of the autonomy of organism. Lysenko's purported findings may, therefore, be taken as the nadir of the history of Lamarckism, and (one hopes) the end of the story. For in his demagoguery the humane view of nature is vulgarized by way of a humanitarian naturalism into the careerist's opportunity. But there is nothing new about it. It is only the most recent expression of that pattern of resistance to science which has attended its entire history in reaction against the objectification of nature.

There was nothing of Christianity in all this, either in Shaw or in the Marxian belief that man does regenerate himself in the revolutionary moment of truth. And surely if one thinks about it, there was never a more unnecessary battle than that between science and theology in the nineteenth century, nor any set of findings more irrelevant to the latter than those of evolutionary theory. In retrospect, it appears that the issue between a naturalistic and providential historiography of nature had been largely worked out on the plane of geology, and needed only to be settled by the culmination of natural history in *The Origin of Species*. This is not to deny the growth of agnosticism in the nineteenth century, and particularly in the world of Anglo-Saxon Protestantism, still living in the aftermath of the Methodist revival. But surely science, as epitomized by the theory of evolution, was at most the emblem or the scapegoat of this movement. It was not the cause.

Darwin's own experience, as he tells about it in his *Autobiography*, was typical of the Victorian intelligentsia. His family background was intermingled with the radical nonconformity of the midlands, where puritanism slowly evolved through Unitarianism to agnosticism. Nor did he lose his little faith as a consequence of the theory of evolution. He became unconvinced for historical and, above all, for moral reasons. Gradually, he saw "that the Old Testament from its manifestly false history of the world, with the Tower of Babel, the rainbow as a sign, etc. etc., and from its attributing to God the feelings of a revengeful tyrant, was no more to be trusted than the sacred books of the Hindoos, or the beliefs of any barbarians." But the ground of the New Testament was no more solid. Discrepancies abound in the Gospels. Those four books are in any case hearsay, whereas "the clearest evidence would be requisite to make any sane man believe in the miracles by which Christianity is supported."

> Thus disbelief crept over me at a very slow rate, but was at last complete. The rate was so slow that I felt no distress, and have never since doubted even for a single second that my conclusion was correct. I can indeed hardly see how anyone ought to wish Christianity to be true; for if so the plain language of the text seems to show that the men who do not believe, and this would include my Father, Brother and almost all my best friends, will be everlastingly punished.
>
> And this is a damnable doctrine.

These meditations were much in Darwin's mind in the years from October 1836 to January 1839. Thus, when he read Malthus and formulated the theory of natural selection in October 1838, he was already far gone in free thought. This is not to argue that Darwin's eminence in biology makes his religious experience—or inexperience—decisive for historical interpretation. Nevertheless, it is interesting that even Darwin's disbelief substantiates a pattern of agnosticism which may equally be exhibited in the repudiation of Christianity by other eminent Victorians—Francis Newman (to name some notable names), James Anthony Froude, George Eliot, John Stuart Mill, and Leslie Stephen. In no one of these cases was the decisive factor the findings of science. In every case it was an ethical revulsion from doctrines of the atonement, everlasting damnation, original sin, and an omnipotent God who permits evil.

We did not, therefore, need Darwin's allusion to "almost all my best friends" to suspect that the decay of theology had left many intelligent Victorians faced with a dangerous choice. Upright, unblinking, committed to truth—TRUTH, come what may—they must repudiate their religion in the name of its ethic. Huxley, it turns out, is the spokesman for Darwin on religion as on science. And Darwin is as vulnerable as Huxley to the very pertinent criticism in A. J. Balfour's *Foundations of Belief*: "Their spiritual life is parasitic: it is sheltered by convictions which belong, not to them, but to the society of which they form a part: it is nourished by processes in which they take no share. And when those convictions decay, and those processes come to an end, the alien life which they have maintained can scarce be expected to outlast them." But science was not the worm in the apple: that worm was manliness, it was honor, it was decency—the Victorian virtues, left defenseless by a theology which Methodism had drowned in rivers of vulgar Evangelical piety, or which the Oxford Movement had blown away on the high, ecclesiastical winds of Tractarian romance.

The *Origin of Species* did cause distress. After all, Christianity does say that man, the possessor of an immortal soul, was specially created in the image of God. Evolutionary theory says that he was evolved by natural processes from other animals. Was it necessary, therefore, to choose between Christianity and science? It was the kind of dilemma which the Victorians loved to face. " 'Tis the crown and glory of organic science," wrote Darwin's old teacher, Sedgwick, acknowledging his presentation copy, "that it *does*, through *final cause*, link material to moral. . . . You have ignored this link; and, if I do not mistake your meaning, you have done your best in one or two pregnant cases to break it. Were it possible (which, thank God, it is not) to break it, humanity, in my mind, would suffer a damage that might brutalize it, and sink the human race into a lower grade of degradation than any into which it has fallen since its written records tell us of its history." In less pensive vein, Gladstone and the Bishop of Oxford made fools of themselves debating the issue with Huxley. One exchange turned on the question whether the destruction of livestock in the miracle of the Gadarene swine had not been a divine invasion of property rights.

Despite all this triviality, it is curious, one hundred years later, that it is the theologians who have learned to live with the theory of evolution. Whereas the ones who have not so learned, and cannot, are those atheists who would substitute nature for God as the source of morality and ethics, private or public. The Scopes Trial was a piece of intellectual buffoonery, after all—the Victorian age expiring in Tennessee. Shaw's preface to *Back to Methuselah* was not. Neither was the Lysenko affair, nor the tedious effusions of Samuel Butler. It is not the intellectual conflict between science and religion which has proved fundamental. It is the conflict between science and naturalistic social or moral philosophy.

If one be clear about the nature of science as a description of the world, declarative but never normative, may not the choice between science and religion be refused? Is it not simply a false problem, arising from a confusion— an ancient confusion going back to the beginning of science—between objects and persons? Science is about nature, after all, not about duties. It is about objects. Christianity is about persons, the relation of the persons of men to the person of God. Biology has found that the human animal is the product of evolution. It has not found —in principle, science (not being omniscience) cannot find—that man is nothing but the product of evolution. Historically speaking, it is precisely those who have said he is nothing but that, nothing but natural, who have found intolerable the meaningless chance which operates under the name of natural selection.

CHAPTER IX

EARLY ENERGETICS

THE HISTORIAN OF SCIENCE approaches nineteenth-century physics in a gingerly spirit. He has before him a great story, perhaps the greatest in his subject. But he is sure neither how to tell it, nor that he always has it right. At what point, he asks himself uneasily, is he beginning to write popular science in retrospect? Compared to conventional fields of historiography, the history of science is but sparsely underpinned by a professional literature of special studies. And here in nineteenth-century physics what is elsewhere a paucity thins to a near vacuum. The historian has to write directly from the sources, and to seek such guidance as he may find from scientists themselves, attributing this or that importance to their predecessors. His situation is like that of a diplomatic historian who should have nothing to work with beyond texts of treaties and official documents, to be studied in awareness of the acts and sayings of practicing diplomatists of his own time. One must respect such men of action, and also recognize that their lights are usually not historical. But an additional and even more serious obstacle lies across the history of science.

Ordinary language always fails in some degree to convey the findings of science. In physics, the measure of this inadequacy curves sharply upward between Carnot and Helmholtz, or between Faraday and Maxwell. After the middle of the century, it mounts exponentially toward the catastrophe of communication which everywhere besets modern learning. Einstein remarks somewhere that as the concepts of science are simplified and become ever more beautiful, the mathematics expressing them grows correspondingly more esoteric. Only the mathematical physicist can follow the lengthening chain of abstraction which connects concepts to human experience, and only he can appreciate the beauty of the simplicity. The rest of us are reduced to silence—to silence and to admiration, but less of physics than of Einstein. Modern physics, indeed, is so many things besides ideas about nature. It is abstraction. It is technique. It is instrumentation, housed in machines complex and expensive beyond the dreams of the most grandiose old engineer. It is power. It is education, diplomacy, and war. Materially at least, it holds all things for all men, the hope or the end of the world. And it is far from clear what degree of entry the history of ideas will give into all this. Techniques crowd out ideas nowadays. They change faster. They have come to have more history—in the last half-century far more. Nevertheless, even though the conscious or assumed structure of ideas about nature occupies a diminishing sector in the expansion of science, it remains a thread. Perhaps it must still be taken as the guiding thread, unless science is to abandon intellectuality altogether for technology at one end or mathematicization at the other, those extremes touching where operationalism meets symbolic logic in the new nominalism.

This is not to apologize, but simply to warn the reader that what follows is rather more provisional than what has gone before. It is a stab in the light. For nineteenth-century physics is to be treated as an end and a beginning. The great themes of classical physics culminated. The great physicists thought of themselves as pressing Newtonian physics into all corners of the structure of nature. Only a few philosophers fretted about the foundations. No one expected the second, the twentieth-century revolution in physics. Nevertheless, that revolution goes far back into the failure of classical physics quite to fit the shape of things. It is impossible to write its history except in the consciousness of that revolution pending, just as no man may write the history of the *ancien régime* except in the knowledge of the French Revolution.

It will be well to recall, therefore, the bold strokes in the world picture of classical physics. So far had Cartesian rationality prevailed over Newtonian caution during the Enlightenment, that the physical universe had come to seem a problem in mechanics. An idealized solar system served as gross model for a world of inertial billiard balls moving and impacting through an infinite void. The eighteenth century brought great triumphs. Lagrange completed the formal development which made analytical mechanics a purely rational subject, abstracted from all particular physical properties of bodies or graphic representations of their motions. In less elegant vein, Laplace demonstrated as instances of mutual gravitation among planets the apparent anomalies of observational astronomy. Benjamin Franklin brought electrostatics within the ambit of conservation laws. Coulomb demonstrated the inverse square relationship in the interaction of electrostatic charges. This was the most dramatic encouragement to belief in the formal universality of Newtonian force law. And surely the faith was worthy that other phenomena would come in, once techniques of measurement were sufficiently refined: the emission and motion of light corpuscles; the bonds between chemical atoms and molecules; the flow and interchange of energies in electricity, magnetism, and heat, following some fluid model of an imponderable hydraulics.

How facile it is to identify in retrospect the two flaws in this structure, the one metaphysical and turning on the Newtonian doctrine of space, the other physical in the excessive commitment to corpuscular mechanism. It will be remembered with what *sang-froid* Newton had incorporated the void of the atomists into the structure of classical physics. The void is what makes inertial motion conceivable, and its introduction is the device by which Newton accommodated the discontinuity of matter to the continuous extension of space. But Newton's void is Epicurean only in function. In form it is Euclidean, not only an arena but a coordinate system for motions of which the equations are written in analytical geometry. Now, the essential point of classical, Galilean kinematics had been to turn motion from metaphysical change into a relational state. Bodies are in motion, not in expression of indwelling essence, but in relation to other bodies. Into this kinematics, Newton inserted his distinction between relative and absolute motion, and made the void of space the reference against which absolute motion occurs.

It was a stroke of brilliant nonsense. As Leibniz very properly objected, the constitutive role of the void posits the existence of the nothing. Rather than define space as a substance consisting of emptiness, Leibniz preferred to consider it as itself a relation, that of simultaneous events. These objections were covered up in the contumely which classical science heaped upon all metaphysics, and were buried under the gathering mass of Newtonian successes. But they were buried alive. For it was to turn out, as often before in the history of thought, that metaphysics has a way of avenging itself. Meanwhile, Newton filled space with aether, weightless, elastic, vibratory, to serve as medium for transmission of gravitational force. But aether aroused no great interest throughout the eighteenth century. Unlike space, it remained a word rather than a working part of physics. It came forward into prominence only in the nineteenth century, as an increasingly important compensation to the inadequacies of a mechanics consisting only in forces defined by the motion and impact of particulate bodies.

These inadequacies began to emerge in phenomena before they did in philosophy. And the movement of nineteenth-century physics may be followed in the continuing dialogue between unity in nature and variety in events, the structure of the whole or the behavior and arrangement of the parts. Any physical science must concern itself with the interaction of matter and energy and with the properties of space—among them, the propagation of physical phenomena. And these alternative and complementary preoccupations may be seen as polarizing respectively the two great aspects of nineteenth-century physics—thermodynamics and mechanics. Both evolved in similar patterns: exorcising by a sophisticated statement a hypostatic fluid constructed for spatial transmission. Both illustrate the dictum that one trouble emerging in classical physics was that its concepts were too complex and its mathematics too elementary. The study of forces inherited such a fluid from Lavoisier. This was caloric, the matter of heat, which had extension, served conservation (supposedly), and was (therefore) mathematicizable. Its effect on particles of matter was taken to be repulsive. Caloric vanished into a more elaborate energetics, of course. Nevertheless, as a construct it was the starting point of thermodynamics. The problems of that science did not easily reduce to kinetics. Was it to be a parallel branch of physics, therefore, or perhaps even an alternative to classical corpuscular mechanics as the route to a unified science? There were those at the end of the century who still thought so, and whose hopes were to prove not so much wrong as irrelevant to the way the world is made.

For in the event, mechanics, the other great branch of physics, proved the more fruitful in its revolution into electromagnetic theory. There aether played the role opposite to caloric. It was a medium for attractive rather than repulsive forces, but it was the same kind of construct, not comparable to the life-forces of vitalism or the sympathies of Goethe, but a mathematicizable notion under conservation. Aether accommodated more fundamental relations than caloric, and was correspondingly more difficult to exorcise; but as caloric disappeared into thermodynamics, so aether did into relativity, the culminating expression of the effort to express the unity of law in a geometric continuum. In the nineteenth century, field phenomena—everything for which the aether was invoked, especially the wave theory of light and electromagnetic induction—represented thinking about spatial problems. And here it is obvious what the opposing terms were: kinetics, statistical mechanics, ultimately radioactivity and the discontinuities which argued quanta.

❖

IN 1824 A YOUNG FRENCH ENGINEER, Sadi Carnot, published a short memoir, *Reflections on the Motive Power of Heat*. It begins as a piece of reasoning about steam engines, and this no doubt was what inspired L. J. Henderson's famous remark that science owes more to the steam engine than does the steam engine to science. Outwardly, Carnot's spirit seems Baconian enough. He sees the implications for all civilization of the permanent revolution in power. "Already the steam-engine works our mines, impels our ships, excavates our ports and our rivers, forges iron, fashions wood, grinds grain, spins and weaves our cloths, transports the heaviest burdens, etc. It appears that it must some day serve as a universal motor, and be substituted for animal power, waterfalls, and air currents." Consider England's industrial leadership: "To take away today from England her steam-engines would be to take away at the same time her coal and iron. It would be to dry up all her sources of wealth, to ruin all on which her prosperity depends, in short, to annihilate that colossal power. The destruction of her navy, which she considers her strongest defense, would perhaps be less fatal." But Carnot was no Watt, and no Brunel. He was the son of Lazare Carnot, the Revolutionary organizer of victory, himself an engineering mathematician. Too, indeed almost inevitably, he was a graduate of *Ecole Polytechnique.* In his rationalism and technocracy he was one of the purest spirits to emerge from that portentous institution, at once Cartesian and positivist, which would assimilate all engineering to science, and all statecraft to social science, the engineering of humanity. For the steam engine only provided Carnot with problems, and the empirical flavor is soon lost in the thought experiment: "Notwithstanding the work of all kinds done by steam-engines, notwithstanding the satisfactory condition to which they have brought today, their theory is very little understood, and the attempts to improve them are still directed almost by chance."

The question as it first presents itself is an old one—whether there is any limit in the nature of things to the "motive power" that can be drawn from heat, and whether any other agent might be found to transmit that power more effectively than steam. *Polytechnique* trained its men to put things in the most general way possible, however, and Carnot immediately abstracted the problem of "the production of motion by heat" from the particular characteristics of steam engines, to the hypothetical properties of "all imaginable heat engines." On this universal basis, the subject becomes something quite other than the magic touch of British enterprise. It becomes a question of the adequacy of a mechanistic physics for the description of nature. So long as engineers had to do only with power that could be levered, pullied, geared, or screwed out of the motions of men or animals, wind or water, the theorems of classical mechanics sufficed. "All cases are foreseen, all imaginable movements are referred to these general principles, firmly established and applicable under all circumstances. This is the character of a complete theory." But as soon as heat became the motor, classical mechanics failed to help. The concept of moment did not connect the thrust of steam against the piston to the fall of temperature within the cylinder. And to arrive at a theory which would help, the laws of physics would need to be "extended enough, generalized enough, to make known beforehand all the effects of heat acting in a determined manner on any body."

Carnot wrote in quiet language. But his proposition was very radical, and must really be allowed to redress any picture which would represent nineteenth-century physics as sunk in naïve mechanism until put straight (or crooked) by Einstein. Carnot did not discuss the inadequacy of mechanism. Rather, he resumed that theoretical study of heat as a physical problem which had been seriously inaugurated by Laplace and Lavoisier. Lavoisier, it will be remembered, had conceived of caloric as a physical agent, a vehicle of expansive force, accompanying chemical transformations but not participating in them. Fourier published his perfected *Analytical Theory of Heat* in 1822. That memoir created the mathematical techniques for describing the conduction of heat in solids and in liquids, but excluded from analytical consideration the "repulsive forces produced by heat" which determine the behavior of gases. Like Lagrange's *Analytical Mechanics*, which was its inspiration, Fourier's was a work of differential equations rather than of physics. It enabled the analyst to calculate the flow of heat, but denied to the imagination any model of wherein heat consisted, any consideration of what might be done with it, or any notion of how its study might lead further into nature. Carnot may have felt the physical sterility of this. He would study power. But he would be no less general—he would abstract from any particular type of engine, but not from the capacity of heat to move bodies. In Carnot's thought, heat is the sensible effect of caloric. And caloric is not intrinsic to steam or the hot body in general, but rather is associated with it, as it is in Lavoisier's concept of chemical reaction.

Since energetics became the opposite pole to mechanics in nineteenth-century physics, it will be best to be explicit about what was to be at stake. Newtonian mechanics knows only the extension, mass, and motion of bodies acted on by the (mathematical) force of gravity. It defines force in terms of those quantities, as the product of mass times change-of-motion. (In the concept of mass lurked another metaphysical difficulty, comparable to that which vitiated the absoluteness of space, but that is not for the moment germane.) Extension, mass, and velocity—even amidst the Newtonian enthusiasm of the eighteenth century, such measurements seemed hardly adequate to embrace all the action in the world. What was the physicist to make of manifestations of action? Of Power? Work? Heat? Fire? Chemical Activity? Magnetism? Electricity? Life? Before Newton, force itself had been just such a vague word. It continued to be used ambiguously through half the nineteenth century. And one task of physics would be to divest some such words of all physical meaning, and to invest the others with positive significance.

The need had been obscurely felt from the beginning. It was a difficult issue between the Cartesians and Leibniz whether momentum or *vis viva* (mv or mv^2) be the quantity conserved in a dynamical situation. The former corresponds to the Newtonian definition of force as mass times acceleration. That question had been resolved analytically rather than physically by Lagrange, after having been dismissed by d'Alembert as semantic, though at the expense of what was to become the concept of work. Finally, of course, when the status of heat as a form of energy was established, the problems would collapse into a single subject. But historically, it was heat rather than *vis viva* (kinetic energy) which brought them before the bar of physics.

In a sense, physics itself was on trial. Were these problems to extend and sophisticate mechanics? Or was physics to enrich itself in them and transcend mechanics? Ultimately, the mechanists, notably Maxwell, Boltzmann, and Gibbs, would make thermodynamics into kinetics, a special case of a statistical mechanics. But it is the enrichers who were always the more verbal and eloquent, and who would treat the phenomena of thermodynamics as manifestations of a deeper level of organization than any to be attained by bodies in motion. And their appeal is always just a touch suspect, in that one senses in it, perhaps, some lingering yearning to transcend a dismalness in science, to replenish the old Democritan death of the soul in things.

This, however, is to go way beyond the implications of the Carnot memoir. Carnot did not labor the inapplicability of the theory of machines to heat. Like the early students of electricity, he simply accepted it, and adopted the hydraulic analogy of heat as caloric, a fluid under conservation. For caloric was not some fuel of motion. A conservative quantity is not consumed. Motion in the steam engine is to be attributed to the flow of caloric, *"its transportation from a warm body to a cold body*, that is, to its re-establishment of equilibrium—an equilibrium considered as destroyed by any cause whatever, by chemical action, such as combustion, or by any other." And it is hardly possible to insist too strongly upon the originality and clarity of this observation: that the marshalling of heat creates a deliberate disturbance in the state of things, and that it is the return to evenness, caloric seeking its own level, from which power may be drawn.

Rather than looking forward to entropy, however, Carnot was certainly looking backward to hydrodynamics, and particularly to a treatise, *General Principles of Equilibrium and Motion*, which his father had published in 1803. Brunold has pointed out the similarity of the language. Thus the father on water power:

> To employ a wheel of which the blades are moved by impact of the water is not the way to produce the greatest possible effect in a hydraulic machine turned by running water. In practice, two factors prevent this means from realizing the greatest effect. First, it is essential to avoid any percussion whatever. Secondly, the fluid after impact would still have a velocity, which would appear as a pure loss, since this residual velocity could then be employed to produce a further effect, augmenting the initial impulse. In order to make a perfect hydraulic machine, therefore, the core of the problem would consist in (1) arranging that the fluid should lose all its motion in its action on the machine, or at least that it should retain only the quantity necessary to get clear of the mechanism; (2) that it should lose all this motion by insensible degrees, without any percussion occurring, either in the fluid, or in the interactions of the solid parts of the machine. Further than this, it would make very little difference what the form of the machine was; for a hydraulic machine which fulfills these two conditions, will always produce the greatest effect possible.

And thus the son on caloric power:

> Since every restoration of equilibrium in caloric may cause the production of motive power, any restoration of equilibrium which takes place without production of that power ought to be considered as a pure loss. Further than this, a moment's thought will show that any change of temperature which is not due to a change in the volume of bodies must be just such a useless restoration of equilibrium in caloric. The necessary condition for maximum

effectiveness, then, is that in bodies used to realize the motive power of heat, no change of temperature should occur which is not due to a change of volume. Reciprocally, whenever that condition is fulfilled, the maximum will be reached.

This principle ought never to be overlooked in the construction of heat engines. It is their fundamental basis. If it cannot be rigorously observed, at least the departures should be as slight as possible.

For this reason alone, it is hopeless to think of using solid or liquid bodies as heat engines. This is a purely theoretical assertion, quite independent of the practical difficulty of devising linkages to harness the expansion and contraction of solids and liquids. It has to do only with the impossibility of getting much motion out of the cooling of a hot bar of metal or whatever, for liquids and solids are cooled not by expansion, but by conduction or radiation. It is an important finding but not surprising (after all, neither Watt nor anyone had ever tried to substitute a thermocouple for a boiler). The same is true of Carnot's other proofs; that the capacity of a heat engine is a function of the absolute temperature differential, that the difference (but not the ratio) of specific heats at constant pressure and constant volume is the same for all gases. In each case, the reasoning rather than the application is what holds the interest.

The reasoning founded the science of thermodynamics. Carnot asks us only to remember what we know from experimental physics of the thermal behavior of gases. Compression heats gases and expansion cools them. Let us suppose that we wish to maintain unchanged the temperature of a gas under compression. We must somehow carry off the caloric. Reciprocally, we must supply caloric to keep the temperature unchanged during an expansion. This in no way implies, however, that the caloric in question pertains to volume as opposed to pressure, which alters inversely in service to Boyle's law. And this is all we know. We do not know what laws relate changes in caloric to changes in volume, or whether its quantity varies with the identity of the gas, or its density or temperature. These relations are what we must establish, and for that Carnot shows us that, in fact, we already have all we need to know.

We are to imagine an idealized system, exhibiting the

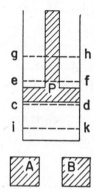

elements of any heat engine. The body A represents the furnace. It is an inexhaustible reservoir of caloric. The body B represents a condenser, or sink for caloric, and stays unchangeably at a lower temperature than A. A quantity of any gas (say air) is confined initially in the cylinder under a weightless, frictionless piston. Arrangements are possible to let caloric flow without leakage from A to the cylinder to B.

A sequence of imaginary operations ensues in six steps: (1) Caloric flows from A to the cylinder. (2) The piston rises from *cd* to *ef*, all the while that A supplies caloric to keep the temperature constant. (3) Caloric from A is cut off, but the piston moves on out (with the gas cooling now) to *gh*, where the temperature has fallen to that of B. (4) Now the gas is compressed and the piston returns to *cd*, but the caloric displaced is permitted to flow into B so that the temperature continues constant at the lower level. (5) The flow of caloric to B is cut, and the compression continues to bring the temperature back to A with the piston at *ik*. (6) The cylinder and A are again connected, and the constant temperature expansion transpires to *ef*. And from here on, the operations may be repeated as often as desired.

Thus did the idea of reversibility spring full cycle from the mind of its creator. Carnot immediately pointed out that these events could equally well be imagined backwards.

> The result of these first operations has been the production of a certain quantity of motive power and the removal of caloric from the body A to the body B. The result of the inverse operation is the consumption of the motive power produced, and the return of caloric from the body B to the body A; so that these two series of operations annul each other, after a fashion, one neutralizing the other.

And it is obvious that the diagram (first imagined by Clapeyron in 1834) by which modern texts render the ideal cycle is simply a more economic representation: the iso-thermal followed by the adiabatic expansion, the system restored to the initial condition by an isothermal followed by an adiabatic compression, the work delivered (or ex-pended) denoted by the area enclosed in the graph.

Nevertheless, Carnot's argument does not quite lead to the conclusion expected by the modern student. It is we, the readers, who have to perceive the proportionality of the segment *cdki* to the area of the pressure-volume graph as a measure of the work performed. Carnot did not define the concept of work. He could not, so long as he held that no caloric was lost. His interests were otherwise directed. In step (6) he returned his piston past *cd* to *ef*, the point of maximum volume at initial temperature. For he wished to reason (quite correctly) that for any given position of the piston, the temperature is higher during expansion than during compression.

> During the former the elastic force of the air is found to be greater, and consequently the quantity of motive power produced by the movements of dilation is more consid-erable than that consumed to produce the movements of compression. Thus we should obtain an excess of motive power—an excess which we could employ for any purpose whatever. The air, then, has served as a heat-engine; we

have, in fact, employed it in the most advantageous manner possible, for no useless re-establishment of equilibrium has been effected in the caloric.

That is to say, all the caloric which has flowed from A to B has been employed in the motions given to the piston. None has been wasted in friction, dissipated by conduction, or employed to change the state of the air in any but a volumetric respect. And even in this ideal arrangement, than which none could be more advantageous, caloric still has to flow from a body at higher temperature to one at lower to produce motion. Nor is there any way to get more power out of the system. If there were, it would be possible to use some of the excess to direct caloric from B back to A, to pump it uphill so to say, "from the refrigerator to the furnace." If this were possible, one could restore the initial conditions without expenditure of caloric (that is to say the position of the piston at terminal point of the cycle would be *cd* instead of *ik*, and the motive power requisite to compress it to *ik* would be disposable).

This would be not only perpetual motion, but an unlimited creation of motive power without consumption either of caloric or of any other agent whatever. Such a creation is entirely contrary to ideas now accepted, to the laws of mechanics and of sound physics. It is inadmissible.

Statements of the second law of thermodynamics sometimes take a form which imply that the impossibility of creating energy, which is to say perpetual motion, were a finding of thermodynamics. Historically at least, this is quite misleading. Historically speaking, the science rather rests upon Carnot's accepting that impossibility. This is more fundamental an assumption than that caloric may be treated under conservation. It is simply an axiom about method on which the very possibility of science was conditional. And its justification was nothing less than the whole experience of "sound physics."

Carnot's is no cautious memoir, therefore. He never wrote the formula, $\dfrac{T-t}{T}$, for the motive power of heat. Nevertheless, his argument does establish that that power depends only on an absolute temperature differential, and not on what vapor fills the heat engine. But even while Carnot proved the indifference of the results to the identity of the gas, he also showed that the expansibility of vapor is the most efficient means of realizing the motive power of heat. He tabulated the specific heats of different gases at constant volume and constant pressure to bring out the invariance of the difference. And only a really daring mind would study the properties of this or that fluid—steam, oxygen, air, or whatever—in order to rise above them, and to found its conclusions rather on the conservation of a hypothetical fluid, which those actual fluids only convey, and ultimately on the impossibility of perpetual motion. For Carnot's was a fertile imagination. He conceived the very categories of thermodynamic reasoning. The ideal heat engine, the isothermal and adiabatic changes, the reversible cycle—these are to thermodynamics what force, velocity, mass, and linear extension have been to mechanics.

It remained to name and to mathematicize them, and especially the notion of reversibility. For therein lies the most intimate dependence of thermodynamics on its historical antecedents in caloric. In the evolving structure of thermodynamic theory, the reversible process may be compared to the concept of inertial motion in seventeenth-century mechanics. No real motion can persist forever in a straight line. No real process is reversible. Nevertheless, it is the ideal datum which gives the hand of mathematics something abstract to lay hold on. Inertial motion, it will be recalled, had entered science as a physical consequence of Descartes' geometrization of space-matter, combined with the immutability of God. Carnot, too, introduced the notion of reversibility as a physical expression of the continuum, of flux as flux was understood in the infinitesimal calculus. Important results depended upon altering the conditions of the imaginary experiment so that the temperatures of A and B should differ indefinitely little. Then we might neglect steps (3) and (5), since these (adiabatic) changes of volume would not sensibly affect the cycle. We would be left with an isothermal expansion and compression. Now, by the gas laws of Mariotte (Boyle's law to an Anglo-Saxon), Gay-Lussac, and Dalton, the pressure-volume relations are invariant for all gases at constant temperature. It follows as a simple consequence that the quantity of caloric involved in an isothermal change is independent of the identity of the gas. Similarly (not to follow the reasoning in detail), the demonstration of the constant difference of specific heats turns on neglecting an infinitesimal term.

For Carnot thought about the fluidity of caloric as one might imagine a materialization of the calculus. The analysis did not require a physicist to follow some actual reversibility to and fro. Rather, its cogency depended on the physicist's being interested only in the initial and the final states of the system. Instead of an actual steam engine, full of friction, leaking heat at every pore, one in which the piston could scarcely tell whether the slam of the steam was adiabatic or isothermal, Carnot would meditate about an ideal engine, departing from and returning to the same initial state. And just so might the physicist always substitute for a real, irreversible change, a theoretical, reversible change—as long as the initial and final states were comparable. And since a reversible change is one in which each state differs infinitesimally from that preceding it, reversibility as a fiction implies continuity and permits the application of the differential calculus. That was the work of

Clapeyron, who published a mathematicization of Carnot's theory in the *Journal de l'École Polytechnique* for 1834. So completely had Carnot's work been lost to view a quarter of a century later, that Clausius knew it only through the medium of Clapeyron's treatment. Reversibility was the essential condition from which Clausius drew the concept of entropy. It is, he wrote, like "a limit which may never be reached, but which may be approached as closely as we wish. In theoretical considerations, therefore, we may speak of the limiting case as if it were realizable, and even as a limit it plays an important role in theory."

Anticipating Clausius will suggest the debt which the science of thermodynamics owes to the caloric theory of heat exchange as fluid flow. For the complementary view of heat as motion in the particles of matter imposed a kinetic model of gases, to be analyzed by statistical techniques which did not yet lie readily to hand. Carnot's results are often said to be independent of his use of caloric. It may be so. He himself qualified his belief in caloric in later jottings, found only after his early death. Certainly one may everywhere substitute the phrase "quantity of heat" for caloric in his memoir without spoiling the reasoning. Even so may one deprive Newton of his aether without spoiling anything about his physics except its intelligibility. But there is another question, more interesting historically. Without the model of caloric, would Carnot ever have arrived at those results? What would he have thought about? Conservation of what? The impossibility of perpetual motion in what? For it is difficult to see how a kinetic theory of heat could have led anyone to the essential and paradoxical notion of reversibility. Lacking a supple statistics, what could reversibility have meant—in a box of giddy molecules? With no sense of direction? In any case, the notion of heat as motion did not lead to reversibility, either in Rumford, who went beyond Lavoisier's calorimetry to found the experimental physics of heat, or in Joule, who determined its mechanical equivalent.

❖

CARNOT'S MEMOIR remained unappreciated and almost unknown until the 1840's, the crucial decade for the foundations of thermodynamics. Otherwise, it might seem as if these problems had been straightened out backwards. For it is the second law of thermodynamics, not the first, which treats of direction and uses the analytical concept of reversibility abstracted from the caloric theory of heat-flow. The first law, on the other hand, the law of the conservation of energy, derived historically from the complementary criticism of the notion of heat as substance, and depended on demonstrations of the convertibility of heat into "motive power." Nor is convertibility at all the same thing as reversibility. For Carnot was wrong about the conservation of heat. It does not just flow. In actual conversions, heat is degraded. What is conserved is energy, a more general quantity, which is not to be measured as a substance, but changes in which are to be expressed as a function. Where thermodynamics obtains, this function is the sum of work and heat, and heat appears as one form of energy.

Historically, the first law of thermodynamics emerged as one of the rewards rendered physical of vague notions of the interchangeability, perhaps the identity, of ultimate forces of nature, possibilities which tantalized the imagination of the early nineteenth-century physicists. Thus, light entails radiant heat, and heat-soaked iron glows first red, then white. Chemical and electrical forces passed one into another through Volta's pile, and in its successor, Humphry Davy's wet cell. Heat disappeared into or boiled out from chemical reactions. The electric current appeared as heat in wires. In 1820 Oersted detected its magnetic effect. Electricity might even have to do with life forces, since it had first betrayed itself in the twitching of Galvani's frog leg. The first experiments (expressly designed) to demonstrate the convertibility of motion through friction into heat were by Benjamin Thompson, a Yankee country boy turned Tory and careerist, who served the King of Bavaria as ordnance expert, was created Count Rumford, and married (unhappily) Lavoisier's widow. He argued that heat consists in motion by reaming a cannon barrel with a blunted borer and flamboyantly boiling water with the heat produced. That was in 1798. Rumford was too early to know other convertibilities. Nor did he quantify that of motion into heat. Forty years later these wider possibilities were the starting point for the man who did, a student of John Dalton, James Prescott Joule of Manchester.

Joule was a no-nonsense physicist, very dexterous and very nice in his experimental conscience. His father, a brewer, set him up snugly in a private laboratory. He married at the age of thirty-nine, chose Chamonix for his wedding trip, and took with him a long thermometer to try the Alpine waterfalls at various altitudes, in the hope of finding the temperature increasing with the motion. The trip held disappointment. There was too much spray. Nor was his belief in some common origin of the forces of nature less matter of fact. Early experiments suggested a constant ratio between the expenditure of mechanical power and the evolution of heat. "I shall lose no time in repeating and extending these experiments," he wrote in 1843, "being satisfied that the grand agents of nature are by the Creator's fiat *indestructible*; and that, wherever mechanical force is expended, an exact equivalent of heat is *always* obtained." And he got on with the job of measuring it, with no further ado about theology or metaphysics. He wrote a summary paper in 1849. It is prefaced by two statements, the first from Locke:

Heat is a very brisk agitation of the insensible parts of the object, which produces in us that sensation from which we denominate the object hot; so what in our sensation is *heat*, in the object is nothing but *motion*.

And the second from Leibniz:

The *force* of a moving body is proportional to the square of its velocity, or to the height to which it would rise against gravity.

Joule was a man of few and simple ideas. He spent the years between 1843 and 1849 demonstrating that the first of these statements is reducible to the second.

Electromagnetism had dominated the physics of the 1830's, and it was through inconclusive experiments on motors that Joule came to the study of heat. He despaired of putting electricity on a footing to compete economically with steam. In 1843 the British Association for the Advancement of Science met in Cork. Joule read a paper "On the Calorific Effects of Magneto-Electricity." He had made a little electromagnet spin in a field under water. Joule measured the current induced, the heat generated, and the energy expended. Each of the last two quantities proved proportional to the square of the current, and was, therefore, equivalent one to the other. By those first determinations 838 foot-pounds corresponded to the amount of heat required to raise one pound of water through one degree Fahrenheit. Then he sought an independent (and direct) confirmation of the same quantity. He set up fluid friction by passing a perforated piston to and fro in a cylinder of water, and found 770 foot-pounds for the mechanical equivalent of the heat produced. At so preliminary a stage, and with such small temperature differences, agreement on the order of magnitude outweighed the discrepancy in the figures, and Joule devised techniques capable of vindicating his views on convertibility with precision.

In the meanwhile, Carnot's work came to Joule's attention, and he put in hand experiments of his own on temperature changes in gases undergoing compression or expansion. He expressed his findings as a dissent from the "opinion of many philosophers" that the mechanical power of the steam engine arises simply from the passage of heat from a hot to a cold body. Joule attributed to Carnot and Clapeyron an opinion which Carnot at least had never held, that the *vis viva* of flowing caloric was the source of power. He dismissed this reasoning "because it leads to the conclusion that *vis viva* may be destroyed." Any theory which "demands the annihilation of force, is necessarily erroneous." Thus by a logic which (like Carnot's) excludes the inadmissible, Joule requires the reader to accept his own hypothesis. On expansion in the cylinder, the steam loses heat *exactly* in proportion to the mechanical force communicated to the piston. No *vis viva* is destroyed, because instead of consisting in the flow of heat as substance, it is conserved in the total situation which transfers *vis viva* from the particles of steam to the outward thrust of the piston, and cools the steam in equivalent degree.

It remained to determine that equivalence directly, and Joule imagined and designed a famous piece of apparatus. He had a small brass paddle-wheel made which turned horizontally in a calorimetric bath. It was set in motion like a pendulum clock by the descent of weights. Thus the work done by the weights descending directly measured the heat consumed in raising the temperature of the water by the friction of the blades. Joule appreciated the importance of reducing turbulence. His vessel contained a system of vanes for baffling the rotation of the liquid. In later determinations he substituted sperm oil and mercury for water. The paper which summarized the series in 1849 states his results in a most restricted fashion:

1st. *That the quantity of heat produced by the friction of bodies, whether solid or liquid, is always proportional to the quantity of force expended.* And,

2nd. *That the quantity of heat capable of increasing the temperature of a pound of water* (weighed in vacuo, and taken at between 55° and 60°) *by 1° Fahr. requires for its evolution the expenditure of a mechanical force represented by the fall of 772 lb. through the space of one foot.*

At the wish of the committee of referees, Joule suppressed a third proposition, to the effect that "friction consisted in the conversion of mechanical power into heat." But he could not quite bear to leave it out of a footnote.

Even though this measures quantities of heat by dimensions drawn from mechanics, the historian must, nevertheless, respect this restraint, and attribute to Joule, not the enunciation of energy conservation in general, but the narrower demonstration of the mechanical equivalent of heat. Its importance is an illustration of the gathering domination of science by the laboratory. Joule could report constant and definitive results only in 1849. Two years earlier Helmholtz in his memoir *On the Conservation of Force*, which did announce the principle in full generality, still regarded Joule's work as peripheral and uncertain. And in an earlier paper Joule himself included more interesting speculative remarks.

A short letter to the *Philosophical Magazine* of 1845 assumes, by way of a postscript on the "absolute quantity of heat in matter," that the expansibility of elastic fluids is a consequence of the centrifugal force of "revolving atmospheres of electricity." Joule seems to mean that these surround each molecule. The pressure will be proportional to the total *vis viva* of these atmospheres. From the ratio 480:481 of gas pressures at 32° and 33°, Joule estimated absolute zero to be 480° below freezing.

We see then what an enormous quantity of *vis viva* exists in matter. A single pound of water at 60° . . . must possess a *vis viva* equal to that acquired by a weight of 415,036 lb. after falling through the perpendicular height of one foot. The velocity with which the atmospheres of electricity must revolve in order to present this enormous amount of *vis viva* must of course be prodigious, and equal probably to the velocity of light in the planetary space, or to that of an electric discharge. . . .

But Joule's was not the mind to pursue these interesting thoughts, and though he dealt from time to time with electrical and chemical forces, no economical statement of conservation beyond that of heat and work escaped his pen.

Very different considerations led J. R. Mayer to similar results in a pair of papers published in Germany in 1842 and 1845. Joule did not then know those essays. Nor, perhaps, would his work have been less independent if he had, so alien was the manner of reasoning, so speculative and metaphysical the thought. For Mayer was a Kantian. He derived the mechanical equivalent of heat algebraically from the gas laws under the inspiration of a highly idealistic metaphysics of causal force. It is scarcely surprising that his work was less appreciated than the mundane measurements of Joule. Mayer's mind collapsed for a time in the 1850's, under the strain (it is said) of non-recognition and after an ignoble dispute with Joule about priorities. What seems more likely to one who has read his correspondence is that an emotionalism in his nature went deeper than chagrin, and that introspective brooding was the source of his entire train of ideas as well as of his momentary madness. Mayer was a medical man, a doctor of Heilbronn, specially interested in physiology. Like Darwin's, his interests were formed by meditations pursued during a tropical expedition which he served as medical officer. A sailor fell ill of some lung disease. Mayer bled him, observed that venous blood was a brighter red in the tropics, much closer in color to arterial, and concluded that metabolism drew less oxygen from the blood in hot climates because maintenance of body temperature required less heat.

In that phlebotomy began a lifelong concern with the conversion of forces, always oriented toward understanding the passage of inorganic into vital forces through metabolism. "Forces are causes," he wrote in good Kantian style at the outset of his first paper, *Notes on the Forces (Kräfte) of Inorganic Nature*. "We may, therefore, apply literally the principle, *causa aequat effectum*. If the cause *c* has the effect *e*, then *e* equals *c*"; and forces may no more be lost than may the sides of a true equation cease to equal one another. For causes are quantitatively indestructible objects, which (unlike matter) have the additional property of qualitative variability; and this is why it is to forces that we must look for that combination of conservation and change which is reality in action, instead of just in being.

Force and matter, action and being—Mayer announces his metaphysical position right from the start. These are the two great categories into which fall all causal entities. Matter has extension and weight. Its conservation no longer arouses doubt. For force, on the other hand, the negative property of imponderability has generally sufficed. But forces, too, must serve the causal principle (which in Mayer is simply sufficient reason by another name), and must, therefore, be "indestructible, variable, imponderable objects." Mechanics offers a familiar special case—the conservation of *vis viva*. Beyond this, chemistry, electrodynamics, and engineering contain many vaguer indications that heat consists in motion. What Mayer calls Fall-Force (*Fallkraft*) and motion (*Bewegung*) seem to disappear, only to reappear as heat. Now, to translate *Fallkraft* at this juncture as potential energy would be like looking up the answers before doing the problem. It would conceal the difficulties that Mayer faced and half-resolved. For though Mayer did express in symbols the quantity now written as $m \times g \times s$, equated to the other quantity

½ *m* *v*² (where *m* is mass; *v*, velocity; *s*, a height; and *g*, the gravitational constant), nevertheless the word he uses for kinetic energy is *not* the restricted *vis viva* of mechanics, but rather *Bewegung*, motion in general considered rather as a quantifiable object than as a state of changing spatial relations among bodies. *Bewegung* is force, he will assert at the outset of the 1845 essay. It is the parent of all forces. And all this seems very obscure, until one appreciates that Mayer was, nevertheless, perfectly clear about the problem—as clear as Joule and far more interesting, however archaic his terms. That problem was to find what amount of heat corresponds to some certain quantity of *Fallkraft* or *Bewegung*, "for example, how high a standard weight must be raised above ground level, so that its *Fallkraft* should be equivalent to the heating of an equal weight of water from 0° to 1°?" And in 1842 he gave the answer as 365 meters.

Mayer derived this result from the difference between the specific heat of gases at constant pressure and at constant volume. The difference had been determined by Gay-Lussac in 1807, and the data refined by Dulong over the next fifteen years. At constant pressure, about forty percent more heat is required to raise the temperature of a gas than at constant volume. Mayer's reasoning was admirably original. The extra increment of heat expands the gas against the pressure of the atmosphere. The 1842 memoir was cryptic in the extreme and withheld the computation. The second paper of 1845 is fuller and more generous with numbers. Mayer performed no experiments himself. But he expressed the ratio of specific heats in general algebraic terms, substituted the best figures from the literature on gases, and arrived at the result that,

$$1° \text{ of Heat} = 1 \text{ gram at} \begin{bmatrix} 367 \text{ m} \\ 1130 \text{ Paris feet} \end{bmatrix} \text{Elevation},$$

to which a footnote compares Joule's results as of 1845,

which in these terms reduced to 425 gram-meters. (A text book of our own day by Gerald Holton, who writes in awareness of the history of physics, converts Mayer's result to 3.6×10^7 ergs for the mechanical equivalent of the calorie, and points out that the modern value is about 4×10^7 ergs.) These were positive results. The expression

$$C - c = \frac{R\sigma}{E}$$

became known as Mayer's relationship in the developing structure of thermodynamics. The left-hand side is the difference between the two specific heats, and on the right, E is the mechanical equivalent of heat, R the c.g.s. quotient of absolute temperature over atmospheric pressure, and σ the volume of one gram of gas, the latter two quantities being taken at 0° centigrade and 760 mm of mercury.

In continuing contrast to Joule, Mayer was drawn on by implications that went deeper than mechanics into the working of nature. He is, indeed, best understood as a medical physicist. The title of his 1845 treatise is *Organic Motion in Relation to Digestion*. Applied mathematics (he begins) has assumed the leading role in the sciences during the last century. Only biological studies have failed to draw profit from the discoveries and methods of a Galileo or a Newton. As yet, no formulas apply to living nature. The old proverb seems to hold good, by which "The letter killeth, but the spirit giveth life." And it would be the study of motions in the organic realm that would close up the gap between physics and physiology.

Mayer might almost be mistaken for a Lamarck of physics. He is concerned with activity, not with things. He adduces chemistry as the most familiar example of constant quantities persisting through qualitative changes in form. He sees the mechanical equivalent of heat as simply an instance of the indestructibility of "force." Given the gas laws and attendant data, it happens to be readily computable. And Mayer went on from there to classify the forms of force. *Fallkraft* and *Bewegung* are mechanical cause-and-effect; heat, magnetism, and electricity are imponderable forms of force. Electricity overlaps the category of chemical force. All pass without loss or gain from one form to another in the processes of the universe, wherein the total reservoir of force is inexhaustible and unchanging. And the argument of the treatise tells how plants draw force from sunlight into the organic order, and how animals in turn feed upon the store of force in plants. What expansibility is to gases, the property of chemically transmuting organic into mechanical motions is to living tissues. Mayer considers individual organisms as living heat engines, so to say. His treatise is an essay in the energetics of metabolism. For this was how he meant it: to quantify that about organisms which can be quantified, balancing intake with expenditure of force. And his physiology is valuable, not as the refuge in science of the inexact and the ineffable, but as an arena of exact processes, where all conversions make a perfect fit. He would enrich physics by addressing dynamics to life processes, and complementing the abstract mechanics of extended bodies with a living science of forms and numbers.

The comparison to Lamarck must not be pressed, therefore. Mayer intended no escape from numbers. This respect for numbers, indeed, is what distinguished Kantian idealism from Goethean romanticism in science. It is a distinction central to the proper appreciation of nineteenth-century science in its cultural implications. For Mayer was an idealist but no romantic. He never retreated into the organismic metaphor. Rather he mounted a counter-offensive of physics against *Naturphilosophie* into the realm of biology. But Mayer's was physics with a difference. For Mayer did found the philosophical protest against mechanics which would find a richer way for physics, not in biology, but in energetics. Mechanics, in Mayer's view, was true, but limited by its own abstractions to objects that are extended. It handles only matter, in place or in motion, and the relevant kind of mathematics is analysis of spatial relations by geometry, or its slippery offspring, calculus. But forces, too, are causes (for in Mayer's world the strictest causalism rules). Forces, too, are objects—imponderable objects, it is true, but objects nonetheless. And where Mayer improves upon others who would handle imponderable objects—electricity, caloric, aether, or whatever—is in distinguishing forces ontologically from matter. No subtle fluids for him: "There is no immaterial matter," he says severely, laying it down as a principle of sound science. In his work (and this is its originality) the quantities that were to become energy differentials part company with the ineffable eighteenth-century fluids imagined to convey different sorts of action. And it is precisely in assigning fundamental ontological status to force, equivalent to that which monistic mechanism vests in matter, that Mayer penetrated beyond classical physics.

Beyond it, or perhaps back to its source, for in a very discerning passage Mayer invokes Newton's distinction between a mechanical and a mathematical force. Like gravity, heat is to be treated as a mathematical, but not a mechanical cause of motion. For Mayer was no reductive philosopher assimilating heat to the corpuscular philosophy. His numbers express equivalences rather than levels of analysis. Unlike other critics of caloric, he seems neither to have studied Carnot, nor to have wished to make heat into vibration in the particles of matter. He located the realm of the imponderable and continuous object which was heat deeper than a hypostatic fluid, and made it more fundamental than oscillations. Heat is an expression of force, to be studied by a new physics of forms. For the numerical correspondences between the forms of force are constant and exact. And in Mayer, mathematical physics was something other than the analytical or geometric resolution of the dimensions of bodies in Euclidean space. For him all space is *Lebensraum* rather than extension. Its mathematicization consists in fixing the numerical equivalences of forces, considered as objects of which the quantities are intensive rather than extensive. In his imagination, indeed, mathematical physics became a numerology of the forms of force, a science at once quantitative (in the amounts) and qualitative (in the manifestations). The mechanical equivalent of heat was only the point of departure. The latter part of his essay computed the consumption of heat and the expenditure of work represented by a variety of objects and events: weights of coal, acts of draft-animals, chemical reactions, electrical occurrences, contractions in muscle-tissue, and incidents depending for power on metabolism. And the very last sentence foretells how energetics, insofar as it followed Mayer in spirit, would seek to inform a physics, fragmented by the atomizing touch of kinetics, with a renewed sense of an ideal unity in the activity of nature: "Only in the cooperation of all instruments lies harmony; in harmony alone lies life."

❖

CONSERVATION of Energy as a principle lurked (to give technology its due) everywhere latent in the steam. It has played a fundamental role, the most fundamental as some would say, in the last half-century of classical physics, and since then in quantum mechanics. Sooner than give it up, Poincaré once remarked, we would imagine new forms of energy to save it. In the neutrino, indeed, nuclear physics did just that. Yet it is a curiosity that this, which is the first law of thermodynamics, should have so fruitfully watered the stonier, the more austere terrain of mechanics. "By what magic," asks Bridgman, "has our stream risen higher than its source?"—in defiance, as it were, of some intellectual application of the second law. The answer may be historical. Historically at least, the growing appreciation of the first law and its scope owed much to the urbanity and economy of Helmholtz's beautiful memoir of 1847, *On the Conservation of Force*. His was perhaps the most gracious personality of nineteenth-century science. Though he devoted much thought and effort to lectures on popular science, his advocacy never transgressed bounds of good taste in the insistence of a Tyndall, the too conscious cleverness of a Huxley, or the romantic brutality of a Haeckel. Nor did his authority chill, like the Gallic hauteur of a Claude Bernard. In Helmholtz, German warmth and naturalness appear at their best, lending a welcome serenity to a fine physical intelligence and civilizing science by passing it through a cultivated mind, alive with a sense of cultural responsibility.

For Helmholtz's was a tranquil talent. It indicates how recently science has become professional that this man, in his thinking the most universal of the nineteenth-century physicists, should have been trained for medicine, and begun his career as a Prussian army doctor. No doubt this unspecialized approach, in which his generation was the last to feel at ease, held its own advantages. The range of Helmholtz's contributions runs from physiology through physics to philosophy. He invented the ophthalmoscope, which reflects the examining physician's beam of light through the pupil onto the retina and permits his own observation through the hole in the center of the parabolic mirror. Helmholtz had been trained under Johannes Müller in Berlin, and had been much impressed by his demonstration that many sensations are specific rather to the nerve than to the stimulus. Studies of the retina led Helmholtz into the physics of perception, first of light and then of sound, and by a kind of German instinct for harmonies (that which Mayer, too, expressed) Helmholtz applied optics to the elucidation of beauty in paintings, and acoustics to the recognition of loveliness in music. The problems of perception, finally, brought him back to epistemology and the philosophy of science, where his education had begun. There his dissatisfaction with Hegelian process, its affirmation of spiritual absolutes as against science, led him to look behind romanticism to rationalism and Kantian idealism as the guide in science itself.

Helmholtz was more explicit than Mayer, and more aware of the difficulties. In conservation of energy, he made his most comprehensive statement. Helmholtz, too, came to it from physiology. His earliest researches investigated the quantity of heat generated by laboratory animals, and found it equal to that produced by burning their food in a calorimeter. Thus metabolism is oxidation rather than soul at work. But this was preaching to the converted, for the most part, and what marks Helmholtz is the comprehensiveness of his reasoning, and the form he gave it. When he published *On the Conservation of Force*, he was twenty-six years old, and still tending the ailments of Prussian conscripts. A memoir of some sixty pages, it is certainly the seminal work of his life. Unlike Mayer, Helmholtz enlisted the forces of nature in the service of mechanistic images. Instead of beginning with heat, or force in general, he went back to classical eighteenth-century dynamics, started with its fundamental principle, the conservation of *vis viva*, and assimilated heat to that by applying conventional analytical mathematics to problems of energy. For though not a creative mathematician, he was gifted with a powerful mathematical grasp, and with the skill, becoming essential to him who would be a creative physicist, to wield the sharpest weapons available in the mathematical armory of his day. And he couched his discussion in the most sophisticated language known to physics, not all weighted down by lumps of data like Joule's heavy-handed laboratory reports, nor confined to the primitive numerical equivalences of Mayer, but in the graceful, taut, and lissome differential equations of classical dynamics.

The introduction purports to eschew philosophy in favor of a "purely physical hypothesis." But in the very act of repudiating philosophy, it becomes clear how Helmholtz (like Mayer) had formed his expectations about science along Kantian lines. Experimental science, he explains, groups phenomena together descriptively under general rules. The law of refraction is one example. The gas laws are another. Theoretical science, for its part, seeks to grasp phenomena in accordance with causality and sufficient reason. It is a study which is, so to say, enjoined on man, a thinking being, by the axiom: "Every change in nature must have a sufficient cause." We are bound to pursue investigations until we discern the invariant cause, which we distinguish from effects precisely by its constancy. Not that science presumes to embrace all facts, or to assert that nature is wholly intelligible. Perhaps there is in things a domain of spontaneity, of liberty. But the limits of causality are limits of science.

Within that domain, science contemplates the objects of the world under two points of view, in their existence and in their activity. Matter is the abstraction which we make in classical mechanics of the existence of bodies. It has the properties of extension and mass. Its quantity is constant through all eternity. But—Helmholtz, too, brings us to the point where energetics penetrates beyond the Democritan world picture—we cannot reduce qualitative differences in objects to changes of position in matter. And force is the abstraction which we make of the cause of events. What with his emphasis on causality, Helmholtz directed theoretical (as contrasted to merely experimental) physics rather toward force than matter. Helmholtz and Mayer were moving under the same inspiration, therefore, but Helmholtz occupied the new ground with far greater assurance. And in the minds of both these German thinkers, it was the Kantian identification of cause and effect in the intellectual effort of theory which raised force to a status equivalent to that of matter—equivalent ontologically but more interesting physically. Thus the types of force (*vis viva*, heat, chemical bonds, electricity, magnetism, gravity, and perhaps others yet unperceived) were to be objects of science equally with the kinds, pieces, and locations of matter. Matter and energy were to be the two aspects of that which has been from everlasting to everlasting, instead of just matter in motion (where force is but the poor thing, mass times acceleration).

The pattern is classic. For no one has ever discovered conservation (of whatever) in some experiment. Rather, conservation has always been assumed as a condition of objective science. Confidence in it rests on the whole experience of science in its history, which may even be seen as the extension of such considerations over ever widening areas of nature. The laboratory may, indeed, provide reassurance that conservation is served in some certain set of experiments, but it is not found there where Priestley found oxygen, Newton the composition of light, and Becquerel radioactivity. Instead, the theorist assumes it, as Lavoisier did, in order to rationalize and control experience. It is meaningless, therefore, to ask who discovered the conservation of energy? No one did. Helmholtz's was a more difficult achievement. He expressed what everyone was vaguely assuming. And the decisive element in his success was that he alone began, not with heat or force (the unknown), but with motion (the known). Motion is the one instance of regular, lawful change of which matter *per se* is capable. The most general principle of dynamics is conservation of *vis viva*. The problem, as Helmholtz saw it, was to extend the laws governing spatial rearrangements to the parallel domain of forces, of assimilating (as he might have said) the activity of nature to the laws that govern its existence, or (as we might say) of embracing energetics in the formalism of mechanics, where conservation already did obtain:

> Unless theoretical science is willing to remain where it is, half-way towards understanding, it must harmonize its views with the requirements of this principle [of conservation of *vis viva*] as to the nature of elementary forces, and also with the consequences. The mission of theoretical science will have been accomplished when it has defined all phenomena in terms of elementary forces, and demonstrated that this definition alone is compatible with the facts. Such a definition should be considered as the necessary form in which to conceive nature. It may be awarded the status of objective truth.

Mechanics offered Helmholtz a sure footing to raise him above the standpoint of Joule and Mayer, who were pulling a science of thermodynamics up from the study of heat by its own bootstraps, the one experimentally and the other metaphysically. And now the threads twisted together by modern physics begin to be discernible, and it becomes evident how it was that conservation of energy, become so fundamental in later mechanics, appeared historically as the first law of Thermodynamics. For the axioms and the equations actually *came* from mechanics. But the object of reasoning was energy, and pursuing that (as yet unnamed) object, Helmholtz looked for guidance right into the heart of the problem itself, neither back to mechanics (which would have led him in a circle to his starting point), nor to Joule or Mayer (confined to their subordinate considerations), but rather to Carnot and Clapeyron, who alone had studied the mechanical power of heat theoretically.

Theirs are the first names to appear in the argument, which starts from the assumption

that it is impossible to create a lasting motive force out of nothing by any combination of bodies. By means of this principle Carnot and Clapeyron have already demonstrated theoretically a series of laws, some already known to science, others not yet confirmed by experiment, concerning the latent and specific heats of the most diverse bodies in nature.

Now, the impossibility of perpetual motion was nothing new. Before the Revolution the French Academy of Sciences had haughtily refused even to notice any more inventions purporting to yield something for nothing. But the despair of perpetual motion rested on nothing deeper than the inevitability of friction. It served scientists only by saving their time. Carnot's theory made it interesting. He employed our incapacity positively, as an instrument of reasoning by exclusion of the inadmissible. He said, not just what we cannot do, but what heat will do, and he laid down the condition of a temperature differential. And now Helmholtz seized on this advantage as on the one device that might lead outward from dynamics to his goal, a generalization of conservation, beginning with *vis viva* but covering heat and all other manifestations of what he (and therefore we for yet a little longer) must still call force.

Thus Helmholtz prepared dynamics to serve his purposes by first importing into it the analytical device which Carnot had fashioned of the concept of reversibility in a conservative situation. By the same token, moreover, he liberated the argument from its dependence on the caloric theory of heat, and clothed it instead in the strong au-

thority of classical mechanics. We are to imagine a
system of mass points acting under reciprocal forces. In
passing from one configuration to a second, the points
acquire velocities which may be exploited to do work. To
draw the same quantity of work a second time from the
system, one would have to restore the initial conditions
by some means, by expending on it, for example, energies
drawn from outside the system itself. Now, it follows from
the axiom that the work consumed in thus going backwards equals that created by the initial process, regardless
of what means are employed, what routes are followed by
the particles, or what the velocity is in either direction.
If it were not so, one could choose one route, say, in
preference to another, in order to profit from the difference. Thus could one create perpetual motion, the inadmissible. The mathematical expression which excludes
that impossibility is simply the law of the conservation of
vis viva, and since we knew this result in principle all
the time, what we have won is confidence in the argument.

Helmholtz now extends that argument. The product
of force into distance had been tacitly adopted as the measure of power in the theory of machines. Helmholtz combined it with the gravitational constant in order to equate
the work expended by a freely falling body with the *vis
viva* acquired. Since the velocity which a body must attain
to reach some height, h, against gravity is $\sqrt{2\,gh}$, then

$$m\,g\,h = \tfrac{1}{2}\,m\,v^2.$$

Next, Helmholtz pauses to show analytically that any
system of mass points in which conservation does hold
must be subject only to radially directed forces among
the particles. Nor is this a digression, for Helmholtz never
let the goal of generality slip from sight. But so tightly
knit is the structure of his thinking that one must look
closely to discern the articulations. What he has done

combines the measure of work taken from the theory of machines with the conservationist reasoning of the Carnot memoir on heat to define force in general as the capacity to do work. But so far the argument reaches only the gravitational case. The equilibrium considerations involve gravitational force as that against which work is done, and make conservation of force identical with the principle of virtual velocities. That is to say, *vis viva* is by now translatable as kinetic energy, but the concept of potential energy had still to be distinguished from the pull of gravity.

The second chapter achieves that abstraction. Now we profit from the demonstration that only central forces operate in conservative systems of point-masses. For gravity, therefore, (which is itself a central force in a conservative system) may be substituted the "tensions (*verbrauchten Spannkräfte*) which correspond to the relative variation of the distance to the center of force." Then, an increase in the *vis viva* of a mass point under any conservative central force equals the decrease in this "tension," or when ϕ is tension, q velocity, and r the radial distance to each center of force,

$$\tfrac{1}{2}\, m\, Q^2 - \tfrac{1}{2}\, mq^2 = -\int_r^R \phi\, dr.$$

The words had still to be attached. "Work" was regularly used for Carnot's motive power only by Clausius, after 1850. Half a century earlier Thomas Young had proposed "energy" as a term less anthropomorphic than *vis viva*, and as denoting a dynamical quantity different from Newtonian force. But the word never caught until revived by Rankine, who borrowed from electronics the qualifier "potential" with which to couple it in replacing *Spannkraft* for the right-hand side of the equation. Kinetic energy, on the other hand, owes its characterization to the determination to keep the left-hand side of the energy equation within the reliable domain of corpuscular mechanics, where it succeeded to *vis viva*.

Helmholtz wrote without benefit of these terms in his first edition, but surely we are by now entitled to anticipate them in order to do justice to the clarity with which he disentangled the quantities concealed amidst the residues of common usage, Newtonian accelerations, and engineering practices, in order, too, to appreciate the finesse with which he wrested a statement of the conservation of energy from the elusive, the almost tautological quality of the notion of force. For this equation is the first expression that contains the full meaning of energy. Kinetic energy is expressed as a difference in levels, and that difference is equated to the definite integral of the "available tensions capable of producing such a certain effect." Nor, though Helmholtz addressed this memoir to the Physical Society of Berlin, was mid-nineteenth-century physics yet so far gone in mathematicization that he did not also express his law in the less economical form of words:

> In all cases of motion of material points under the influence of their attractive or repulsive forces, of which the intensity depends only on distance, the decrease in tension [potential energy] always equals the gain in *vis viva* [kinetic energy]; and contrariwise, an increase in the former equals a loss of the latter. In other words, the sum of *vis viva* and tension is always constant. In this its most general form, we may designate our proposition as the law of conservation of force [energy].

One feature of the history of concepts on force and energy is the narrowness of the area for maneuver between assumptions and conclusions, and the mind unused to these reflections sometimes finds it difficult to perceive in what respect the one goes beyond the other. The decisive element in Helmholtz's argument is the assurance with which he built out from axioms of dynamics, reasoned by analogy about heat, electricity, and chemical energy, and then re-imported his findings back into the sure subject of mechanics to verify their validity. Thus, it is no tautology when the first use that Helmholtz makes of his powerful new law of conservation of energy is to derive from it, as a special case, the principle of virtual velocities, perhaps the best established and the simplest proposition, one will not say of mechanics, but of statics. For he is not in fact thinking in a circle. Rather, he is verifying his law by a thought experiment, showing that what we cannot doubt is a special instance of a new hold on a wider reality. And armed with this reassurance, he restates the conclusions about work which Carnot had rested upon conservation of caloric. But now he puts them in terms of rational mechanics, and derives them from the conservation of energy instead of the indestructibility of heat as a fluid: 1) In a system where energy is conserved, the maximum work obtainable is finite and determinate; 2) If non-conservative forces existed, which did depend on time or velocity, or which acted other than radially, then combinations of bodies could create or destroy energy, which result is inadmissible—ergo such forces do not exist; 3) A system in equilibrium under central forces could never be set in motion relative to other systems by internal, but only by external forces—in other words, nothing is self-starting.

The remaining chapters read the law in analytical detail into the several departments of physics. First, theorems of mechanics are made to follow: the inverse-square relationship of gravity; the popular rule that loss of force in simple machines is proportional to velocity acquired; the kinematics of elastic bodies—wave mechanics, impact, the reflection and refraction of light, the velocity of sound. Next, heat is treated explicitly as a form of energy, and its mechanical equivalent is established theoretically.

And in this, the most important connection historically, it is quite clear from the first edition that the historian is to present Helmholtz rather as parting company from Carnot on the nature of heat, than as conceptualizing the measurements of Joule. Joule's results were too imprecise to inspire confidence in 1847, and it is not because of them that Helmholtz, too, defines "Quantity of Heat" as only a means of expressing "the quantity of the *vis viva* of the calorific movement, or on the other hand, of the *Spannkraft* of the atoms." And he goes on to sketch out qualitatively the way in which such energy is distributed among the three degrees of freedom which, in a moment of profound insight, Ampère had imagined to exist for molecules.

The adaptability of concepts of potential to energetics suggests the importance which electricity was assuming in all of physics. Helmholtz very deftly reduced the motion of electrical and magnetic charges to the formalism which mechanics had developed for material masses of corpuscular texture. It was less fortunate, perhaps, that he also organized his electrodynamical discussion in a fashion reminiscent of the eighteenth century. Thus, electricity has two modes of motion. It may be transported along with supporting bodies (static electricity). Or it may move through such bodies (galvanic electricity). In the latter case, the electric current may be conducted either metallically or chemically in electrolytic solutions. In all cases, Helmholtz balanced the heat generated by currents against its mechanical equivalent, and compared chemical and electrical energies by reference to their thermal equivalents. The induced current was still too novel to be handled with familiarity, and Helmholtz treated it separately, less for its own sake than as a means of including magnetism within energetics through its convertibility to electricity. There remained, finally, the whole of living nature. Plants accumulate large stores of chemical energy, converting it (so far as one can tell) from no source other than the *vis viva* of rays of sunlight. Animals have this to draw upon to fuel their own metabolism. Like Mayer before him, Helmholtz ends as he began, with the energetics of physiology. But Helmholtz only gestures toward biology, for science as yet furnished few data from which to compute equivalents.

Years later, and in popular lectures, Helmholtz became one of the most eloquent preachers of the gospel of energy. For the first law became the favorite text for the sermons on science which fill a large, and in the case of Helmholtz a not too desperately unreadable, literature in the late nineteenth century:

> From a similar investigation of all the other known physical and chemical processes, we arrive at the conclusion that nature as a whole possesses a store of force which cannot in any way be either increased or diminished; and that, therefore, the quantity of force in nature is just as eternal and unalterable as the quantity of matter. Expressed in this form, I have named the general law "The principle of the conservation of force."

> We cannot create mechanical force, but we may help ourselves from the general store-house of nature. The brook and the wind, which drive our mills, the forest and the coal-bed, which supply our steam engines and warm our rooms, are to us the bearers of a small portion of the great natural supply which we draw upon for our purposes, and the actions of which we can apply as we think fit. The possessor of a mill claims the gravity of the descending rivulet, or the living force of the moving wind, as his possession. These portions of the store of nature are what give his property its chief value.

In a curious way, the central ideas of thermodynamics when traced historically lead, not to each other, but back to Carnot's memoir. It is a science shaped, as compared to others, more by intellectual heredity than by experimental environment. Thus, Joule and Mayer established the convertibility of heat and work quite independently of each other, in quite different ways, but both of them in the course of arguments intended to be crucial against the materiality of heat. Helmholtz began with neither argument, but rather with Carnot's own proof of the necessity for a temperature differential, and with the exclusion of perpetual motion. In one respect, indeed, Helmholtz did less with that analytical device than had Carnot. Helmholtz transposed the impossibility of perpetual motion into terms of mechanics, where it had long been assumed, in order to invest the new study of heat with the authority of mechanics. Nor so long as it was a question of work into heat could thermodynamics embarrass a mechanistic science. The kinetic energy of a piston could plausibly be imagined to impart equivalent energy of motion by impact to molecules of a gas. Helmholtz was quite indifferent to questions of direction. He reasoned rather about work into heat than heat into work. Nor did he at first remark that the case for mechanism is far more parlous when the latter is the direction of conversion. For by Carnot's own principle, not all the heat reappears as motion, but some of it is carried down from the higher to the lower temperature, where like a Cheshire cat it simply disappears. In a word it is consumed. And though it may thus serve conservation within some wider set of boundaries, vanishing does not on the face of it conform to corpuscular mechanism.

Two new quantities follow historically from the Carnot temperature differential—absolute zero and entropy. The existence of a bottom to temperature had been implied in Carnot's discussion, as in Mayer's, of the difference between specific heats at constant pressure and volume. But William Thomson first investigated the data experimentally, and fixed the value at minus 273°, where molecules cease to move. He became Lord Kelvin in honor of his distinction in the laboratory—hence the name of the absolute temperature scale. Thomson did for temperature what Carnot had done for heat. He considered its quantity independently of the physical properties of mercury, alcohol, or any particular substance of which the expansion might be measured on a thermometer. He too began directly from the Carnot memoir, of which he seems to have thought himself in some sense the discoverer. Moreover, he saw the difficulty introduced into Carnot's conservation of caloric by the Joule-Mayer demonstration that heat is lost to mechanical power. Nor does the flow of heat necessarily do work. It does none, for example, in simple conduction. Nevertheless, Thomson was unwilling to do without the temperature differential as the source of motive power. "If we abandon this principle, we meet with innumerable other difficulties—insuperable without further experimental investigation and an entire reconstruction of the theory of heat from its foundations." And rather than accept such a setback, he preferred to despair for the moment of resolving the difficulty between the convertibility of heat into work and its transportation and loss to a lower temperature, which appeared as necessary and mutually exclusive conditions for realizing its motive power.

Rudolf Clausius firmly grasped both horns of this dilemma, refused the alternative Carnot *or* Joule, and incorporated the essential principle of each into the second law of thermodynamics. "It is not at all necessary," he wrote in his memoir of 1850, *On the Motive Power of Heat*,

to discard Carnot's theory entirely, a step which we certainly would find it hard to take, since it has to some extent been conspicuously verified by experience. A careful examination shows that the new method does not stand in contradiction to the essential principle of Carnot, but only to the subsidiary statement *that no heat is lost,* since in the production of work it may very well be the case that at the same time a certain quantity of heat is consumed and another quantity is transferred from a hotter to a colder body, and both quantities of heat stand in a definite relation to the work that is done.

The title is a translation of Carnot's, no doubt designedly. For though Clausius knew Carnot only through Clapeyron's mathematical treatment, he falls into the pattern by which he and his contemporaries went rather to Carnot than to one another for the ideas they developed. Clausius seems not to have studied Helmholtz closely at this time, nor to have been preoccupied with the work-plus-heat function in which energy is conserved. Rather, he addressed himself to the conversion of heat into work, and its transportation in the process to a lower temperature. Nor did he find the case so desperate as Thomson, who drew from it nothing deeper than absolute zero. Clausius's was a subtler mind, less concrete than Joule's or Thomson's and less comprehensive, perhaps, than Helmholtz's, but the supplest of their generation in the abstract reaches of physics. And the enigmatic quantity, entropy, awaited his discernment—the second great principle to follow out of Carnot's reasoning and the complement to conservation of energy. Time's Arrow, the second law has been called, in the lapidary phrase imagined by Sir Arthur Eddington to evoke the consequence that in this principle physics first found a difference between time past and time future. In it the history as well as the structure of nature becomes capable of physical description.

Clausius seized upon the point that Carnot's essential principle was not the conservation of caloric, but the unique dependence of the motive power of heat on T_2 and T_1. His was a different problem, therefore, from Helmholtz's. He analyzed not what was conserved, but what was lost, thus distinguishing between that portion of heat which disappeared at the lower temperature and that which converted into mechanical work according to Joule's ratio. It would be tedious to travel isothermally and adiabatically around the Carnot cycle once again, even to survey the changes through Clausius's more sophisticated eyes. He took full advantage of the cyclic device, which cancels out the effects of intermediate internal changes, so that all the work done or heat consumed represents external effects. And he observed that the necessity to consume heat or expend work introduces an asymmetry into the restoration of the initial conditions. The system is back in the same state at the end of the cycle, but something has changed, something not material has disappeared, something inaccessible to conservation laws. That something is amenable, however, to Carnot's principles. Its quantity depends only on the initial and final states of the system, and not on the intervening routes or rates. What has occurred has done so by means of a temperature differential.

The measure of what has changed is entropy. Clausius identified that quantity only later, in a paper of 1854. It will be well to indicate the derivation. Clausius denotes by Q_1 the quantity of heat drawn from the heat reservoir at T_1; by Q, the amount converted into work; and by Q_2, the amount which passes down to the sink at T_2. Then

$$Q_1 = Q_2 + Q,$$

and

$$\frac{Q_1}{Q_2} = \phi\ (T_1, T_2),$$

where $\phi\ (T_1, T_2)$ is some function of the temperatures, which is independent of the nature of the body. By the latter condition, what is true of any body must be true of all, and it is permissible, therefore, to consider the case of an ideal gas. It can be shown from the gas laws that

$$\frac{Q_1}{T_1} - \frac{Q_2}{T_2} = 0.$$

Now, we may adopt the convention that heat taken in is positive, and heat given out, negative; then the sign of the second term changes, and

$$\frac{Q_1}{T_1} + \frac{Q_2}{T_2} = 0.$$

The same reasoning applies to a reversible cycle of any number of stages and temperatures:

$$\frac{Q_1}{T_1} + \frac{Q_2}{T_2} + \frac{Q_3}{T_3} + \ldots = 0;$$

or,

$$\sum \frac{Q}{T} = 0.$$

Finally, we may consider cyclical processes of general form. Even though not consisting of successive isothermal and adiabatic stages, they may nevertheless be analyzed in terms of infinitesimal elements of such lines, so that

$$\int \frac{dQ}{T} = 0.$$

In words, this relation would be:

If in a reversible Cyclical Process every element of heat taken in (positive or negative) be divided by the absolute temperature at which it is taken in, and the differential so formed be integrated for the whole course of the process, the integral so obtained is equal to zero.

From the conditions of a cyclical reversible process, it follows that the expression $\dfrac{dQ}{T}$ must be the differential of a quantity which depends on the configuration of the system, but which is altogether independent of the manner in which that configuration has come about. If we call this quantity S, then

$$\frac{dQ}{T} = dS,$$

or

$$dQ = T\,dS.$$

Clausius coined the word entropy for S: "I prefer going to the ancient languages for the names of important scientific quantities, so that they may mean the same thing in all living tongues. I propose, accordingly, to call S the *entropy* of a body, after the Greek word 'transformation.' I have designedly coined the word *entropy* to be similar to 'energy,' for these two quantities are so analogous in their physical significance, that an analogy of denomination seemed to me helpful."

Two laws now describe the energetic state of a system. The first $(dQ = dU + dW)$ says that energy is conserved. The second $(dQ = T\,dS)$ says that in a real—i.e. irreversible—process, entropy increases; or that energy (however faithful to conservation) becomes increasingly unavailable in the way in which nature does in fact move. And the idea of entropy has proved a very acute instrument of analysis. For although thermodynamics *per se* knows only reversible processes, the second law evades this limitation by measuring the unavailability of energy, the departure of the real from the ideal, the brake (so to say) of friction between things, rather than the inertial persistence in them. This back-handed approach made it possible to consider irreversible and reversible processes in the same formalism. Nor need a reversible process be cyclical for the second law to obtain. Two particular configurations of a system might be considered as a "state-couple," in which case one need not return the system to the starting point to analyze it, provided one accept the wider condition of considering only the initial and final states, and not the manner of the transition.

❖

IN THE ANALOGY between energy and entropy, Clausius hoped to save the mechanical theory of heat from the contradiction between its conservation as energy and its consumption in work. It was a plausible hope. His enthusiasm is infectious and persuasive. Both energy and entropy are functions rather of the state or organization of things than of their mass, location, or extension. Neither one is a measure of materials. Neither one is a point function. Both have significance only in comparison to a previous value, since both are defined with reference to the initial and final configurations of a system, instead of to its instantaneous condition. Both of them, moreover, are highly sophisticated and abstract representations of certain elementary experiences of the world, certain serious intuitions: Energy, of the intuition that there is an activity, a "force" in things beyond matter in motion, that something real makes nature go (and all history teaches that if this be so, then the assumption of conservation is a pre-condition to embracing it in measurement, which is to say in science); Entropy, on the other hand, of the complementary experience of water seeking its own level, of hot bodies cooling, of springs untensing, of magnetism wearing off and electrical charges leaking away, of a destiny such

> That no life lives forever;
> That dead men rise up never;

That even the weariest river
Winds somewhere safe to sea;

of a world getting old and running down.

Nevertheless, the analogy between energy and entropy has not always sustained the weight of Clausius's hopes. It should by now be evident that a certain philosophical confusion hovered over the enunciation of those quantities, symptomatic (it may be) of the formalism of the likeness. There was Clausius himself, for example, saving the mechanistic theory of heat in a concept derived from the unmechanizable fact of the necessary loss of heat on conversion into work. There were Mayer and Helmholtz, conserving "force" on the grounds of strict causalism, which was to say sufficient reason, and to that end promoting energy to ontological partnership with matter— and assigning it a status opposite atoms comparable to that occupied by the void in the old corpuscular philosophy. Across the Channel, on the other hand, where that philosophy still informed the style of science, Joule instinctively conserved energy as *vis viva*—not as metaphysical cause, but on the classical grounds of dynamics.

Moreover, whatever standing one might assign to energy, surely entropy could never be ontologically equivalent. It is too arbitrary and unreasonable a concept. It is an abstraction, not from the measurement of objects of some sort, but from the inability of nature to swim upstream in time. Whether entropy be called positive or negative is a mere convention. Like the working of the steam engine from which the Carnot cycle was abstracted, entropy has reference only to gross processes and never to deep structural relations in nature. It is no measure of particular things, but an index to the condition of a crowd of things, or rather to how its condition has compared or will compare. The reasoning is all about reversible processes, but the finding is that no real process is reversible: we learn about the real by contemplation of the impossible. It seems odd that a fundamental law of nature should rest upon the exclusion of what man cannot do, namely create perpetual motion. Yet there are those (and foremost among them Eddington) who have claimed for the second law the status of the most fundamental among the laws of nature.

The difficulties in the logical relations of the first and second laws go back to a deep inconsequence in the historical order. For the laws of thermodynamics are not like Newton's laws, or Kepler's. They are not reducible one to the other. They are not about the same thing. The first law states an ideal equality, and the second law a real inequality. The first law says that the amount of energy in a system is constant, and it emerged historically from discussions of the convertibility of forms of energy, discussions now mechanistic and again idealistic. The second law, on the other hand, says that the availability of energy in a real process decreases, that energy is degraded in nature, that heat is a wasting asset, and that the world is living on its capital. And this finding emerged historically from the fallacy that heat is conserved, which is to say from the experience that it is lost, even though it is a form of energy. By the second law, nature follows a determined direction, or at least a direction so highly probable that were it violated, all our sense of order in things would go with it. And this is irreconcilable with the great concepts of mechanism, in which time is a geometric parameter. In thermodynamics, time was the course of a process once again, as in quite extraneous biological notions of progress and evolution.

Even there, however, superficial conflicts arose to mar a rapprochement which might have lent a new though false vitality to biological romanticism. Uniformitarian geology and Darwinian biology required hundreds of millions of years, perhaps even billions, for the evolution of species by natural selection. In principle, indeed, the nineteenth-century conceptions of earth history followed Hutton in presupposing infinity in time. Not so the physicists, very severe in their new thermodynamics, who in the person of Kelvin would allow the geologists only ten million years since the earth had been molten and a few million more before it would cool to lifelessness under a waning sun. "I take the sun much to heart," wrote Darwin in dismay, more alarmed for his theory than for his descendants. The second law was even more damaging to the cosmic historicism of Herbert Spencer, who had assimilated the idea of progress to a supposed movement from the homogeneous and undifferentiated to the individualized and the highly organized. For the ineluctable increase of entropy in the universe precisely contradicts this liberal tendency in things.

It skirts tautology to say (as often is said) that entropy is a measure of the disorder or randomness of a system. Nor were all the uses of this quantity imagined until the later decades of the nineteenth century. Since then, entropy has become a protean concept, no doubt because science and philosophy have become more and more preoccupied with problems of organization and information, and less and less interested in classical models of matter, motion, and space. The status, so to say, of energy has suffered from the decline of causalism (true in this retreat to the historical emergence of the first law from the idealist identification of cause and effect in the real), whereas the prestige of entropy as an organizing idea has risen correspondingly. Does it advance upon us along with the disintegration which it indicates? In any case, the fortunes of entropy have expanded along with the probabilistic analysis. Entropy becomes, for example, a measure of the probability of finding a system in such a certain state. And all this decline of dogmatism in classical physics seems curiously reminiscent of the collapse of dogmatism two centuries before, under the erosion of the very Newtonian critique which had established that physics. Indeed, modern philosophy of science sometimes seems like a reprise of the main themes of the Enlightenment. Once again philosophers study the implications of science for the operations of the mind that creates it, for language and communication. In information theory, entropy becomes a measure of the uncertainty of our knowledge. Everything is reminiscent except the mood. For now philosophy teaches us of the difficulty of communication, and not of the educability of man.

But the idea of entropy had already been generalized in the vaguer, the more cosmic pessimism of the late nineteenth-century prophets of despair, for whom the idea of progress had turned to clinkers, and who sat among the ashes contemplating the heat death of the universe. It is odd that this should have been so prevalent a posture at the end of the century. To us it is bound to seem a century which had scarcely known disaster. Indeed, no problem more pressingly invites study than that of the cultural factors which determine what use is made of the ideas and findings of science. And surely the answer will be found in the world of men, and not in the world of nature. For it can scarcely have been science which changed the style from the optimism of the eighteenth century to the callow pessimism of the nineteenth.

Certainly evolution and entropy were the leading novelties which nineteenth-century science offered to the pundit. Exponents of Darwinism, as is well known, interpreted competition in the most ferocious fashion, making strife the law of life, and progress the defeat of the miserable and incompetent instead of the harmonious advancement of mankind along the paths of nature. But Darwin had only found his language in the literature of political economy, where the mood was already fatalistic. Nature holds no such message. Nor, perhaps, is entropy really much help to the understanding of history or the social process. Nevertheless, if the robber baron and the strong man armed took license to ruthlessness from the theory of natural selection, the intellectual who shrank from such successes found in entropy the excuse to indulge his *mal du siècle*. To Americans, the voice of Henry Adams comes as the most familiar of those speaking out disillusionment from the soured dream of progress. Instead of moving toward the perfection of the race, history runs downhill toward a miscellany of the second-rate, a society without classes, structures, gradations, duties, responsibilities, nobility, or culture, without Mont-Saint-Michel or Chartres. The future, in other words, holds only equality and social randomness. But surely Adams had despaired of liberalism, and of his education, before he lost faith in the universe? Surely he, and his literary contemporaries, accepted with something very like irresponsibility and relief the imminent prospect of an end to time, when though the masses would be one with the fastidious, none would perceive the universal degradation. This, indeed, is the release for which Swinburne praises "whatever gods there be:"

> Then star nor sun shall waken,
>   Nor any change of light:
> Nor sound of waters shaken,
>   Nor any sound or sight:
> Nor wintry leaves nor vernal,
>   Nor days nor things diurnal
> Only the sleep eternal
>   In an eternal night.

CHAPTER X

FIELD PHYSICS

YET, AFTER ALL, the history of physics holds less heat than light. Nor could a relation go unsuspected by anyone who had ever reflected on the implication of a burning lens for the interaction of radiation and matter. In 1800 the King's Astronomer, Sir William Herschel, published in *Philosophical Transactions* a series of "Experiments on the solar, and on the terrestrial rays that occasion heat." He showed (what was not entirely novel) that heat is radiant; that "the prismatic colours, if they are not themselves the heat-making rays, are at least accompanied by such as have a power of occasioning heat"; that these are subject to laws of reflection and refraction; that heat rays are of differing refrangibility, the strongest associated with the red; and finally, that rays of heat inhabit a portion of the unseen spectrum "beyond the confines of the red." A few years later J. W. Ritter darkened silver chloride with ultraviolet light. Thus, the Newtonian spectrum found itself bracketed by invisible radiation.

The movement of criticism which now began to study light, not as a stream of luminous corpuscles but as the effect of vibration in a continuous medium, could hardly have failed to carry over into heat. That criticism directed attention rather to the properties of the medium than to the nature of light. And physics had dismissed the "emission" theory of light before the caloric theory of heat began to go the same way—toward its destiny in energetics. The analogy, indeed, seems to have been what suggested to Carnot those rising doubts about the substantiality of

caloric which he confided to his notebook after publishing
The Motive Power of Heat:

> Light is nowadays regarded as the result of vibratory mo-
> tion in an aethereal fluid. Light produces heat, or at least
> it accompanies radiant heat and travels at the same velocity.
> It would be ridiculous to suppose that it is an emission
> from bodies, while the light that attends it is only motion.

It is time, therefore, to turn back and take up the thread
of optics.

The nineteenth-century wave theory of light is a classic
case of independent discovery, and of the hurts which that
recurrent coincidence inflicts on the discoverers. "When
one thinks one has made a discovery," wrote Augustin
Fresnel in 1816 to Thomas Young, who had published the
same interference phenomena in 1802 and 1804, "one does
not learn without regret that one has been anticipated; and
I shall admit to you frankly, Monsieur, that regret was my
sentiment when M. Arago made me see that there were
only a few really new observations in the Memoir which
I had presented to the Institute. But if anything could
console me for not having the advantage of priority, it
would be to have encountered a scientist who has enriched
physics with so great a number of important discoveries.
At the same time, that experience has contributed not a
little to increasing my confidence in the theory which I had
adopted." Perhaps, however, the relationship of Young and
Fresnel should be described as one of independent redis-
covery, and thereby as an illustration of the influence that
style and cultural tradition do have in science. For Young
regarded himself as one who was renewing the undulatory
aspects of Newton's own optical views. He sought to recom-
mend his theory by associating it with Newton's authority.
At the same time, he would deliver the master from the
hands of the epigones of analytical mechanics and restore
the full Newtonian science of light with its original sophis-

tication enhanced by new empirical detail. Fresnel, on the other hand, though his theory differed from Young's only in superior mathematical expression, thought himself to be applying a principle from Huygens to "a system of undulations, where light is nothing but the vibrations of a universal fluid agitated by the rapid movements of the particles of luminous bodies"—applying it, that is to say, to a model of space-matter drawn from Cartesian physics.

In the realm of intellectual habit, national styles do certainly persist. It is hardly too much to say that at the outset of the nineteenth century, aether was an English medium, and caloric a French one; and it is a curious byplay of the history of ideas that Young began by assimilating caloric to aether, or heat to light, and Fresnel the contrary (curious but unimportant, since it came to the same thing, and in recounting his actual experiments, Fresnel tended toward employing aether, the traditional luminiferous medium). Young's method in his first memoir consisted in reprinting long passages from Newton's optical writings in order to bring out their consistency with undulatory views. His first and fundamental hypothesis: "A luminiferous ether pervades the Universe, rare and elastic in a high degree," is elaborated from the query in which Newton asks: "Is not the heat conveyed through the vacuum by the vibration of a much subtiler medium than air? And is not this medium the same with that medium by which light is refracted and reflected, and by whose vibration light communicates heat to bodies, and is put into fits of easy reflection, and easy transmission?"

Fresnel for his part, true to the provincial universalism of French learning, never knew English and could never read Newton's *Optics* except in the translation by Marat, who was hardly equipped to give entry into the fullness of Newton's meaning. For Fresnel the Newtonian theory of light consisted only in the emission of particles, which were in light what atoms were in matter. He began his first memoir by saying that Newton had been obliged to imagine them streaming through space unobstructed by the caloric that fills it. Fresnel refutes this view (which Newton had never held, of course) in pointing out that black bodies under illumination would increase indefinitely in temperature since they absorb all the light that falls on them and convert it into caloric. Similarly, if refraction in (say) glass were explicable by the greater attraction exerted by molecules of glass over corpuscles of light, then a refracting surface should not only bend light but draw heat and thus be warmer than the adjacent air. Finally, Fresnel thought it implausible that particles responsible for the various colors should travel each species at its own velocity. "Periodic variations in the affections of light are much better conceived as produced by vibrations of caloric." Heat itself, suggested Fresnel at the end of his earliest memoir, may more probably manifest vibration in caloric than the emission of a substance. That the latter view prevailed he attributed, acutely enough, to the importance of chemical interests in the study of heat. But in explosions heat and light appear at once; it is reasonable to think that both manifest vibration in caloric; and however that may be, "the continual vibration of caloric and of the particles of bodies cannot be doubted: the force and nature of these vibrations ought to have a great influence on all the phenomena embraced by physics and chemistry, and it seems to me that up till now the study of these two sciences has been overly abstracted one from the other."

So different, then, were the standpoints—one ostensibly Newtonian, the other counter-Newtonian—from which Fresnel and Young reached the same conclusion. The arguments, too, display the strength and weaknesses of the French and English minds in science: the one systematic, rigorous, theoretical, and formal; the other ingenious, in-

ventive, concrete, and physical; the one too elegant to dare
the brusquer innovations, the other too deficient in taste
to feel the force of elegance. For Young could never understand that Fresnel's analytical elaboration of wave mechanics improved upon the ideas that he had struck out,
nor that it was Fresnel's mathematical demonstration and
formalization which wrought the conversion to the wave
theory of light.

"I am sincerely delighted," he wrote to Arago,

> with the success which has attended Mr. Fresnel's efforts,
> as I beg you will tell him; and I think some of his proofs
> and illustrations very distinctly stated; but I cannot fully
> adopt your expression in the letter you wrote by Mr. Dupin,
> that his memoir may be "considéré comme la démonstration
> de la doctrine des interférences"; for neither I nor any of
> those few who were acquainted with what I had written
> can find a single *new* fact in it of the least importance. . . .

An element of injustice, therefore, attends the equal eminence which history has accorded to their pedestals in refusing to distinguish between their claims. Certainly the
priorities were Young's. But it was the work of Fresnel
which commanded assent, nor would he have missed any
essential element if Young had never been there first.

❖

BECAUSE YOUNG WAS THERE FIRST, the historian will begin
with him. Nor must one seem to denigrate the brilliance
and ingenuity of this, perhaps the most various scholar to
emerge from the traditions enjoined by English Dissent,
and from the disabilities which it enjoyed a little. Young
came from a Quaker family of mercers and bankers in
Somerset. As a young man he rebelled against the Friends
to become outwardly an Anglican. The Quaker way never
deserted him, however, plain but of the best sort. When
on his deathbed, he continued working on a dictionary of

Egyptian languages, using a pencil after he could no longer handle a pen. To a friend who remonstrated he answered that the work would be a satisfaction if he finished it, and if he did not, it would still be a satisfaction never to have spent an idle day in his life. Young intended himself for the profession of medicine and practiced for a time. But these qualities of rigidity and austerity, this high sense of self-reliance, made for prickly comfort. He failed on the personal and sympathetic side of medicine. His real interest was natural philosophy, and his patients felt it. In 1801 he succeeded Humphry Davy as Professor at Rumford's Royal Institution. Afterwards he gathered his *Lectures on Natural Philosophy* into a vast compendium, which remains the most valuable single source for the whole state of physics in the first decade of the nineteenth century. But neither were his qualities suited to ingratiating the subject with fashionable ladies or unfashionable working men, and he did not teach for long.

Few could ever have met the standards he set himself. He was manually skilled and knew how to grind lenses and turn a lathe. He had a great bent for grammar, learned Greek, Latin, Hebrew, Chaldee, Syriac, and Samaritan, and read the classics of Italian and French literature together with English. Perhaps, however, languages confronted his intelligence mainly as another set of structures to be resolved into elements. Certainly his mastery lent Young no saving touch of literary grace. In the sciences he put himself through the entire works of Newton, Linnaeus, Lavoisier, Black, and Boerhaave. He studied in London, Edinburgh, Göttingen, and Cambridge. Discouraged at the incomprehension which greeted his wave theory of light, he turned away from physics and became interested in the Rosetta Stone, which in 1799 had been unearthed near the Nile by soldiers attached to Napoleon's Egyptian expedition, appropriated on its surrender by the British, and puzzled over for fifteen years by archaeologists to whom the message remained inscrutable. Young brought to the inscription his linguistic virtuosity, and seized upon the principle of hieroglyphic writing. In this discovery, too, his priorities were dimmed in luster by a Frenchman, Champollion, who carried the work through to a systematic conclusion. Though Young refused a post as editor of the fourth edition of the *Encyclopedia Britannica*, he was its most prolific contributor. He became interested, finally, in problems of the early insurance industry and is counted a founder of actuarial science. Thomas Young, in short, was a prodigy, and if he had had a more systematic training, if there had been such a thing as a scientific profession in the England of self-help, private endeavor, and new wine in old bottles, there is no telling what his energies might have accomplished. He even proposed the word "energy" for *vis viva* fifty years before Helmholtz.

As Helmholtz too would do, Young came to physics from physiological studies. He measured the accommodation of the eye to varying distances, and identified astigmatism and color-blindness as specific conditions. His discussion of colors as modes of perception is in the most sophisticated Newtonian tradition. In 1801 he gave the annual Bakerian lecture before the Royal Society, and published it in *Philosophical Transactions* in 1802. An experimental paper followed in the same year, and a second Bakerian lecture in 1804. Young was too preoccupied with developing his views to distinguish quite as explicitly as we may do between the elements of his work: the general hypothesis that light manifests waves, the law of interference of light, the experimental demonstration of that law, and its success in predictions and computations. Nevertheless, thus to analyze his contribution will do no violence to the order of his thoughts, though it may somewhat subdue their press and hurry. He advanced the hypothesis in passages chosen from

Newton: "That fundamental supposition is, that the parts of bodies, when briskly agitated, do excite vibrations in the ether, which are propagated every way from those bodies in straight lines, and cause a sensation of light by beating and dashing against the bottom of the eye, something after the manner that vibrations in the air cause a sensation of sound by beating against the organs of hearing." Nor was the "prepossession" which Young felt for this notion a question of new evidence, but only of reflections neglected in the nearly universal assent commanded by optical atomism. And it is intriguing that this breach, which was to widen and bring the properties of aether into the arena of physical discussion, should have begun with the consideration which a century later would dissipate the aether in relativity: "How happens it," inquired Young, "that, whether the projecting force is the slightest transmission of electricity, the friction of two pebbles, the lowest degree of ignition, the white heat of a wind furnace, or the intense heat of the sun itself, these wonderful corpuscles are always propelled with one uniform velocity?" Moreover, without admitting an element of periodicity (what Newton himself had called fits of easy transmission and easy reflection), it was quite impossible to explain why a refracting surface should always reflect some portion of an incident ray.

Moving beyond these generalities, Young made a close study of Newton's experiments on the colors of "thin plates" which appear in oil films or soap bubbles. They converted him to the "truth and sufficiency" of the undulatory system. Newton had produced them (it may be remembered) by pressing a plane surface of optical glass to another, ground ever so slightly convex. The circular wedge of air then gave "Newton's rings," a concentric spectrum, and he measured how deep the "thin plate" was at the point corresponding to each refracted band. Young discussed these appearances only in a qualitative way in his first Bakerian lecture, rather to remove objections to the wave theory than to insist upon proofs. The edge of shadow was the gravest. Shadows had prevented Newton himself from making the wave theory more than an auxiliary notion brought in to help save the evidence for periodicity. For light does certainly appear to travel in straight lines. Apart from the difficulties of the case, important circumstances would incline Newton away from the wave theory. In his eyes, it was not that alone. It was also a *pressure* theory of light, which is to say a feature of Cartesian cosmology. Query twenty-eight in the *Optics* had in view less Huygens than Descartes: "Are not all hypotheses erroneous, in which light is supposed to consist in pression or motion propagated through a fluid medium?—If it consisted in pression or motion, propagated either in an instant, or in time, it would bend into the shadow." But other reasons may have hidden the merit of Huygens' waves from Newton, factors deeper than his failure to abstract from Cartesian vortices. It was a defect of his great mathematical qualities that he adhered to the purest Euclidean formalism. Newton had been brought up, moreover, in the tradition by which light, even the light that he had fragmented, was taken to be luminous geometry. And the geometric constructions in which Huygens clothed his more physical notion of an undulating medium were more complicated than the linear tracing of rays obeying the equal law of reflection and the sine law of refraction.

Young, in his first memoir, developed Huygens' demonstrations verbally, showing that the radial development of a wavefront does satisfy those laws of reflection and refraction. In a general way he suggested how the mutual interference of waves of differing frequency or phase might make Newton's rings appear. To introduce this idea in so offhand a fashion was characteristic both of the resourcefulness of Young's temper, and of its limitations. For in it he makes the very important distinction between the motion of a wave system and the motions of the particles it sets to oscillating. But Young leaves this all implicit: "It is well known that a similar cause produces in sound, that effect which is called a beat; two series of undulations of nearly equal magnitude cooperating and destroying each other alternately, as they coincide more or less perfectly in the times of performing their respective motions." Young imagined an experiment for breaking light into colors in a way that supported the analogy. He scratched a polished surface to a depth of the order of magnitude of the wave lengths concerned, and represented the reflected hues as component vibrations of the original beam thus put out of phase.

Thereafter, Young worked systematically to develop and demonstrate this principle of interference, which he now began to distinguish from the more hypothetical question of the undulatory system. Interference at least admitted of demonstration in the laboratory, as a "simple and general law, capable of explaining a number of the phenomena of coloured light, which, without this law, would remain insulated and unintelligible." And Young's experimental papers set this more positive tone. The first of them, presented to the Royal Society later in 1802, opens with an explicit statement of his law of interference:

> Wherever two portions of the same light arrive at the eye by different routes, either exactly or very nearly in the same direction, the light becomes most intense when the difference of the routes is any multiple of a certain length, and least intense in the intermediate state of the interfering portions; and this length is different for light of different colours.

And this Young demonstrated with a series of simple and forceful experiments which opened to observation two new classes of phenomena. First, he examined the fringes of darkness to bring out the facts of "inflection" into the geometric shadow and diffraction by tiny orifices. This led him secondly, to "mixed plate" effects, which he created by varying the sequence between media of different optical properties and refractive indices.

Young began at the edges of shadows, particularly those cast by objects so narrow that their shadows were, in fact, mostly edge. He tried splitting the sight of a candle by threads stretched close across his eye. First he used a horse hair. It was too coarse. A single strand of woollen fiber did better, and a span of silken floss best of all. Fringes of color bisected the flame. They appeared capable of measurement.

I therefore made a rectangular hole in a card, and bent its ends so as to support a hair parallel to the sides of the hole: then, upon applying the eye near the hole, the hair of course appeared dilated by indistinct vision into a surface, of which the breadth was determined by the distance of the hair and the magnitude of the hole, independently of the temporary aperture of the pupil. When the hair approached so near to the direction of the margin of a candle that the inflected light was sufficiently copious to produce a sensible effect, the fringes began to appear; and it was easy to estimate the proportion of their breadth to the apparent breadth of the hair, across the image of which they extended. I found that six of the brightest red fringes, nearly at equal distances, occupied the whole of that image. The breadth of the aperture was $\frac{66}{1000}$, and its distance from the hair $\frac{8}{10}$ of an inch: the diameter of the hair was less than $\frac{1}{300}$ of an inch; as nearly as I could ascertain, it was $\frac{1}{600}$. Hence, we have $\frac{11}{1000}$ for the deviation of the first red fringe at the distance $\frac{8}{10}$; and, as $\frac{8}{10} : \frac{11}{1000} :: \frac{1}{600} : \frac{11}{480000}$, or $\frac{1}{43636}$ for the difference of the routes of red light where it was most intense. The measure deduced from Newton's experiments is $\frac{1}{39200}$. I thought this coincidence, with only an error of one-ninth of so minute a quantity, sufficiently perfect to warrant completely the explanation of the phenomenon. . . .

A later experiment reduced the light source to a pin-prick admitting a sunbeam through a screen. In place of the hair, Young held a slip of a card edge-on in the ray. He blocked one side. The "internal" fringes disappeared on both sides of the shadow; thus, he could conclude that fine objects produce internal fringes by interference between rays inflected round either side, and external fringes (which had remained unaffected on the open side), by interference between direct rays and those reflected (or deflected) by the edge. (Unfortunately this was not correct, since Young was still thinking about longitudinal waves.)

Next, he substituted a lock of woollen hairs for the single strand between his eye and the candle, and he saw a halo around the flame, not unlike Newton's rings. He tried the comparison with "mixed plates," in which substances of different refractive indices were substituted for the "plate" of air between the glass surfaces. He moistened and pressed together two pieces of plate glass so that droplets of water mixed with air would perform the office of the hairs. The effect led him to consider the old question: whether the velocity of light is greater in a rare or a dense medium. Wave theory required that it be the rarer substance, but direct measurements of so small a difference still eluded technique. Fortunately for the theory, however, Newton's rings become smaller as the density of the medium increases, which fact would imply that the wave length is shorter and the velocity less.

Moreover, the wave theory saved an otherwise very puzzling phenomenon of "mixed plates." In Newton's rings the central spot corresponds to the zone where the glass surfaces touch. That spot is black. One sees it under other circumstances, notably on the breaking of a soap bubble. It appears at the moment of vanishing, just as the liquid thins until the tension is unbearable. Young thought it odd that the least possible depth of the medium should black out light: "The actual lengths of the paths very nearly coincide, but the effect is the same as if one of the portions had been so retarded as to destroy the other." Since the difference in trajectories was infinitesimal, it must (Young supposed) be reflection at the surface of an optically rarer substance which retards the wave by half an interval. But this led to verifiable consequences. Suppose a beam be passed through two progressively rarer refracting substances. "The effect would be reversed, and the central spot, instead of black, would become white; and I have now the pleasure of stating, that I have fully verified this prediction, by interposing a drop of oil of sassafras between a prism of flint-glass and a lens of crown glass: the central spot seen by reflected light was white, and surrounded by a dark ring."

It was one of those satisfactions given only by science and the more precise kinds of scholarship, in which a prediction deduced from pure principle is realized in hard fact—a small triumph, but within its compass absolute, thus ministering innocently enough to the expert's self-esteem. By the time Young gave his lectures at the Royal Institution, he had devised a fine demonstration of interference effects or "beats" in light. He made a grating consisting of two minute apertures which diffracted elements of a monochromatic wave front. Bands of light and dark alternated on the screen, and from their dimensions Young recalculated, more closely now, the wave lengths of the different colors that he tried:

The middle of the two portions is always light, and the bright stripes on each side are at such distances, that the

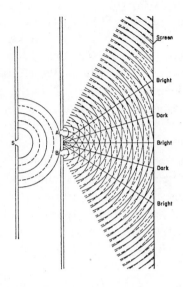

light coming to them from one of the apertures must have
passed through a longer space than that which comes from
the other, by an interval which is equal to the breadth of
one, two, three or more, of the supposed undulations, while
the intervening dark spaces correspond to a difference of
half a supposed undulation, of one and a half, of two and
a half, or more.

From a comparison of various experiments, it appears that
the breadth of the undulations constituting the extreme red
light must be supposed to be, in air, about one 36-thousandth
of an inch, and those of the extreme violet about one 60-
thousandth: the mean of the whole spectrum, with respect
to the intensity of light, being about one 45-thousandth.

All this seems persuasive in retrospect. Later in the
nineteenth century, it seemed more than that: it seemed
conclusive, for the community of physics experienced a
conversion and swung from one pole to the other—from
particles to waves. Young himself made few converts, however, and the reserve which his work evoked has a bearing

on the grammar of dissent in science. Personal factors no doubt weakened his effect. There hung about him, even in his Anglicanism, an invidious air of Quaker self-righteousness in argument. He would return a sharp answer and draw wrath. Moreover, he wrote badly. His papers are short but wordy, achieving cogency mainly in the quotations and in occasional flashes of scorn. Consider, for example, the unlucky clarity of this statement, made when he was still a young man, in which he dismissed a treatise on harmonics by an aged, much respected mathematician: "Dr. Smith has written a large and obscure volume, which, for every purpose but for the use of an impracticable instrument, leaves the whole subject precisely where it found it." The republic of letters was not sorry, therefore, when Young's own papers on light were pilloried in the *Edinburgh Review* by another whom Young had insulted, Henry Brougham, a shallow Scotch rhetorician and pseudo-Bacon, whose later political career is the most difficult to admire in the annals of Victorian opportunism.

But an important vein of resistance went deep into the very structure of classical physics. It is comparable to Newton's own inability to feel the full force of wave theory, and has to do with the influence on science of canons of style and language, which is to say of mathematical taste and technique. Mathematicians are bound to hold an influential voice in any scientific situation, and they can be reasoned with only in their own tongue. Moreover, neither simple conservatism nor Newtonian idolatry was what kept Laplace and Poisson and the French analytical school faithful to a corpuscular model of light. For they were accustomed to think no longer in Euclidean geometry, to be sure, but in the formalism which they had developed for the abstract dynamics of point masses, and into which they and their predecessors of the Enlightenment had cast the whole science of mechanics. It remains the great, the peculiar merit of Young's law of interference that it distinguishes between the motion of the waves (which are continuous) and of the particles (which are not), and the actual displacements of which compound the influence of the several systems of waves traversing the medium. The distinction invited creation of a dynamics of waves, instead of a dynamics in which the elements are point-masses, now idealized and again concretized. This was all unwitting in Young. He did suggest that aether waves might convey electricity along with heat and light—three manifestations, it may be, of one physical reality. But Young was not the man to develop wave dynamics theoretically. He only exhibited the physical evidence, experiments to exemplify and extend the unreasonable proposition that light on light gives bands of darkness. And he illuminated those bands with insights only, and not with theory.

❖

FRESNEL'S ADVANTAGE lay in his education at *Polytechnique*, where students began with theory. An early discovery, one of his happiest, determined that external fringes of shadow develop along a hyperbolic trajectory as the screen is moved away from the object. Arago had just informed him of Young's work, and Fresnel was writing to inquire precisely how Young had forestalled him. If Young had really done so, then he should have reached the same formulations, and particularly this of the hyperbolic path: "For I must say that it was in no sense observation, but rather theory, that led me to this result, which experiment subsequently confirmed." The ultimate problem of that theory was very serious, and Fresnel soon saw its scope. It was to reconcile the specificity of color, and indeed the whole ontology of matter, with the permeation of space by light and heat. Atomism served the former purpose. But for the latter, vibrations in an elastic medium afforded the simpler hypothesis. And Fresnel's rebuke to the classical mechanists may be read as a prophecy of the course which physics would be bound to traverse across the nineteenth century:

> The first [corpuscular] hypothesis has the advantage of leading to more obvious consequences, because analytical mechanics applies to it more easily. The second [undulatory] hypothesis presents, on the contrary, great difficulty on this score. But in choosing a theory, one should pay attention to simplicity in the hypothesis only. Simplicity in computation can be of no weight in the balance of probabilities. Nature is not embarrassed by difficulties with the calculus. She avoids complication only in means.

Fresnel was of that generation of Frenchmen which Stendhal understood, soaring into youth amid the expanding Napoleonic universe, only to confront maturity in the closed world of bourgeois France. Fresnel himself, the founder of wave mechanics; Sadi Carnot, the formulator of thermodynamics; Evariste Galois, an Einstein *avant la lettre*—they were like Julien Sorel, eaglets who did not long survive the eagle. Galois died at twenty-one, Carnot at thirty-five, Fresnel at thirty-nine. By profession Fresnel was a civil engineer serving in the *Corps des Ponts et Chaussées*, responsible since the mid-eighteenth century for the finest highway system in the world. Nor did the tradition of French engineering science imply any scorn of innovation or application. Fresnel carried his studies of light into the design of the lenticular light-house beacon. Its echelon lens replaced the feeble torches and lanterns for which ship captains had anxiously had to peer. Those studies began, indeed, as the distractions of isolation, when Fresnel was at work upon the Napoleonic roads in remote corners of France.

The materiality of light and heat had been presented to students at *Polytechnique* as an example of a subject in an unsatisfactory condition. Political mistakes during the hundred days gave Fresnel an involuntary leave at his family's house at Mathieu in Normandy, quite near to Caen. There with homemade apparatus he found the same phenomena which Young had made appear. He put himself into touch with Arago and the Institute, into which the Academy of Sciences had been incorporated. Some members encouraged him. Others criticized. He got occasional leave for visits to Paris and to proper laboratories. Though Laplace and his circle never adopted his views, they did support his election to the Institute in 1823. That recognition was unanimous. His health was already ruined, and he died in four years of tuberculosis, the disease of nineteenth-century genius, after only six or seven years spent in creative work. But it must not be supposed that Fresnel's contribution consists only of deep intuitions embodied in a few equations. His memoirs on optics, together with discussion of them by the referees of the Institute, occupy two immense volumes of his *Oeuvres*, and his engineering writings fill a third. If, therefore, we adopt Young's distinction between the wave hypothesis of light and the (phenomenalistic) law of interference, and assign the latter to Young, it was Fresnel who created the truly mathematical theory of light as waves, and who thereby converted the scientific community to his model of things.

Rather than follow Fresnel across the ground already traversed by Young, it will advance the subject faster to specify the breaches he exploited in the theoretical attack. What was decisive in Fresnel's tactics was the device of considering each point in an oscillating medium as a center of propagation of spherical wavelets. The idea was in Huygens, who for lack of the calculus could not develop it. Fresnel did dispose of the infinitesimal analysis, and improved his opportunity by combining it with the principle of interference. By his argument, the only detectable result of the composition of motions within the medium is the advancing wave front. All but the radially directed elementary oscillations destroy each other, and the resultant motion of the wave, an envelope of elementary wavelets, contains, therefore, rectilinear propagation of its effects as a special case. From this it was predictable (and Fresnel did predict) that diffracted waves would appear on a screen as alternating bands of light and dark, according as the coincident rays differed by an even or odd number of half wave-lengths. And this, of course, simply embedded Young's law of interference more intimately into the wave theory than had his own unadorned statement of the facts.

Fresnel's further prediction of the hyperbolic trajectory of external fringes illustrated the power of his method in more arresting fashion. And though Young had seen this evidence experimentally (as Newton had, indeed, and before him, Grimaldi), he had not appreciated what a very interesting result it was, nor made more of it than other more trifling appearances. Fresnel illustrated the phenomenon with a diagram. S is a light source, and AB a narrow object in the shadow of which fringe patterns are to form. Fresnel treats the ends A and B as centers of diffraction. The solid lines represent the enveloping wave fronts of the three systems emanating from S, A, and B and taken at the nodes of compression. (For Fresnel is still supposing longitudinal percussion waves on the model of those which stand in an organ pipe.) The dotted lines, on the other hand, represent the foci of the dilated phase. Then the trajectories of the dark bands are F_1, F_2 externally, and f_1, f_2 internally. Geometric analysis shows that F_1, F_2, etc. are neither circular nor rectilinear but hyperbolic. And experiment answers perfectly to this prediction. Moreover, the theory explained how it is that the number of internal fringes is greater nearer the object.

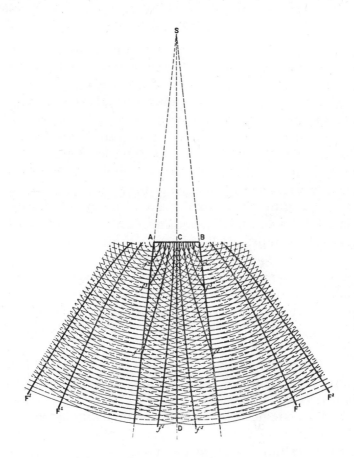

Impressed by the importance of these optical questions, the Academy proposed as its annual competition in 1818:

1° To determine by exact experiments all the effects of diffraction of luminous rays, direct and reflected, when they pass either separately or simultaneously close to the extremities of one body or of several bodies of limited or indefinite extension, with special attention to the intervals between these bodies, as well as to the distance from the source of light rays.

2° To conclude by means of mathematical inductions from these experiments what the motion of rays is in their passage close to bodies.

Fresnel assembled his theories and experiments into the winning memoir. Laplace, Poisson, and Biot sat on the Commission, all three skeptical if not downright hostile, along with Arago, who was enthusiastic, and Gay-Lussac, who was impartial. Speaking as a mathematician, Poisson observed that Fresnel's integrals entrained a paradox. At certain distances the intensity of diffracted light at the center of the geometric shadow of a tiny disc should be identical with the illumination at the apex of the conical projection of an aperture of the same diameter. An obstacle and a hole would thus be indistinguishable by diffracted light. They challenged Fresnel to test this alarmingly implausible consequence. He performed the experiment—with perfect success. And he later read a little lesson to the mathematicians who had preferred to tighten the grasp which their accustomed equations had given them on corpuscles. Already the wave hypothesis "furnishes much more extensive methods of computation. That is one of the least equivocal characteristics of the reality of a theory. When a hypothesis is true, it ought to lead to the discovery of numerical relations. . . ."

Nor was Fresnel indifferent to ampler implications. His equations are of the same type as those which Maxwell would later throw over the electromagnetic field of force. They are second-order partial differential equations which contain both the propagation and refraction of light. Was anything implied, then, about the interaction between aether and the ordinary stuff of physics, ponderable matter in motion? On the one hand, the fact of aberration relates the motion of the earth to the light from the stars. On the other hand, Arago had just shown in a very celebrated experiment that the motion of the earth has no influence on prismatic refraction of rays from the fixed stars. Does an atmosphere of aether move with bodies, therefore? Or do they sail through it? Fresnel never solved that problem. But to raise it was a great and a portentous thing, and raise it he did in a most intriguing speculation. Ponderable bodies, he suggested, convey in their interstices only a portion of the aether with which they coincide in space. That amount, a kind of specific aethericity (the phrase is not his), represents the excess over some hypothetical minimum which any body of equal volume will contain. Now, considering the luminiferous function of aether, Fresnel assumed that the total aether in a transparent body is proportional to the square of its refractive index. That, in turn, is the ratio between the velocity of light *in vacuo* and in the body. Finally, the velocities of the body and of light add vectorially, so that the absolute velocity of light within the body becomes a measure of the quantities of aether which pertain to space and to the body. Thus, thought Fresnel, the theory of propagation of light will contain whatever results would flow from the motion of the aether.

Years later, Hippolyte Fizeau tested this hypothesis in a fine experiment which he published in 1859 in *Annales de chimie et de physique*. It is less famous, for some reason, than the failure of Michelson and Morley to detect the aether drag, but no less significant. For it showed that the velocity of light increases in a medium according to the formula, $v \left(1 - \dfrac{1}{n^2}\right)$, where v is the velocity of the medium, and n its refractive index. His apparatus consisted of parallel tubes through which he sent water at high speeds in opposite directions. He split a single beam of light into two fractions, and with a mirror he made each traverse the tubes in opposite directions. He then compared the interference fringes produced by light moving with and against the current and found them displaced in opposite senses by measurable and reproducible amounts. If, therefore, the aether functions as a medium, its state of motion should also have a detectable influence on the velocity of light. But the existence of the aether was not then, of course, the point at issue. Fizeau designed his experiment to choose between three alternatives: first, that aether accompanies bodies in their motion; second, that bodies move through aether; or third, that the truth lies in between, so that the velocity of light should be increased by some specific amount less than the velocity of the medium. This last was the finding which Fizeau thought he had established, all alien though it was to the either-or outlook of nineteenth-century physics.

Some years earlier, indeed, a different experiment by Fizeau had sealed the victory of the wave theory. This one he performed in 1849 with Foucault, who also mounted the great pendulum in the dome of the Pantheon and for the first time demonstrated the rotation of the earth. The wave theory of light was a belief less well established, perhaps. But in this case, too, the experiment preached to the converted. Most physicists had already adopted the wave picture by mid-century. The old emission theory explained refraction by the superior attraction of the denser medium for the corpuscles of light. Its velocity should therefore be greater in water than in air. Only instruments more precise than any of Fresnel's generation could detect the difference. Fizeau and Foucault disposed of such instruments and showed the contrary to be true. It seemed a *coup de grâce* to particles of light.

This is to anticipate, however, for Fresnel's work on diffraction did not necessarily transcend the formal. For the study of interference, analogy with sound waves sufficed, wherein vibrations pulse back and forth in the direction of propagation of the wave. Hereafter, however, in the final stages of his work, Fresnel moved beyond Young in conceptualization as well as in formalization, and introduced the notion of transverse light waves. This, if we may not quite call it the optical revolution against mechanics, was certainly an optical rebellion against classical dynamics. Polarization led him on—polarization and the special case of it in double refraction. Iceland spar had aroused optical curiosity ever since Newton's generation. A crystal of that mineral will split a ray into two refracted beams, one obeying the ordinary sine law, and the other straying off at an idiosyncratic angle. Huygens made room for double refraction by supposing that the crystal put the aether into two wave systems: one expanding spherically to convey the ordinary ray, the other ellipsoidally for the extraordinary.

The gem tourmaline complicated the explanation, and indeed escaped it. It, too, shows double refraction. If a crystal be split and one slice rotated at right angles, the pair becomes opaque. Tried with Iceland spar, moreover, tourmaline will transmit the ordinary ray alone in one position, and the extraordinary ray when turned through ninety degrees. That light has "sides" was a proposition to be explained in the corpuscular theory only by supposing its particles to be either unsymmetrical or possessed of a qualitative polarity akin to magnetism. Neither prospect pleased. The wave theory did not help, on the other hand, so long as it imagined vibrations normal to the front. Young (once again) saw the problem and set a friend to working on polarization. He even suggested that waves oscillating transversely would yield such asymmetries. But this was a passing fancy, to which Young recurred as a fiction on occasion; he thus left this notion, too, to Fresnel to incorporate integrally into theory.

He came to it gradually. In 1809 Malus found that light which is reflected at angles within a range characteristic for certain surfaces—glass, water, polished metals—will traverse a crystal of Iceland spar as if it were one of the two fractions produced by a preliminary refraction through another crystal. That reflection also polarizes light greatly widened the interest of the phenomenon. From the beginning Fresnel had hoped that his own researches might bear on polarization. He discussed with Ampère how it might happen that rays polarized in different senses will never interfere. And Ampère, always ingenious and imaginative, threw out the notion that two wave systems might be compounded of transverse oscillations perpendicular to each other and of equivalent frequency and amplitude. Then if they were out of phase and destroyed one another, the forward motions alone might appear—hence the apparent independence of polarized rays. Fresnel took this idea and gradually qualified the notion of a forward component until ultimately he dropped it altogether and made the discreteness consist in the inclination of the planes of vibration.

It took determination, for the mechanical objections seemed insuperable to his colleagues. The elasticity of a solid might support such shear waves. The force in which rigidity consists would serve to bring each particle back across the point of equilibrium as gravity returns the pendulum after every upswing. In a fluid, on the other hand, only a harmonic sequence of impacts would send each oscillating particle back from the extreme amplitude of the wave across the mean. But what in a fluid would perform this office if it were oscillating at right angles to the direction of impact? Even Arago refused to follow Fresnel in these final "acrobatics." For it required little less to compose the wave motion of a beam in the round from an infinity of coaxial plane vibrations, each oscillating independently of the others in the same tube of space and in the same fluid medium, the nature of which did not admit of such motion. In deference to Arago, Fresnel withheld this feature of his theory until 1821.

He also deferred to his own conscience. He had thought of transverse vibrations before Young, always "more daring in his conjectures," but he would not rush the idea into print before satisfying himself that it could march with the principles of mechanics. And he had to satisfy himself, for as his thoughts progressed, he came to see transverse vibration as a necessary part of his theory. Otherwise, the undulatory system would fail in its claim to conceptual simplicity and coherence. There was one special case of double refraction which it could contain on no other terms. This was the action of biaxial crystals. It might seem a detail, even a trivial detail. But given Fresnel's temperament, progressive resolution of new and more difficult problems replenished his courage. It will be best to abstract an account of his views from the two lengthy memoirs on double refraction in which he developed and perfected them. In the second of these Fresnel squarely faced Laplace and the corpuscular school:

> The theory which we combat, and against which other objections may be raised, has led to no discovery. The learned computations of M. de Laplace, remarkable though they may be as elegant applications of mechanics, have taught us nothing new on the laws of double refraction. We do not think, however, that the advantage to be drawn from good theory ought to be confined to computing forces, when the laws of the phenomena are already known. That would contribute too little to the advancement of science. Certain laws there are, so complicated or so singular, that observation assisted only by analogy could never discover them. In order to divine these enigmas, it is necessary to be guided by a theory resting on a *true* hypothesis. The theory of luminous vibrations has that character and affords these precious advantages. We owe to it the discovery of the most complicated laws of optics, and the most difficult to divine; whereas all the other discoveries, numerous and important though they are, which have been made in that science by physicists partial to the emission system, beginning with those of Newton, are rather the fruit of their observations or their sagacity than they are mathematical consequences deduced from the Newtonian system.

What gave Fresnel such confidence was the success of his prediction of the refractive effects in biaxial crystals.

In crystals symmetrical around a single axis, one would still expect the ordinary ray to follow the usual laws of refraction. But in biaxial crystals this should no longer be the case. The molecules should be subject to asymmetrical tensions in all directions, and the elastic effects should, therefore, be different from those of mediums isotropic even in one plane. "This is precisely what I verified by experiment, a month after having communicated it to M. Arago: it is true that I did not present him this result as a certain fact, but as so necessary a consequence of my theoretical ideas that I should be obliged to abandon them if experiment had failed to confirm this singular character of double refraction in biaxial crystals." For in Fresnel, theory itself takes on an almost crystalline structure, so that the pattern of ideas must touch the facts not just here and here and there, but at every point, while the failure of new facts to appear where called for would not just mar one facet but would demolish the whole shape of thought.

An earlier note argues in qualitative terms the mechanizability of a transverse wave picture. Analytical mechanics, Fresnel points out, has admitted as the force of propagation only the differential state of dilation or condensation between successive phases. Their equations compose elastic fluids of points which may be crowded together or separated according as impacts are more or less frequent. But these equations can scarcely be supposed to express the true state of things. In reality the particles of a fluid are separated by distances that are very great compared to their diameters. In reality (it may be) they never touch, and true equations of fluid dynamics must, therefore, be concerned with such spatial relations, rather than with Newton's laws. No terms in the classical equations of the "geometers" allow for the motions that will be set up, should one layer of a fluid be supposed to slip between others. And Fresnel asks us to imagine three parallel rows of particles in a state of equilibrium between their mutually repulsive forces. If the molecules of the outside rows are opposite one another, those of the middle row will be staggered at the intervals. Suppose the molecules of this row to be displaced longitudinally. Since equilibrium obtained, the repulsive forces will tend to restore each to its initial spot. Vibration will be set up, therefore, compensated according to the law of action and reaction by vibrations in the outside rows. And these last displacements will have both a longitudinal and a lateral component. This is a development, evidently, of Ampère's idea. And it is not, perhaps, worth repeating the turn Fresnel gives to the argument to show that under aethereal conditions only the lateral component will be sensible. The point is only to picture the possibility of transverse waves in a fluid as a mechanism in which polarization could, and indeed must, consist.

> Thus, direct light may be considered as the reunion, or more exactly as the rapid succession of systems of waves polarized in every direction. In this way of looking at things, the act of polarization no longer consists in creating transverse motions, but in decomposing them along two invariant directions at right angles, and in separating the two components one from the other. For then the oscillations in each will always occur in the same plane.

The memoirs on double refraction develop this picture mathematically. Fresnel would analyze a three-dimensional bundle of radiation by resolving the actual oscillations in a ray into components projected upon two planes at right angles through the axis. This was simply an application of geometric composition and decomposition of motions to his model, followed by a derivation of the laws of optics expressed so as to include (for the first time) the phenomena of polarization. For his analysis of double refraction, Fresnel went back to the geometric constructions of Huygens, who had made the ordinary ray develop as a spherical wave surface, and the extraordinary ray as elliptical. This fitted the shape of things but had found no basis in the corpuscular constitution of light. And by a most elegant transformation into analytical terms, which he himself regarded as his most signal achievement, Fresnel made that missing connection between the evidence from interference for the constitution of light as vibrations in aether, the inference from polarization that those vibrations are transverse, and the prediction therefrom of idiosyncratic refractions in particular species of crystal. His equations related the two refractive indices of biaxial crystals to the proportions of their major and minor axes. Thus, he contained all the phenomena of optics in laws of the propagation of waves.

By those laws every point in a medium traversed by a vibration becomes itself a center of infinitesimal waves. The wave front is an envelope. In an isotropic medium the waves are spherical, and in an anisotropic medium they have the form of a more complicated surface of the fourth degree, depending on what axes of symmetry the medium may display.

The full generality of this synthesis lay beyond the power of Fresnel's own generation to appreciate. Nor, indeed, will his mechanical models of shear waves, his details of crystalline molecules tensed into unsymmetric patterns and setting up corresponding off-beats in the aether, quite bear the theoretical weight he put on them. Nevertheless, the generality and power of his own discussion forced the reorientation of physics toward the periodic aspects of radiation. He recaptured optics from the mechanists, turned light from corpuscles back to waves, and made this study an arena for the physics of the continuum instead of the physics of particles. And if this last, of the transversity of vibrations, was less crucial to his system than he fondly believed, at least it did set physicists upon what must seem in retrospect the most portentous train of thought in nineteenth-century physics: the mechanical properties of the aether—the medium for action at a distance—for the propagation of phenomena through space. So disconcerting were the requirements of transverse vibrations that Fresnel forebore to dwell upon anything beyond the evidence. Not so Young, who had no such stake in theory to lose, and who, since he had had the first word, may also be allowed the last:

> This hypothesis of Mr. Fresnel is at least very ingenious, and may lead us to some satisfactory computations: but it is attended by one circumstance which is perfectly *appalling* in its consequences. . . . It is only to solids that such a *lateral* resistance has ever been attributed: so that if we adopted the distinctions laid down by the reviver of the undulatory system himself [i.e. Young] in his *Lectures*, it might be inferred that the luminiferous ether, pervading all space, and penetrating almost all substances, is not only elastic, but absolutely solid!!!

❖

SCIENCE HAS KNOWN no tidier investigator than Michael Faraday. The papers collected in *Experimental Researches in Electricity* report researches extending over a quarter of a century, from 1831 until 1855. He numbered the paragraphs consecutively throughout the sequence. On November 20, 1845, he read before the Royal Society the nineteenth series. It announces magnetic rotation of the plane of polarized light, and paragraph two thousand, two hundred and twenty-two reads:

> The relation existing between *polarized* light and magnetism and electricity is even more interesting than if it had been shown to exist with common light only. It cannot but extend to common light; and, as it belongs to light made,

in a certain respect, more precise in its character and properties by polarization, it collates and connects it with these powers, in that duality of character which they possess, and yields an opening, which before was wanting to us, for the appliance of these powers to the investigation of the nature of this and other radiant agencies.

And he goes on to say in the next paragraph but one:

The magnetic forces do not act on the ray of light directly and without the intervention of matter, but through the mediation of the substance in which they and the ray have a simultaneous existence; the substances and the forces giving to and receiving from each other the power of acting on the light.

Writers of science fiction, and of science for young people, might reflect on their responsibility, and their opportunity. It was one of their predecessors who drew Faraday toward science, a Mrs. Marcet, author of *Conversations on Chemistry*. "Do not suppose," wrote Faraday of his childhood, spent in poverty, "that I was a very deep thinker, or was marked as a precocious person. I was a very lively imaginative person, and could believe in the *Arabian Nights* as easily as in the *Encyclopaedia*. But facts were important to me, and saved me. I could trust a fact, and always cross-examined an assertion. So when I questioned Mrs. Marcet's book by such little experiments as I could find means to perform, and found it true to the facts as I could understand them, I felt that I had got hold of an anchor in chemical knowledge, and clung fast to it. Thence my deep veneration for Mrs. Marcet—first as one who had conferred great personal good and pleasure on me; and then as one able to convey the truth and principle of those boundless fields of knowledge which concern natural things, to the young, untaught, and inquiring mind." Faraday's father was a blacksmith who could not provide the boy with education or surround him with external graces. All his grace, and it was lovely, was native to his mind and to his hands. The history of science has known no sweeter disposition, nor any gentler spirit, qualities not often associated with the urge toward discovery, or the ambition to be recognized. His ambition, or rather passion, was different. It was to lead the life of a philosopher. For he preferred the term "philosophy" to science.

For a man of no education, no other route than the experimental was open. At the age of thirteen Faraday was apprenticed to a bookseller. The business then included the binding of sheets received from the printer. For the statutory seven years Faraday bent to the task, hating trade all the while and losing himself in chemistry and physics in the hours after work. In 1812 he managed to attend lectures by Humphry Davy at the Royal Institution. He wrote out a fair and formal copy of his notes, bound them in the shop, and sent them to Davy, with a letter asking for a job in any scientific capacity. Davy was a skillful scientist, who had brought the first measure of order into electrochemistry and made it the most active sector of physical research in the previous decade. He advised Faraday against abandoning the security of his craft for the chances and limited opportunities of science, and then when he persisted, gave him a post as laboratory assistant. There is a charming memoir, *Faraday as a Discoverer*, by his younger colleague (disciple is perhaps the word) John Tyndall, who writes, "Davy was helpful to the young man, and this should never be forgotten." Neither, unfortunately, can a disfiguring careerism in Davy's conduct be forgotten, nor the unhappy jealousy of a more brilliant junior which in 1823 led him to oppose (unsuccessfully) his own protégé's election to the Royal Society. But Faraday never allowed this episode to spoil the regard he felt for his patron.

Faraday's researches of the 1820's were mostly chemical,

worthy enough but not yet seminal. He had still to develop, in Tyndall's words, "the power which Faraday possessed in an extraordinary degree. He united vast strength with perfect flexibility. His momentum was that of a river, which combines weight and directness with the ability to yield to the flexures of its bed. The intentness of his vision in any direction did not apparently diminish his power of perception in other directions; and when he attacked a subject, expecting results, he had the faculty of keeping his mind alert, so that results different from these which he expected should not escape him through preoccupation." Alone among the great scientists, Faraday might be taken in his success as justifying the suspicion which Diderot had once expressed for mathematics with its haughty spirit, and the democratic preference for the common touch of craftsmanship as conferring on the humble soul a power of divination, the ability to sense by manual inspiration how it must be with nature. Indeed, a German colleague once hit on the same figure as Diderot in evoking this artisan-like ability to "subodorer" the truth. "Er riecht die Wahrheit—he smells the truth," said Kuhlrausch of Faraday.

A philosopher of science who should study Faraday might well be led to write of prophetic rather than simply predictive validation of theories. For the sobriety with which he curbed his imagination in the laboratory never stultified him. Tyndall writes, too, of his ability to exalt a subject out of the microcosm of the laboratory into the macrocosm of nature; of how he would move out in his mind's eye from the little magnet in the Royal Institution attended by its curvilinear pattern of iron filings to the earth as a great magnet, with lines of force running through the atmosphere and through the seas, continually cut (it might be) by the flowing and ebbing of the tides, and inducing (if so) electrical currents in the oceans and the air; and too, of how difficult he was to follow for those who had been trained to express theoretical ideas in the mathematical formalism of conventional science.

> He does not know the reader's needs, and he therefore does not meet them. For instance, he speaks over and over again of the impossibility of charging a body with one electricity, though the impossibility is by no means evident. The key to the difficulty is this. He looks upon every insulated conductor as the inner coating of a Leyden jar. An insulated sphere in the middle of a room is to his mind such a coating; the walls are the outer coating, while the air between both is the insulator, across which the charge acts by induction. Without this reaction of the walls upon the sphere you could no more, according to Faraday, charge it with electricity than you could charge a Leyden jar, if its outer coating were removed. Distance with him is immaterial. His strength as a generalizer enables him to dissolve the idea of magnitude; and if you abolished the walls of the room—even the earth itself—he would make the sun and planets the outer coating of his jar.

Criticism can only speculate about the influence on his work of his entire ignorance of mathematics. Certainly he was the last physicist who could have borne up under such a handicap, and those not gifted mathematically would, no doubt, like to argue that it was a positive advantage to him, that it threw him back upon the experimental way he trod with a success unmatched before or since. Faraday himself liked to tick off the mathematicians on occasion, though always very gently. As against Arago, Babbage, and Herschel, his first paper on electromagnetic induction demonstrated (among other things) that the force whirling the copper plate was tangential. "It is quite comfortable to me," he wrote in a private letter, "to find that experiment need not quail before mathematics, but is quite competent to rival it in discovery." Nevertheless, it is difficult to think that a command of mathematics would not have advanced him further. As it was, he had to rely upon analogy rather than abstraction as an instrument of ordering and a guide to fruitful experiment. He wielded it with all possible finesse and achieved real elegance, the first physicist since Newton to do so in the British experimental tradition, but at a fearful cost in efficiency. Analogy, after all, depends upon a kind of linear transfer of ideas from one area to another, while abstraction frees ideas from the physical and poises the mind for the thought experiment. Only a mind as distinguished as Faraday's could have kept its bearings amid the mass of his experiments. "Faraday's resources as an experimentalist were so wonderful," admitted Tyndall, "and his delight in experiment was so great, that he sometimes almost ran into excess in this direction. I have heard him say that his paper on vibrating surfaces was too heavily laden with experiments." No scientist except Kepler has left a fuller record of his thoughts and trials. His *Diary*, published recently in some seven volumes, records his private thinking. But the papers he printed are themselves almost transcriptions of a laboratory notebook, full of weights, lengths, circumstances, results, discrepancies, false trails, failure, and success. They are not good reading. Already the literature of science wilts under the blight of the passive voice. Only occasionally did he loose his pen publicly to write down the speculations about how the world is made that filled his mind and guided his hand through all this mass of fact.

It is, moreover, very curious, and very wonderful, that he could distinguish by a kind of instinct those of his ideas which were subsidiary, from those which were fundamental and which guided the whole course of experiment through the forty years or more of his career. The former he would abandon without a sign of regretting the fate that his brainchildren suffered at his own hands. The latter illustrate rather the special faith that does animate science:

I have long held an opinion, almost amounting to conviction, in common I believe with many other lovers of natural knowledge, that the various forms under which the forces of matter are made manifest have one common origin; or, in other words, are so directly related and mutually dependent, that they are convertible, as it were, one into another, and possess equivalents of power in their action.

Instead of leading Faraday toward heat, however, and thence into thermodynamics, this conviction drew him to the deeper relations, as he believed the ultimate unity, of electricity, magnetism, and gravity. For there was given to him as to few scientists a sense of the spatial. He would almost see the moving wire slice through the lines of force and the current stir within. Perhaps, after all, it was the reward of his incapacity for abstraction, this vision of nature in the round, and in depth—deeper even than Tyndall said. He was a Victorian Kepler following on no Pythagoras and knowing no geometry—a laboratory Leonardo who could see but could not draw. His passion for knowing nature transcended the modes of expression open to a man of the nineteenth century. And always there was innocence, and that saving modesty. In the world he deported himself like the elder that he was of his strict, Nonconformist congregation. To the historian reading through his papers the speculative passages come as welcome respites from the interminable experimental detail to which Faraday's conscience condemns him. But the reprieves are brief: "I shall do better to refrain from giving expression to these vague thoughts (though they will press in upon the mind), and first submitting them to rigid investigation by experiment, if they prove worthy, then present them hereafter to the Royal Society."

The fact of electromagnetic interaction was already ten years old in August 1831, when Faraday wound two coils of copper wire onto opposite sides of an iron ring, completed one circuit by a wire passing by a magnetic needle and the other through a battery of wet cells, and found that the pointer would kick one way at make and the other at break. It was the first transformer. The obvious similarities, formal and physical, between magnetism and static electricity fed the general suspicion of some connection between these manifestations. In 1820 Hans Christian Oersted succeeded in the search. His is perhaps the only major discovery ever to come out of one of those lecture demonstrations in which professors carefully contrive the surprises they practice on nature and their students. Nothing happened when he tried the compass needle at right angles to a wire bearing a current. Then he turned it parallel and found that the wire, when placed below the magnetic pole, would "drive it toward the east, and when placed above it, toward the west." Thus, in this situation the current acts in its conductor like a magnet.

Word of his effect reached Paris, where its interest was immediately apparent. Ampère took up the investigation and immediately established the mutual influences of currents over each other, the conductors attracting each other if the currents are in the same direction and repelling in the contrary case. But Ampère was a polytechnician, and true to his education he named the new subject electrodynamics and concentrated on embracing induction in the formalism of analytical mechanics. That he seemed to do so was a triumph of mathematical virtuosity. For he had to treat the elements of each current infinitesimally and assume that the force joining any two is radial. He announced himself at the outset as one for whom scientific explanation consisted in reducing phenomena to the description of equal and opposite forces between pairs of particles. He brought great clarification and abolished the distinction between frictional electricity and galvanic. By analogy with the history of mechanics, the term statical applied to the former, once it was seen to be a special case, a kind of arrested dynamics. That appeared as "tension" (or potential) in the one case which causes continuous flow in the other. Only electricity in motion, Ampère pointed out, wraps a magnetic influence around itself, or affects another current across the space between conductors. And it was left to Faraday to find the missing piece in the puzzle of induction, the inverse creation by magnetism of an electrical current, and to meditate about the space where these effects transpire, in configurations quite inescapably curvilinear, and by forces which took their purchase on bodies tangentially and not centrally. There lay the difficulty. Since Kepler there had been no tangential drag in physics.

Reciprocity required that if electricity is convertible into magnetism, magnetism ought to be convertible into electricity. Faraday began systematic experiments on his return to the Royal Institution from a summer vacation in 1831. Nor did he come to the problem as an electrical novice. Though he had until then worked primarily on chemistry, he had given thought to the new science of induction. As early as 1821, he had made a bar magnet revolve around a wire by weighting one end with platinum and floating it upright in a mercury bath into which the wire dipped, thus completing a voltaic circuit. Reciprocally, he made a wire revolve around a fixed magnet. Logically enough, he next tried to make the wire or magnet rotate in place in the center of the cup of mercury. He failed and—this was how his mind always worked—suspected that perhaps the current turns within the conductor. So he bent it into a crank shape. He was right. Now it did turn. And he devised an arrangement delicate enough to show the rotation of a conductor in the magnetic field of the earth.

His first attempts to induce the electrical current held similar disappointment, which he exploited with the same importunate instinct. He had expected to generate a continuous flow of galvanic electricity in his second circuit, comparable to the continuing magnetism set up by electricity in motion. Instead, he got only that quick flick of the needle, at the moment of connecting the battery, and then inert quiescence until he broke the circuit, which expired with an equally feeble twitch the other way. Einstein once said that experiments are dull to read about and exciting only to do. Certainly no summary can do justice to the patience and ingenuity with which Faraday substituted helices for simple coils, with and without iron cores, altered his connection, tried its parts in every possible plane vis-à-vis the others, and then bethought himself of an experiment by Arago, in which a permanent magnet hanging over a copper disc has no effect until it begins to rotate, when the disc is constrained to follow, and vice versa. May not induced currents be the agent? And if so, it is the motion of magnetized metal relative to the copper which creates them. Faraday had, in any case, hoped to create electricity from magnetism, rather than simply to provoke one current by another. And now he dispensed with the battery, connected a galvanometer between the ends of a wire wound into a tight coil or helix around an open iron cylinder, and tried thrusting a bar-magnet into the hole. Everyone knows of his success: "A powerful pull *whirling the galvanometer needle* round many times was given," says the *Diary*. Still it was only on inserting or withdrawing the magnet that the current stirred. But now Faraday saw that he must exploit the fact rather than overcome or evade its limitations, and thus find a way to enlarge and continue it. This simply required ingenuity. He mounted a copper plate to rotate between the poles of a great magnet belonging to the Royal Society, and drew from this the first electrical current to be sustained otherwise than chemically. It needed only the substitution for the disc of a coil wound round an armature to become the magneto; and what followed was in all the long history of science the first truly portentous application of a major piece of basic research (as opposed to rational method) to the occasions of industry.

Faraday's own imagination moved in another direction, outward as was his wont toward the great magnet of the earth, and then inward toward the simplest and most elegant representation of the contrivances of nature. First he devised various arrangements for inducing a current by the action of the magnetic field of the earth alone. The most economical consisted only of a copper plate rotated at right angles to the line of a dipping needle: "The effect at the needle was slight but very distinct and could be accumulated upon the needle by reversion and reiteration of motion." It was always Faraday's hope to detect the grand analogue of such artificial local currents, and to observe electrical currents induced between Dover and Calais in the tides of the Channel. In that he was disappointed. But he did use the magnetism of the earth in the Second Series of his *Experimental Researches*. He formed a single strand of copper wire into a rectangle, with a galvanometer included in one of the long sides. Then when the wire was rotated around the galvanometer, the needle would swing in angles up to 90°. This with a single wire, he noted in his *Diary*, was a "truly elementary experiment," the results "beautiful," and (as he remarked in the published paper), "The exclusion of all extraneous circumstances and complexity of arrangement, and the distinct character of the indications afforded, render this single experiment an epitome of nearly all the facts of magneto-electric induction." An entry in his *Diary* of March 26th contains a most characteristic instance of the working of his mind, and of the solid sense of the spatial in which his thinking would issue. In one short statement, and a notebook sketch, he relates the new, electro-magnetic dimension of physics to motion, the classic preoccupation of science:

> The mutual relation of electricity, magnetism, and motion may be represented by three lines at right angles to each other, any one of which may represent any one of these points and the other two lines the other points. Then if electricity be determined in one line and motion in another, magnetism will be developed in the third; or if electricity be determined in one line and magnetism in another, motion will occur in a third. Or if magnetism be determined first then motion will produce electricity or electricity motion. Or if motion be the first point determined, magnetism will evolve electricity or electricity magnetism.

From these, his truly seminal discoveries, Faraday turned to an exploitation of the vast terrain of electromagnetic happenings, and first of all to establishing the identity of what some still took to be different kinds of electricity, "common, animal, and voltaic," for otherwise there could be no confidence in the uniformity of his results. And he imagined a series of experiments to exhibit identical effects physically, thermally, physiologically, and chemically, whether his electricity was produced by friction, a battery, or a fish. This held neither surprise nor disappointment, but in the course of these rather routine demonstrations, Faraday did think to notice what was not entirely novel: that when water turns to ice, it ceases to conduct. Immediately he turned his interests towards the physical state of conductors, and found it to be a general law that solutions cease to pass the electrical current when they congeal. The contrary is true of heat, he remarked, which is conducted more readily across ice and other solids than through liquids. The difference might open some insight into the "corpuscular condition" of the substances concerned, but Faraday left the structure of matter in abeyance for the nonce to concentrate on the mechanism of electrolysis, as he was the first to call it. For it seemed that substances convey currents chemically, and batteries produce them, at the price of their own decomposition.

The researches on electrolysis were a brilliant and characteristic digression, displaying Faraday's imagination to the very best advantage—and his clarity of mind. A curious feature of his scientific personality here emerged into peculiar prominence. Though he had no mathematics, he chafed, nevertheless, in the prison house of vulgar symbols and naïve models. And perhaps it was the combination of these qualities and disqualities, so to say, which led him beyond the abstractions he could not express into a kind of intimacy with nature to which no mathematician might attain. Neither, unhappily, might Faraday share it. The hydraulic image suggested by the phrase "electric current" aroused his special impatience, and he proposed to liberate physics from bondage to all terms which pre-judged an investigation. Poles imply attraction. He coined in their place the word "electrode," paired off as anode and cathode. A material which conducts in decomposing is an electrolyte, and its constituent "ions" are cations and anions. Poles imply attraction or repulsion. But "according to my view the determining force is not at the poles, but *within* the body under decomposition; and the oxygen and acids are rendered at the *negative* extremity of that body, whilst hydrogen, metals, etc., are evolved at the *positive* extremity. . . . The *poles*, as they are usually called, are only the doors or ways by which the electric current passes into and out of the decomposing body; and they of course, when in contact with that body, are the limits of its extent in the direction of the current." In place of poles, therefore, he coined "electrode" as a word implying nothing inconsistent with this or any picture.

How many physicists, one wonders, know why anode stands for positive and cathode for negative? "Wishing for a natural standard of electric direction to which I might refer these, expressive of their difference and at the same time free from all theory, I have thought it might be found in the earth. If the magnetism of the earth be due to electric currents passing round it, the latter must be in a constant direction, which, according to present usage of speech would be from east to west, or, which will strengthen this help to the memory, that in which the sun appears to move." In any electrolysis, therefore, Faraday imagines the solution placed so that the current through it will be parallel to that around the earth and in the same direction; then the electrodes will have an invariant reference, and that towards the east will be called the *anode* ("*ana*," upward, toward the rising sun; and "*hodos*," way) and its opposite the *cathode* ("*kata*," downward, toward the setting sun). Analogously, a material which does conduct in decomposing is an electrolyte, and its constituent "ions" are anions or cations. "These terms being once well-defined, will, I hope, in their use enable me to avoid much periphrasis and ambiguity of expression. I do not mean to press them into service more frequently than will be required, for I am fully aware that names are one thing and science another."

So far are we from Condillac, and Faraday went on immediately to substantive matters, determining by electrochemistry the equivalent weights of electrolytic substances. The evidence was, of course, a powerful confirmation of Dalton's law of definite proportions, and therefore of the atomic theory of matter. And there is a certain irony in this. For in Faraday's determination to transcend a fluid conception of electricity, he imagined his ions ferrying that charge in motion which was the current across the decomposing mass. In disembodying the current, therefore, he added body to the corpuscular image of matter, and so successfully that a century later his model of electrolysis still served the needs of textbooks. But he himself, the more he thought on the structure of matter and its relation to electromagnetism, the more he despaired of the corpuscular account as a sufficient representation of the texture of reality. Once again his inquiry turned out of the common path he had just widened toward the mechanism of induction and conduction, which was to say the relationship of force to matter. This was the preoccupation which would engulf his later years. For him, all untrained in academic philosophy, it could be no metaphysical problem, as it was for Mayer and for Helmholtz. It was altogether physical. No more than Newton could he accept the notion of action at a distance. But neither could he bring to a happy issue the attempt which he began in 1837, and continued for several series of experiments, to see induction and conduction as the action of "contiguous particles," passing on or delaying the electrical impulse according to their specific properties, or perhaps their freedom to move. The analogy contrasted the behavior of an electrolyte in the liquid and in the solid state, and it failed. He could never escape consideration of the distances between bodies. If action or induction at a distance does occur, it must be in straight lines between centers of force, like gravity. But magnetic and electrical forces are not radial; they seem to curve through space; and it was now that Faraday began to write of lines of force, generalizing the possibilities apparent in the patterns assumed by iron filings round a magnet. But then it must be the medium, somehow, and not the mechanistic jostlings of the particles which will lift the curse of unintelligibility from action at an apparent distance. Tyndall tells of the difficulty of following Faraday's thinking through the experiments which lodged these musings in the laboratory.

It would, however, be easy to criticize these researches, easy to show the looseness, and sometimes the inaccuracy, of the phraseology employed; but this critical spirit will get little good out of Faraday. Rather let those who ponder his works seek to realize the object he set before him, not permitting his occasional vagueness to interfere with their appreciation of his speculations. We may see the ripples, and eddies, and vortices of a flowing stream, without being able to resolve all these motions into their constituent elements; and so it sometimes strikes me that Faraday clearly saw the play of fluids and ethers and atoms, though his previous training did not enable him to resolve what he saw into its constituents, or describe it in a manner satisfactory to a mind versed in mechanics. And then again occur, I confess, dark sayings, difficult to be understood, which disturb my confidence in this conclusion. It must, however, always be remembered that he works at the very boundaries of our knowledge, and that his mind dwells in the "boundless contiguity of shade" by which that knowledge is surrounded.

Such, in any case, were his dilemmas and preoccupations, and in 1841 Faraday collapsed under their weight. He was just fifty years of age. His prostration is always attributed to overwork, and indeed no scientist in all history comes to mind who ever spent such long and faithful hours in the laboratory. Nor is historical understanding much enriched by retrospective and amateur psychoanalysis. Nevertheless, it is moving to think for a moment of the tensions under which Faraday lived: a great physicist, ignorant of the language of his science; a daringly speculative thinker, constrained, both by his own deficiencies and the intellectual fashion of the time, into the paths of the most dutiful, painstaking, and arduous experimentalism; a proud and passionate nature, under the dual code of the Nonconformist conscience and the Victorian way of life. Generosity came naturally to him. But it may be that moderation did not. They were terrible problems on which he brooded, and if one turns from these to one of the many likenesses that survive, they make a painful contrast with an element of childlike simplicity, almost of beauty, in his face. Could it have hurt him that he was of such very short stature, not much over five feet tall? In any case, his collapse was one of those nervous catastrophes to which the Victorians were prone. Similar crises, to recall two examples, beset John Stuart Mill and Florence Nightingale. All his life Darwin suffered from prostration. For two years Faraday could do no science and see no company. Ultimately the classic nineteenth-century remedies worked their slow magic—the loving care of a devoted helpmate, the soothing influence of Alpine scenery. And Faraday resumed his studies.

But with a difference. Speculation and theory about the structure of things occupied an increasing sector of his attention. They became, one begins to feel, the object of his search instead of the recreations of his imagination. Not that he gave over spending his days in the laboratory, or ceased to make discoveries great and small. In the hope of getting closer in to the actual contiguity of particles, he tried examining with polarized light—and discovered its rotation under magnetism. He entitled the paper reporting these researches "The Magnetization of Light, and the Illumination of the Lines of Magnetic Force." "A few years ago," runs paragraph 2,614, "magnetism was to us an occult power, affecting only a few bodies; now it is found to influence all bodies, and to possess the most intimate relations with electricity, heat, chemical action, light, crystallization, and through it, with the forces concerned in cohesion; and we may, in the present state of things, well feel urged to continue in our labours, encouraged by the hope of bringing it into a bond of union with gravity itself." And a little later, in words which might have been written very much later and by a greater thinker, "Here end my trials for the present. The results are negative. They do not shake my strong feeling of the existence of a relation between gravity and electricity, though they give no proof that such a relation exists."

The inclusion of crystallization refers to Faraday's investigation of the magnetic properties of crystals, no doubt on the analogy which had led from double refraction to an understanding of polarized light. He, too, found strange effects, having to do with directions of force. Certain crystals—metallic bismuth, for example—experienced neither attraction nor repulsion in the grip of magnetic forces, but simply aligned themselves with their magnetic axes tangent to lines of force, so that they might offer the least resistance. "I do not remember," he wrote, "heretofore such a case of force as the present one, where a body is brought into position only, without attraction or repulsion." And it is time to introduce the meditations of his later years, which consisted in the gradual possession of his mind by belief in the physical existence of lines of force.

In earlier years, in the 1830's and 1840's, Faraday treated lines of force only as representations of direction. It is not clear at what point he began to think of them as really existent in the physical condition of space. Indeed, the germ of belief must have been in his first observation of the pattern assumed by iron filings around a magnet. What is clear is that the wish to believe grew upon him in the 1840's. Experiments multiplied in which he investigated magnetic and "diamagnetic," conducting and insulating properties of bodies. He would represent his findings in ever more explicit drawings of lines of force streaming concavely together through those bodies which support magnetism or electricity, or bending convexly outward to strain away from bodies repellant to those forces.

The graphic advantage was great—so great that for a century teachers taught that the quantity of induced electricity is proportional to the number of lines of force cut by the armature. Surely the majority of persons who remember anything about induction still remember this. Certainly Faraday himself saw his magnets as loci of a host of lines

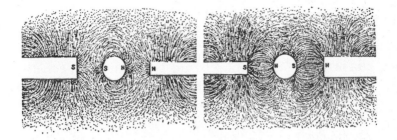

of force, pinching and clustering together as they streamed in from magnetically neutral space to pass through the magnet, spreading and streaming away, on the other hand, to avoid as far as possible inhospitable "diamagnetic" bodies. And whatever the psychological explanation of Faraday's gradual surrender to the wish to believe, it is clear what the physical phenomena were that fed it. To begin with, the curvilinear form of electromagnetic lines of force argued their physical existence. Sometimes Faraday seems to betray an instinct that only the mathematical may follow the straight Euclidean line, and that his lines of force are what constrain the physical away from the abstract. It is true that his hope of penetrating beneath the abstractions of classical dynamics into a real physics of unified force was frustrated by the radial direction of gravity, and the radical incompatibility of tangential and central forces. But neither could Ampère's equations from classical dynamics adequately describe the condition of the electromagnetic medium. To Faraday, therefore, belongs the honor of identifying that incompatibility which he could not resolve. Ampère and the analytical school, for their part, sought rather to preserve appearances in mathematics. Moreover, in thinking on the physical foundations of force, Faraday felt gathering dismay at action-at-a-distance, across a space which entered science only as a permission to ignore fundamental difficulties. All the physics which had supervened since Newton did not armor over that Achilles' heel.

In 1844 the *Philosophical Magazine* published an essay in the form of a letter to Richard Taylor in which Faraday freely and explicitly discussed the perplexities that afflicted him at the very foundations of physics. These, it must be remembered, were the decades of mounting triumph for the atomic theory in chemistry, and also in crystallography. And yet, Faraday points out, the implications of atomism as a hypothesis of the constitution of matter go beyond the facts of definite proportions, equivalents, and constant composition. In the minds of those whom the atomic theory pleased there was surely something more than the sum of those phenomena. Faraday himself had often sought to imagine how the corpuscles of bodies convey or repel charge and current. But ultimately, atomism could not please him as a representation of reality: "Light and electricity are two great and searching investigators of the molecular structure of bodies, and it was whilst considering the probable nature of conduction and insulation in bodies not decomposable by the electricity to which they were subject, and the relation of electricity to space contemplated as void of that which the atomists called matter, that considerations something like those which follow were presented to my mind."

If the common atomic view of the constitution of matter were correct, the particles of matter are one kind of thing, and space quite another. They are raisins and the cake, such that "space must be taken as the only continuous part, for the particles are considered as separated by space from each other. Space will permeate all masses of matter in every direction like a net, except that in place of meshes it will form cells, isolating each atom from its neighbours, and itself only being continuous." What, then, of a piece of shellac, a non-conductor? It must follow from the atomic theory that space is an insulator, for if it were atoms rather than space which insulates, the space in the most resistant of bodies would still conduct around the atoms. But what of metals? What of platinum or copper? For by the atomic theory again, space alone is the continuous aspect of matter, and therefore, space is a conductor. Thus, "the reasoning ends in this, a subversion of that theory altogether; for if space be an insulator it cannot exist in conducting bodies, and if it be a conductor it cannot exist in insulating bodies. Any ground of reasoning which tends to such conclusions as these must in itself be false."

Such were the contradictions which led Faraday to repudiate atoms-and-the-void as a model of reality. He did recognize how naturally the phenomena of crystallization, and of chemistry and physics generally, grouped themselves around the acknowledgment of centers of force. "I feel myself constrained, for the present hypothetically, to admit them." But he would admit as little as possible. In place of the conventional picture he preferred the atoms of Boscovich, an eighteenth-century Jesuit philosopher somewhat out of the main stream of science, who had defined atoms only as centers of force, and not as particles of matter in which powers somehow inhere. To blur the ultimate massy atom of chemistry into a vaguer focus of such manifestations as we can observe would in no way injure its capacity to order that science. But it would immensely ease the task of imagining what happens in the case of electrical conduction, chemical reaction, and the interactions of heat, electricity, and magnetism with matter. What Faraday pleads for, indeed, is nothing less than abolition of the boundary between matter and space in order to assimilate both to the manifestations of energy in extension. If we consider a conductor, say potassium, we can scarcely imagine that its conducting power belongs to it "any otherwise than as a consequence of the properties of space." So, too, its properties in relation to light, or magnetism, or even solidity, specific gravity, and hardness: "For where is the least ground (except in a gratuitous assumption) for imagining a difference in kind between the centres of two contiguous atoms and any other spot between those centres? A difference in degree, or even in the nature of the power consistent with the law of continuity, I can admit, but the difference between a supposed little hard particle and the powers around it, I cannot imagine."

And it is indeed interesting, and to anyone with a sense of the drama of culture across the centuries, it is moving to read Faraday, whose own untutored, self-taught state surely left him the nearest thing to a child of nature, not to say a noble savage, that the history of science can show, coming, after a lifetime spent on doing science with his own hands, to that same dilemma between atoms and the continuum which has given structure to the history of science since its opening in Greece. By now, indeed, what with the success of analytical mechanics, it was a turn-about tale, with the mathematicians enlisted in the service of atoms. And it was rather Faraday who came down, the greatest of the empiricists and the first of them to do so, on the side of the continuum and the unity of nature. "Now the powers we know and recognize in every phenomenon of the creation, the abstract matter in none; why then assume the existence of that of which we are ignorant, which we cannot conceive, and for which there is no philosophical necessity?" For in Faraday it is not geometry, as it had been in Descartes and would be again in Einstein, which encompasses matter and energy in extension. It is imagination. "To my mind, therefore, the . . . nucleus vanishes, and the substance consists of the powers; . . . and indeed, what notion can we form of the nucleus independent of its powers?" In this view, atoms will be highly amorphous, instead of being hard little balls bounding and rebounding. They will no longer have to be assigned any determinate shape, nor any exclusive location. "If an atom be conceived to be a centre of power, that which is ordinarily referred to under the term shape would now be referred to the disposition and relative intensity of the forces." And finally,

> The view now stated of the constitution of matter would seem to involve necessarily the conclusion that matter fills all space, or, at least all space to which gravitation extends (including the sun and its system); for gravitation is a property of matter dependent on a certain force, and it is this force which constitutes the matter. In that view matter is not merely mutually penetrable, but each atom extends, so to say, throughout the whole of the solar system, yet always retaining its own center of force.

Thus would Faraday save the "old adage, 'matter cannot act where it is not,'" and these were the dark views to which Tyndall alluded, these the "flashes of wondrous insight and utterances which seem less the product of reasoning than of revelation." These, too, were the "theoretic views" of which Faraday persuaded very few. Nor was Tyndall, for all his sympathy and admiration, among them.

A final word of caution must be said against the allusion in many books of physics to Faraday as the founder of field physics. In the most important sense, no doubt, he was, in that he re-addressed the science to the properties of space, and required it to pay attention to the propagation of phenomena instead of simply to the dimensions of bodies, serving as parameters of ever more analytical and elegant equations. But he did not write of fields. His space is full of tubes of force. Nor was he the one to identify the structure of space with the mechanical properties being visited by wave theory upon the aether. About the aether, indeed, Faraday maintained that reserve which forsook him when it came to his own tubes of force. Even about space and matter he retreated in practice back into that necessity which he had admitted for centers of force in his own thinking, and for some distinction. He was not the man to plunge physics back into Cartesianism. In Series 25 of his *Researches* appears an experimental paper addressed to the Royal Society in 1850, and paragraph 2,777 contains these observations:

> It seems manifest that the lines of magnetic force can traverse pure space, just as gravitating force does, and as static electrical forces do; and therefore space has a magnetic relation of its own, and one that we shall probably find hereafter to be of the utmost importance in natural phenomena. But this character of space is not of the same kind as that which, in relation to matter, we endeavour to express by the terms magnetic and dia-[non-]magnetic. To confuse them together would be to confound space with matter, and to trouble all the conceptions by which we endeavour to understand and work out a progressively clearer view of the mode of action and the laws of natural forces. It would be as if, in gravitation or electric forces, one were to confound the particles acting on each other with the space across which they are acting, and would, I think, shut the door to advancement. Mere space cannot act as matter acts, even though the utmost latitude be allowed to the hypothesis of an ether; and admitting that hypothesis, it would be a large additional assumption to suppose that the lines of magnetic force are vibrations carried on by it; whilst as yet, we have no proof or indication that time is required for their propagation, or in what respect they may in general character assimilate to, or differ from, the respective lines of gravitating, luminiferous, or electric forces.

For Faraday's thinking about these problems contains no refuge for small minds.

❖

To MOVE without preliminaries from Faraday's reflections about the electromagnetic relations of space into the field theory of James Clerk Maxwell will convey the intimacy in which the physical discoveries of the one merged into the theoretical formulations of the other, so that something like an intellectual continuum joined their thinking. Certainly Maxwell himself conceived the problem of electrodynamics from the very outset to be the mathematical, and not simply the intuitional description of the tensions and convolutions in continuous media.

Unrepentant over the ignominious decline of the corpuscular theory of light, the continental analysts had proceeded in the wake of Coulomb and Ampère to express electromagnetic forces in the differential formalism which had served classical mechanics, and just failed in optics. They resolved electrical and magnetic charges into infinitesimal elements of electricity and magnetism, and currents into point-masses in conservative motions serving Newton's laws. They displayed the effects of attraction, repulsion, and induction as interaction of particles at a distance, taking their ultimate license to do so, of course, from the Newtonian law of gravity and the extension of the inverse-square relationship to electromagnetic forces. It was, as Maxwell admitted, a powerful method, "warranted by the universal consent of men of science." Faraday's lines of force had evoked no such consent. Indeed, few mathematical physicists paid them serious attention, and most regarded them with either indulgence or a touch of scorn as another instance of the mathematical incapacity and consequent theoretical immaturity or barbarism of the English experimental tradition.

Not so Maxwell, to whom it seemed that Faraday had stated the laws he discovered "in terms as unambiguous as those of pure mathematics," and for whom the mathematician's part was to receive physical truths, deduce other laws capable of being tested by experiment, and thus to assist the physicist in arranging his own ideas. Those truths were that currents are produced by changes in the magnetic or electrical state of the medium surrounding a conductor, and that no complete description could be yielded, therefore, by any action-at-a-distance analysis which ignored the reality and efficacy of the strains and stresses in the medium. He would be Faraday's mathematicus.

> In the following investigation, therefore, the laws established by Faraday will be assumed as true, and it will be shown that by following out his speculations other and more general laws can be deduced from them. If it should then appear that these laws, originally devised to include one set of phenomena, may be generalized so as to extend to phenomena of a different class, these mathematical connexions may suggest to physicists the means of establishing physical connexions; and thus mere speculation may be turned to account in physical science.

And this was the program of the charming and ingenious memoir, "On Faraday's Lines of Force," which Maxwell published in 1856 in the *Transactions of the Cambridge Philosophical Society*. Faraday himself was astonished and delighted. "I was at first almost frightened," he wrote, "when I saw such mathematical force made to bear upon the subject, and then wondered to see that the subject stood it so well."

But how might anything so nebulous in the minds of others as Faraday's lines of force be quantified? And it was the peculiar turn, one almost wrote the Scotch turn, of his ingenuity, at once rational and handy, which made Maxwell precisely the right man in the right place to answer to that opportunity. Physics must steer, he said in effect at the outset of his paper, between the Scylla of the abstract and the Charybdis of the concrete.

> We must therefore discover some method of investigation which allows the mind at every step to lay hold of a clear physical conception, without being committed to any theory founded on the physical science from which that conception is borrowed, so that it is neither drawn aside from the subject

in pursuit of analytical subtleties, nor carried beyond the truth by a favourite hypothesis.

In order to obtain physical ideas without adopting a physical theory, we must make ourselves familiar with the existence of physical analogies. By a physical analogy I mean that partial similarity between the laws of one science and those of another which makes each of them illustrate the other. Thus all the mathematical sciences are founded on relations between physical laws and laws of numbers, so that the aim of exact science is to reduce the problems of nature to the determination of quantities by operations with numbers. Passing from the most universal of all analogies to a very partial one, we find the same resemblance in mathematical form between two different phenomena giving rise to a physical theory of light.

Thus, Maxwell specified, there had been two alternative models for light. That of particles in inertial motion did (and still does) account for the direction of light beams, and was long thought to be the true explanation of refraction. That of waves, on the other hand, though it carries science much farther, must not simply for that reason be confused with the truth of things. For it rests on a resemblance *"in form* only" (the italics are Maxwell's) between the laws of light and those of vibrations. Maxwell introduced the warning preliminary to cautions he addressed to attraction-at-a-distance and gravity as a model for electromagnetism. Nor can he then have foreseen to what lengths he would push comparisons. But a sound intuition had planted in this, his first important memoir, the seeds of a relation between the wave theory of light and the electromagnetism which was his subject.

For the moment, however, his concern was the continuity of those lines of force in spaces and situations where electromagnetic effects obtained, and he turned, not yet to waves, but to a more primitive manifestation of the continuum—to the behavior of a fluid so defined as to be the physical expression of mathematical relations.

He reminds us what Faraday meant by a line of force: a line passing through any point of space so that it represents the direction of the force exerted by a surface electrified positively on a particle of like sign, or on an elementary north pole. We might fill all space with such lines and thus have a geometric model indicating the direction of some force at every point. But still we should need to know its intensity, and for this we body the image into three dimensions and consider these curving forms "as fine tubes of variable section carrying an incompressible fluid." Now we can have an expression of intensity, since velocity will vary inversely as the cross-section of the tube. Moreover, electrical and magnetic forces have properties which allow us an immense simplification of the picture. We can so reduce the tubes in imaginary diameter as to leave no gaps. "The tubes will then be mere surfaces, directing the motion of a fluid filling up the whole space." And what these images express, of course, is a return from the algebraic or analytical to the palpable and the geometric imagination.

> By referring everything to the purely geometrical idea of the motion of an imaginary fluid, I hope to attain generality and precision, and to avoid the dangers arising from a premature theory professing to explain the cause of phenomena. If the results of mere speculation which I have collected are found to be of any use to experimental philosophers, in arranging and interpreting their results, they will have served their purpose, and a mature theory, in which physical facts will be physically explained, will be formed by those who by interrogating Nature herself, can obtain the only true solution of the questions which the mathematical theory suggests.

For Maxwell's was an extraordinary fluid, not at all like Carnot's caloric. "It is not even," he immediately pointed out, "a hypothetical fluid which is introduced to explain actual phenomena. It is merely a collection of imaginary properties which may be employed for establishing certain theorems in pure mathematics in a way more intelligible to many minds and more applicable to physical problems than that in which algebraic symbols alone are used." Thus, it was to resemble ordinary fluids only in its perfect flowiness and incompressibility. It was to have no mass and to be capable of inertial motion only in direction, so to say, but not in acquiring momentum. Those were its physical properties, or disproperties. Its advantages lay in its mathematical properties, for such a fluid became accessible to the mathematical operations developed for hydrodynamics. Electrical or magnetic intensity might be assimilated to the velocity of the fluid, and the lines of force to flow-lines. Any consistent system of units might apply, by so defining the tubes of flow (force) that a unit of volume passed any section in a unit of time. By assuming the unit as small as desired, one could reduce the tubes to infinitesimal proportions, fill the whole of the space under consideration with unit tubes, define thereby the motion of the whole quantity of fluid, and even ascertain the state of its motion at any given point of space. Nor need one confine one's image to the description of closed circuits. The tube might end in the boundary to the space of the experiment, in which case what lies beyond is like Carnot's inexhaustible reservoir or unfillable sink for caloric. But even this is not necessary. The fluid might be supplied or swallowed up within the space. We need suppose only a source or a sink capable of yielding or disposing of unit fluid in unit time: "There is nothing self-contradictory in the conception of these sources where the fluid is created, and sinks where it is annihilated. The properties of the fluid are at our disposal, we have made it incompressible, and now we suppose it produced from nothing at certain points and reduced to nothing at others." Sources and sinks will be assigned a value equal to the number of units which they emit or absorb in unit time, and Maxwell will in fact use the two the same way, simply assigning to the one a positive value and to the other a negative.

His fluid defined with such *sang froid*, and in greater detail than this summary indicates, Maxwell put it before the reader as the object of a mathematical analysis. He derived the relations of velocity, pressure, and work which would obtain in its uniform motion through a resisting medium. And with these he had expressions of flux for Faraday's lines of force, considered as the flow lines of his fluid. And he proceeded to apply them to major instances of electromagnetic action: static electricity, permanent magnets, induced magnetism, and electrical currents— leaving for future consideration the analysis of what he had not yet named the electromagnetic field.

Paradoxes keep rising out of Maxwell's originality of mind as if to protect his privacy from the intrusion which interpretation is bound to perpetrate. On the one hand, he took as his mission the mathematicization of Faraday's discoveries. But on the other hand, mathematicization was to him no end in itself, was not at all the desideratum of a science that it was to the continental school. "My aim," he writes after having applied his mathematical fluid to these first manifestations of electromagnetism, "has been to present the mathematical ideas to the mind in an embodied form, as systems of lines or surfaces, and not as mere symbols, which neither convey the same ideas, nor readily adapt themselves to the phenomena to be explained." For it is not to mathematics that he looks for clarity: it is to physics. He himself committed no impertinence of interpretation against Faraday, his master: "The conjecture of a philosopher so familiar with nature may sometimes be more pregnant with truth than the best established experimental law discovered by empirical inquirers, and though not bound to admit it as a physical truth, we may accept it as a new idea by which our mathematical conceptions may be rendered clearer."

Maxwell was referring to Faraday's temporary hypothesis of an "electrotonic state," some special condition into which bodies are put by the mere presence of magnets or currents. It is undetectable so long as undisturbed, but it betrays any change by the appearance of a current or a magnetic impulse, and vice versa. Faraday himself had abandoned the notion as superfluous and given his spatial hostages to fortune rather in the form of lines of force. Maxwell recurred to it in the latter part of his paper—in a much more tentative way, however, than he had handled electrostatics, magnetism, interacting closed currents, and other specific effects. His fluid no longer served, and as always when this became true of one of his models, he dropped it: "The idea of the electrotonic state . . . has not yet presented itself to my mind in such a form that its nature and properties may be clearly explained without reference to mere symbols." He did derive mathematical expressions to show that the phenomena of induction are not explained by an account of the currents alone, but that the effects presupposed a contribution from configurations in the medium.

The second of his mentors, William Thomson, had already investigated formal analogies between electrical phenomena and elasticity, and had compared the displacement from equilibrium in an incompressible solid under strain to the distribution of forces in an electrostatic system. Maxwell may have been guided, too, by the thought that the transversity of light waves required an aether mathematically identical with an elastic solid. In any case, he looked to a study of the laws of elastic solids and viscous fluids to furnish him with "a mechanical conception of this electrotonic state adapted to general reasoning." And though he did not arrive at that conception in this first paper, he did state in verbal form a first approximation to what would become Maxwell's laws of the field, once he had conceived the field more clearly in his mind and named it. One may, perhaps, quote the third of his laws as an illustration of the stage his thought had reached, "The entire magnetic intensity round the boundary of any surface measures the quantity of electric current which passes through that surface." Already he was thinking of directions and intensities in space, of the "electro-tonic state at any point of space as a quantity determinate in magnitude and direction." And it was not some lag in his mathematicization which left Maxwell dissatisfied with the state to which this paper brought his subject. He stood at the opposite pole from Lagrange, whose goal had been to free mechanics of all graphic elements, and held rather that

> The discussion of these functions would involve us in mathematical formulae, of which this paper is already too full. It is only on account of their physical importance as the mathematical expression of one of Faraday's conjectures that I have been induced to consider them at all in their present form. By a more patient consideration of their relations, and with the help of those who are engaged in physical inquiries both in this subject and in others not obviously connected with it, I hope to exhibit the theory of the electro-tonic state in a form in which all its relations may be distinctly conceived without reference to analytical calculations.

As the form did take shape in Maxwell's mind, it became, in its detail and in its ingenuity, in its fertility and in a kind of mechanistic wit, the most engaging (in every sense) of all the imaginary constructs by which Maxwell vivified mathematical analysis in his peculiar pictorial physics. He published the second of his seminal memoirs, "On Physical Lines of Force," in installments in the *Philosophical Magazine* in 1861 and 1862. There the word "field" entered into physics. Now he treated not just of the streaming of his lines, their number and direction, but of the whole state of the medium. He went beyond flux to the field and studied its electromagnetic effects considered mechanically. What are the tensions in a medium, and what the motions, which will connect magnetic attraction with electromagnetism and induction? If that can be answered, "We shall have found a theory which, if not true, can only be proved to be erroneous by experiments which will greatly enlarge our knowledge of this part of physics."

Maxwell generously indicated the course his thinking had taken since his earlier memoir. He had been meditating upon the remarkable formal analogies between the mathematical laws of attraction, conduction, and elasticity in rigid bodies—phenomena which otherwise seem to have little relation one to the other. In 1847 William Thomson had shown that electromagnetic forces might be represented as displacements of particles of a rigid solid under strain. If the angular displacement at every point in the solid is made proportional to the electromagnetic force at a comparable point in the field, then the absolute displacement of the particle will correspond to that of the electrotonic state in the field, and the displacement relative to a neighboring particle to the quantity of the electrical current. The problem, therefore, was to find a physical picture which would make sense of this mathematical assimilation of effects in an electromagnetic field to stresses and strains in an elastic solid.

Magnetism, however, presents certain features that differentiate its effects from electricity, and Maxwell began with magnetic force. These phenomena appear to exhibit two fundamental characters: first, a stress along the line of force, which must by the mutual behavior of magnets be a tension along the line, like that in a rope, rather than a pressure, as in the case of gravitational force, which acts in the opposite direction when the configuration of lines is the same. (That is to say, if we calculate the lines between two north poles and two gravitating bodies, they avoid each other and spread out into space in both cases, but the magnetic effect is repulsion and the gravitational is attraction.) So magnetism argues a tensional strain along each line of force. But it displays a second, and perhaps more interesting configuration. Magnetic tubes of force tend to contract longitudinally under axial tension, but to spread out equatorially into space. And "the next question is, what mechanical explanation can we give of this inequality of pressures in a fluid or mobile medium? The explanation which most readily occurs to the mind is that the excess of pressure in the equatorial direction arises from the centrifugal force of vortices or eddies in the medium having their axes in directions parallel to the lines of force." Moreover, when one compares magnetism to electricity, the connection "has the same mathematical form as that between certain pairs of phenomena, of which one has a *linear* and the other a *rotatory* character." Faraday had already discovered the latter quality experimentally in the magnetic rotation of the plane of polarized light. But Maxwell deferred these last comparisons to the end of his memoir, since they might lead even deeper into nature than the model under construction.

With that model, Maxwell's imagination was off into what may well be described as his neo-Cartesian fantasy— a new physics of the vortex. Now, Faraday's tubes of force became not the stream lines of a mathematical fluid, but rather a system of roller bearings filling all space. Maxwell now made each magnetic tube of force rotate on its axis instead of flow along it. Investigating the mechanical conditions of equilibrium in such a state of affairs, Maxwell showed that at every point in the medium the pressures will be different, that the direction of least pressure will follow that of the lines of force, and that the ratio of greatest and least pressures will be the square of the intensity of the force at any point. This, in turn, will yield the ordinary laws by which magnets act on currents. But this is only the beginning. Our model does not yet tell us why the vortices producing magnetism should rotate at all, or why they should be arranged to exhibit the laws of force in a magnetic field. "We have, in fact"—and this is the second part of the memoir, which applies the theory of magnetic vortices to the electric current—"We have, in fact, now come to inquire into the physical connexion of these vortices with electric currents, while we are still in doubt as to the nature of electricity, whether it is one substance, two substances, or not a substance at all, or in what way it differs from matter, and how it is connected with it." Truly, as Maxwell observed in passing, these problems are "of a higher order of difficulty." For to know why a particular distribution of vortices indicates an electric current would lead a long way toward answering a very important question indeed, namely: "What is an electric current?"

A preliminary mechanical difficulty had to be resolved. The theory required that all the vortices revolve in the same direction. But how could this be, if they lay side by side? One gear-wheel engaging another turns it in the opposite direction. And Maxwell turned to the real engineers in his extremity, and imported into the aether the device of the "idle" wheel: "The hypothesis about the vortices which I have to suggest is that a layer of particles, acting as idle wheels, is interposed between each vortex and the next, so that each vortex has a tendency to make the neighbouring vortices revolve in the same direction with itself." But Maxwell needed play in his medium and could not do with fixed axles for his aetherial idle wheels;

he pursued the engineers into even more intimate detail and adopted specifically the analogy of Siemens's governor on steam-engines, which employed idle wheels whereof the centers were capable of translation. In such a case the motion of the center is half the sum of the circumferential motion of the driving wheels on either side. "Let us examine," writes Maxwell, "the relations between the motions of our vortices and those of the layer of particles interposed as idle wheels between them."

That examination yielded other restrictions which must be imposed, chief among them that the idle tubes must be held to a thickness of one molecule, and that they must rotate between the vortical tubes of force without slippage. But perhaps this much will exemplify the capacity of Maxwell's imagination for literal detail. It makes a truly startling combination with the grandiosity of the scheme as a whole, which—it must be remembered—was no less than a mechanical explanation, in principle though not in actual fact, of magnetism and electricity. Thus, magnetism becomes the kinetic energy of rotation of the vortices of a medium filling all space; the tangential pressures called into play by the transmission of rotation from one part of the field to another then constitute electromotive force; next, the electric current becomes the translation of the particles of the medium under the influence of that force directed along the idling layer, which is turning between the vortices without developing angular momentum since it is unimolecular in depth; finally, the resistance to this translation transforms energy into heat, and is the only occasion by which electromagnetic energy is degraded in the operations of the field.

All this can be calculated. And we are justified, therefore, in our initial assumption that, "Magneto-electric phenomena are due to the existence of matter under certain conditions of motion or of pressure in every part of the magnetic field, and not to direct action at a distance between the magnets or currents." It remained to cover electrostatics with the model, which Maxwell accomplished by considering the statical electricity accumulated on condensing surfaces as the potential energy of strain stored up in physical displacements of the medium.

Thence flowed consequences far more interesting than any following out of magnetism as the kinetic energy of aethereal rotation. The first was specific and had to do with a new theory of the dielectric; the second, universal and assimilated electromagnetism to light. Throughout the entire analysis of the model, Maxwell had to assume that the substance of the medium possesses the perfect elasticity of rigid bodies. He saw conduction of the current as the flux of particles. Insulators would not, by definition, transmit that flux. But while barring the current, they did transmit electrical effects. And Maxwell hit on one of his nicest analogies in illustrating what the physicist means by a dielectric: "A conducting body may be compared to a porous membrane which opposes more or less resistance to the passage of a fluid, while a dielectric is like an elastic membrane which may be impervious to the fluid, but transmits the pressure of the fluid on one side to that on the other."

Thus, there is no flow of particles in the walls of a Leyden jar. What induction, or charging, may be thought to do is to polarize each molecule *in situ*. "We may conceive that the electricity in each molecule is so displaced that one side is rendered positively, and the other negatively electrical, but that the electricity remains entirely connected with the molecule, and does not pass from one molecule to another." Across the whole surface the effect will be, therefore, a linear displacement, which though it is not a current, and is indeed rather a sudden strain, may nevertheless be considered as the beginning of a current and be analyzed accordingly: "We cannot help regarding the phenomena as those of an elastic body, yielding to a pressure, and recovering its form when the pressure is removed." And of all Maxwell's figures, this displacement current in a dielectric was the most puzzling to his contemporaries. It was as if he had tried to blur the clear distinction which allowed them to say categorically when a current is not a current. It did not help towards clarity that in later discussion he abandoned the physical lurch of the electrified molecule into a position of polarized strain and substituted an abstract change in the structure of aether. It forced, or would have forced had it been accepted, a change in the conception of condensers. Instead of considering their coatings as termini where currents ended and charge accumulated, it would have closed all currents through the body of the dielectric. But important though that is for sophisticated thinking about the nature of electrical energy, it was the second consequence of his model as an electrostatic medium which (for a welcome once) combined drama with depth.

Maxwell had seen immediately that the wave theory of light required of a medium just the same sort of elasticity as did his field theory of electromagnetism. The equations took the same form. It was possible from his elastic-solid model of the medium to compute the velocity with which a shear wave would be propagated, which on the analogy is the same as an electromagnetic wave. In 1856 two German physicists, Weber and Kohlrausch, working on quite a different hypothesis, had determined the velocity with which an electrical impulse will be propagated along a wire. They were concerned with the proportionality of units between electrodynamics and electrostatics. That factor must have the dimensions length over time, which is to say of velocity, since electrostatic repulsion is the same type of quantity as electrodynamic repulsion between two lengths of wire carrying a certain charge in a certain time. They measured the velocity experimentally and found the value to be almost 3.1×10^{10} centimeters per second. But in 1849 Fizeau had already refined the determination of the speed of light by rotating a cut-out perforated wheel at such a speed that a ray of light passed through a gap between the teeth was reflected and arrested on the return by one of the teeth. The dimensions and angular velocity permitted calculation of the speed of light. His figure (an improvement on that known from eclipses of the moons of Jupiter) was 3.15×10^{10} centimeters per second.

"The velocity," wrote Maxwell, "of transverse undulations in our hypothetical medium, calculated from the electromagnetic experiments of MM. Kohlrausch and Weber, agrees so exactly with the velocity of light calculated from the optical experiments of M. Fizeau, that we can scarcely avoid the inference that *light consists in the transverse undulations of the same medium which is the cause of electric and magnetic phenomena.*"

Now it is not only the form of equations which are the same. The italics are Maxwell's—and well they might be!

Finally, in his third and definitive paper, "A Dynamical Theory of the Electromagnetic Field," Maxwell sloughed off all this structure of imaginary mechanical detail. But out of sight was not out of mind, either for him or for his science, for the crucial terms and the ideas they express remain: the electromagnetic flux from his first paper, displacements and fields from the second, and from the third the very conception of a marriage of electricity and dynamics in the theory of the medium. Thus electrodynamics remains the science of the energy of the field: "The theory I propose may therefore be called a theory of the *Electromagnetic Field,* because it has to do with the space in the neighbourhood of the electric or magnetic bodies, and it may be called a *Dynamical Theory*, because it assumes that in that space there is matter in motion, by which the observed electromagnetic phenomena are produced." And though he abandoned all specific models, he must not be supposed to have abandoned the principle of mechanism. That is to say, electromagnetic phenomena are clearly expressions of motions of some kind, causally interrelated and communicated by forces. They were subject, therefore, to the general laws of dynamics, and specifically served the conservation of energy. Only gravity escaped the unification which Maxwell's Laws of the field imposed on magnetism, electricity, and light. It still seemed to act at a distance between like bodies. Nor was Maxwell optimistic: "As I am unable to understand in what way a medium can possess such properties, I cannot go any further in this direction in searching for the cause of gravitation."

Whoever understands those laws will already know them as familiarly as the law of falling bodies. Whoever does not, will scarcely appreciate their symmetry. Still, there is an astringency about the mere appearance of the field equations which may be salutary as the distillation of an imagination so various as Maxwell's

1) $\operatorname{div} E = 0$ 3) $\operatorname{div} H = 0$

2) $\operatorname{curl} E = -\dfrac{1}{c}\dfrac{dH}{dt}$ 4) $\operatorname{curl} H = \dfrac{1}{c}\dfrac{dE}{dt},$

where E is the intensity of the electrical field; H, of the magnetic field; t is time; c, the velocity of light; and div and curl, certain mathematical operations in vectorial analysis, i.e. a way of specifying a directional aspect in the effective quantities.

Einstein once observed that the statement of those laws had been the most important event in physics since Newton. And from having followed the route of Maxwell's mind we may, perhaps, appreciate what it is that these laws do. They connect in every respect the amount, the flux, and the interrelation of an electrical field with the magnetism it induces, and vice versa. The velocity of propagation appears as a constant, time as a parameter, and space by virtue of the definition of the field. These equations connect phenomena in a different way from Newton's. Newton's laws conserve energy in the motions of systems of bodies and take no account of the space in between; Maxwell's laws conserve energy in its gradations in space and take no account of bodies as such.

Nevertheless, Maxwell had pressed himself more closely into the problem of describing electromagnetic effects than relativity theorists have supposed who would represent his attention spread out over all space, or his soul committed to the aether. His equations in the form just given are abstracted into free space, or free aether, where there are fields but no charges or currents. In fact, he was much concerned with both, and his own expression of his equations is correspondingly less laconic. A surprise may await the physicist who would pursue Maxwell's laws to their first statement. The vectorial notation does not appear. In another memoir, however, Maxwell did coin the wonderfully graphic term "curl" for the operation that mathematicizes the behavior of lines of force wrapping around a wire. It is like Newton inventing "fluxions" out of the necessities of his physical ideas, and then not using the calculus to explain them. Instead, therefore, of the four equations now usually given as his laws, there are twenty. Figuring therein are as many variables which he called electromagnetic momentum, magnetic intensity, electromotive force, true current, displacement current, total current—each of which has three components along the x, y, and z axes—and finally two undirected (or scalar) quantities, free electricity and electric potential. For the relations among any of these quantities and those of the others which might concern a particular problem, were what interested Maxwell. He was concerned with conservation of energy, not with aether or the structure of space. He might, indeed he would, abstract with the greatest generality from particular mechanical images, and warn that he used terms like momentum and elasticity in an illustrative sense only.

> In speaking of the Energy of the field, however, I wish to be understood literally. All energy is the same as mechanical energy, whether it exists in the form of motion or in that of elasticity, or in any other form. The energy in electromagnetic phenomena is mechanical energy. The only question is, Where does it reside? On the old theories, it resides in the electrified bodies, conducting circuits, and magnets, in the form of an unknown quality called potential energy, or the power of producing certain effects at a distance. On our theory it resides in the electromagnetic field, in the space surrounding the electrified and magnetic bodies, as well as in those bodies themselves, and is in two different forms, which may be described without hypothesis as magnetic polarization and electric polarization, or, according to a very probable hypothesis, as the motion and strain of one and the same medium.

For Maxwell was a nineteenth-century physicist, after all, and the prince of nineteenth-century physicists.

❖

But field physics tells only half the story, nor has physics yet decided which was the more important half. "Modern physics," wrote Max Planck in the centennial symposium, *Clerk Maxwell, 1831-1931*, "recognized two main conceptual schemes, the physics of discrete particles and the physics of continuous media, and it is since Maxwell's time that the distinction between them first became more apparent. These schemes correspond nearly but not quite to the physics of Matter and the physics of the Aether. In both regions Maxwell introduced new and fruitful ideas." Planck did not specify the exceptions contained in that "not quite." Perhaps he had in mind that Maxwell's kinetics arrested the anti-mechanistic tenor of energetics, hitherto the physics rather of force than of matter, by reducing thermodynamics to the behavior of particles. Maxwell himself was a better historian of science than Planck—as a scientist, indeed, he took the subject very seriously. He knew how old the issue was, and how modern, and once lectured to the British Association on the subject of molecules:

> The mind of man has perplexed itself with many hard questions. Is space infinite, and in what sense? Is the material world infinite in extent, and are all places within that extent equally full of matter? Do atoms exist, or is matter infinitely divisible?
>
> The discussion of questions of this kind has been going on ever since men began to reason, and to each of us, as soon as we obtain the use of our faculties, the same old questions arise as fresh as ever. They form as essential a part of the science of the nineteenth century of our era, as of that of the fifth century before it.

In the nineteenth century the discussion carried reason into the laboratory to deal with atoms, leaving the geometers' imagination outside to deal with continuity. For particle physics has been peculiarly associated with the laboratory tradition, to which Maxwell equally belongs by virtue of having planted his feet so firmly in both camps. This is a corporate tradition, a little sectarian perhaps, but not lonely like that of mathematical idealism, into which a Descartes or an Einstein withdraws to take thought. Maxwell, indeed, was the founder of that laboratory which was to the second scientific revolution in physics almost what Padua had been to the first—the

Cavendish Laboratory in Cambridge. There he held the first chair of experimental physics in the university.

A Cambridge man himself, he had gone back to Scotland as professor of natural philosophy at Aberdeen, and thence to King's College, London, in 1860. His own most creative work occupied the five years he spent in London. As a teacher of general physics he was not a success. Ideas would occur as he lectured to divert him from the path he had intended. Bending over their notebooks, his students could not follow. He retired to his family property in 1865. It is a pity that the Cambridge post was not established until 1871, nor the laboratory under way until 1874, because Maxwell died in 1879 at the age of forty-nine; and it was only as the director of a laboratory that he came into his own in the communication of his science. There he was at his best, in the inner circle atmosphere which a great laboratory creates. His charm and his wit, his quizzical turn of mind, needed intimacy to flourish. His verses still form part of the oral tradition of physics, the personal lore which physicists hand on from one generation to the next in a kind of pale Pythagoreanism—pale but a little heavy, for to the outsider the verses do not seem very good.

The quizzical quality emerges in his scientific work, in the fun he evidently had with the mechanistic detail of his electrodynamic aether models, in experiments to determine how a cat lands on its feet, and in turns of phrase: "The top which I have the honour to spin before the Society . . ."—so he introduces a paper on "the motions of a body of invariable form about a fixed point, with some suggestions as to the Earth's motion." His first paper on particle physics, the paper which won him reputation, was a dynamical study of the conditions of equilibrium in Saturn's rings. From there to the kinetic theory of gases was a reduction in scale, but not in importance. For

Maxwell imported into molecular science that technique of analysis upon which it has thriven ever since, statistical mechanics. He resigned himself to statements of probability, and made the most of this inevitable rape upon the reason.

The kinetic theory of gases started in the attempt to extend dynamical considerations over phenomena which normally preoccupied the chemists. What we mean by explanation, Maxwell pointed out in an address "On the Dynamical Evidence of the Molecular Constitution of Bodies," is just that we should extend over some set of phenomena a principle applicable to others. No chemist would have presumed that the various geometrical constructions he imagined for his compound molecules were anything more than a symbolic shorthand for the physical relations of the constituents. Most physical theories had been constructed by just such methods of hypothesis— calculating what would happen if some hypothesis were true, and then comparing to actual events. The difficulty in chemistry, as in most branches of physics at similar stages, was that speculators lacked terms or methods of sufficient generality at the early stages of induction. But help was at hand: "In the meantime the mathematicians, guided by that instinct which teaches them to store up for others the irrepressible secretions of their own minds, had developed with the utmost generality the dynamical theory of a material system." And surely the most unobjectionable hypothesis about the constitution of bodies is that which assumes no more than that they are material systems serving the laws of mechanics, the most general yet established. But only in the case of gases can we reasonably expect the particles of matter to enjoy sufficient freedom from undetectable and indistinguishable relations to reward dynamical analysis.

The first kinetic model of gases was imagined by Daniel Bernoulli in a treatise of *Hydrodynamica* in 1738. Applying the corpuscular philosophy literally enough to confined airs, he considered pressure as the effect of bombardment of the container by particles travelling in paths resolved normally to the sides. In the early nineteenth century one Le Sage, another Genevan, radically enlarged the model to cover the entire universe, and in *Lucrèce newtonien* he imagined gravity to be caused by the impact of atoms streaming at random through infinite space. If there were only one body in the universe, there would be no effect. But since there are more, each masks the others from a certain proportion of the corpuscular bombardment, and the effect is a diminution of pressure between bodies which gives the appearance of gravitational attraction. Maxwell devoted some pages to this fancy in his article "Atom" in the ninth edition of the *Encyclopedia Britannica*; nor though he saw the grave objections, chiefly that it fails to permit a steady state of temperature in bodies undergoing bombardment, did he read it right out of court.

A little-studied English physicist of the same generation, John Herapath, recurred during his lifetime to a favorite hypothesis which made Boyle's law a consequence of molecular kinetics. He followed Bernoulli in making the pressure on any face of a cubical container equal to $\frac{1}{3} MV^2$. He also assumed, less fortunately, that temperature is measured by the momentum of the particles. This gave him pressure as proportional to the square of the absolute temperature, so that his computation of absolute zero, the rate of diffusion, the value of Avogadro's number, and the relation of specific heat to atomic weight would have been wrong even if based on sufficient data. But it was a brave attempt, and permitted computation in cases where temperature was held constant. For example, Herapath gave the relation

$$p = 1/3 \, \rho \, v^2,$$

where p is pressure and ρ density, from which Joule calculated in 1848 the average velocity of the hydrogen molecule as 6,055 feet per second at 32 ° F and atmospheric pressure.

Clausius took up the subject in 1857, and so deepened what had hitherto been vulnerable to criticism as the occasional resort of the mechanistic enthusiast or crank that kinetics carried off for a time the main stream of thermodynamic reasoning. His paper "On the Kind of Motion We Call Heat" inaugurated the dynamical study of gases as an actual program of analysis and not just an assertion of particulate policy. He grappled with the problem that others had evaded in their postulates of ideal gases, consisting of point-masses. Arguing that the total energy of a gas must be divided between the various modes of motion of its molecules, he developed a partition expression for that portion of the total kinetic energy due to the motion of molecules in translation and that due to their spin about their own axes. This explained why the specific heat of gases is greater than what had been predicted so long as all the heat, like all the pressure, was supposed to manifest translational momentum. The pressure, too, would be affected by whatever mutual attractions or repulsions molecules might exert over one another. And on purely analytical grounds Clausius developed approximations to diminish these discrepancies.

In a second paper of 1859 Clausius addressed himself to a further embarrassment with which experience confronted the kinetic theory of gases. Gases do not diffuse with nearly the velocity which the theory expected of the average molecules. The deficiency Clausius attributed to the short path each can follow before it hits another and is deflected in some new direction. Thus, as in a crowd dispersing, each individual bounces this way and that, making much less time than if he could step straight out, and the rate of progress is a function both of velocity and the mean free path.

With these refinements, theoretical predictions of the pressure, volume, density, and temperature relations of actual gases came closer to laboratory measurements than before, but agreement was still far from reliable; and it was at this point that Maxwell stepped into the kinetic scene with a new and a portentous analytical idea, which he had employed in the management of his computations of the motion of particles in the rings of Saturn. Clausius and the others had considered the molecules of a gas as if all moved at the same velocity. It was, of course, an unreal assumption, but Clausius seems to have seen no alternative except the desperate, indeed the impossible, expedient of following every molecule throughout its whole career in the experiment. Maxwell published his first paper on the subject, "Illustrations of the Dynamical Theory of Gases," in 1860, the year before his second paper on lines of force. They both display the same imaginative use of models. Clausius had determined no parameters for his algebraic expression relating the mean free path to the distance between the centers of mass of particles at collision. It seemed to Maxwell that experimental data for the viscosity, heat conductivity, and diffusion of gases would permit such computations: "In order to lay the foundation of such investigations on strict mechanical principles, I shall demonstrate the laws of motion of an indefinite number of small, hard, and perfectly elastic spheres acting on one another only during impact." And if the results turned out to be consistent with known gas laws, then they might lead in to more accurate knowledge of the constitution of matter: they might, in other words, bear upon the fundamental discussion of atomism.

But first he must grapple with the distribution of velocities, and to this end Maxwell introduced into a fundamental physical problem the calculus of probability, that
mode of analysis which had hitherto seemed natural when
applied to games or affairs, but an admission of our inability to know nature when things were the object of
study. Not that Maxwell himself regarded its rigor as
equal to that of strict dynamics, but no other method
presented itself, either experimental or mathematical.
Thus, Maxwell inaugurated in this memoir the science of
statistical mechanics. Except under very rare conditions of
impact, collisions would be bound to alter the velocity of
particular molecules. But in a steady state, distribution of
velocities from zero to infinity must follow some definite
law. One might, therefore, describe the system by computing what portion of the molecules would be at each velocity.

Though he did not then draw the famous bell-shaped
curve, and though the later refinements were serious by
which Boltzmann would win the honor of co-namesake,
nevertheless Maxwell did ascertain that distribution in
principle. And the most interesting thing about the pattern is that it was not new. The velocities might be plotted
on a diagram in which each is measured by a vector drawn
from any fixed point. Then if a dot be recorded at the
far end of each vector, the same diagram could represent
bullet-holes spattered around the bull's eye of a target, or
an astronomer's record of successive observations of the
same fixed star. It was Gauss's law of error, turning up
thus unexpectedly in the hitherto deterministic physics of
particles.

In retrospect, the method will seem the most significant
contribution of this paper, but the immediate results were
startling enough to speak for its value. With the results
of experiments on air by Stokes, Maxwell's first kinetic
equations gave $\frac{1}{447,000}$ of an inch as the mean free path,

and an average of 8,077,200,000 impacts per particle per second at standard conditions. As a surprising consequence of one equation, viscosity turned out to be independent of density in a gas—a result which no experiment then confirmed, but which did establish itself as the theory took on sophistication and refinement. But for Maxwell perhaps the most encouraging success was his confirmation of Gay-Lussac's law of combining volumes. Early in the nineteenth century Gay-Lussac had argued on strictly chemical grounds that different gases at the same temperature and pressure must contain the same number of molecules in unit volume. Now Maxwell found the identical result by a kinetic consideration of the simple diffusion of one gas into another. When diffusion is complete, the average kinetic energy is the same for a particular molecule in either gas. This will also be a state of equal temperature. It follows that the condition of equal temperature is that the average kinetic energy of a single molecule should be the same. It had been established that the pressure of a gas is given by two-thirds the kinetic energy in unit volume. Hence, if pressure and temperature are the same, then total kinetic energy as well as average kinetic energy of each molecule are equal, and hence unit volume must contain equal numbers of molecules. It made a precise example of what Maxwell meant by explanation, gathering chemical and dynamical considerations under the one principle of atomism, and thus strengthening the principle in these mutual supports.

The kinetic theory of gases remained far from perfect. Its great difficulty, which it did not overcome in Maxwell's lifetime, was to divide the total energy of the molecules between translation, rotation, and the various modes of vibration of which the atoms might be capable. Pressure had to depend on the kinetic energy of translation alone. Specific heat would depend on the rate at which total energy would rise with temperature. Theoretically the ratio of specific heats at constant pressure and constant volume would yield the ratio of the increments of the whole energy to that of translation. But except for the infrequent monatomic gases, the ratio predicted by theory was always too high for the observed value. The more complicated and various the modes of vibration which do occur in the poly-atomic molecule, the worse the disagreement with gases, which obstinately refused to absorb as much heat as they seemed to need.

In his first paper Maxwell proved that the known relation between specific heats ruled out the assumption (with which he had himself begun) of hard, spherical particles. And in 1866 Maxwell published his greatest memoir on the subject, "On the Dynamical Theory of Gases." Therein he abandoned his model to the facts as cheerfully as when he discarded the mathematical and dynamical fluids on which he had forged his field equations. In lieu of the description of the molecule as minuscule, impenetrable, elastic, round, and free flying, he substituted other mathematically more convenient characteristics. He had confirmed by experiments of his own the most surprising, if not the most seminal, consequence of his earlier formulations—that viscosity is independent of density. Now he invented a molecule that would yield that fact with a certain facility.

This idea too he may have owed to Faraday, for his second paper fuzzes the effective molecule from an impacting sphere into a body which exerts intermolecular forces to some certain distance beyond its physical surface. Since it is still a question of molecular motion, the force must be repulsive. And he makes the new molecule one that repels its kind with a force inversely as the fifth power of the distance between them. The advantage is mathematical. Maxwell needed to assume no particular conditions in molecular collisions beyond the conservation of energy and momentum. In particular, he need not concern himself with the distribution of velocities in a gas which, instead of being in a steady state, is streaming or diffusing. Assuming billiard-ball molecules, Boltzmann labored for years over that intractable problem, employing one horrible approximation after another. Maxwell simply abolished the difficulty by altering the nature of the encounter. With a repulsive force of high enough inverse power, molecules at any considerable distance from each other will move at almost constant velocity. Only when they come close do they change speed or direction. But if we are concerned with conservation and not with summing up impacts, we need not know how much of the total velocity pertained to each. With a fifth-power law, the closest approach in a head-on collision is inversely proportional to the fourth root of the relative velocity; and the relative velocity (which is what no man could determine or postulate in a streaming gas) disappeared from the final equation of viscosity. So Maxwell used a fifth-power law, and then compared the consequences to nature. Boltzmann was overwhelmed with magniloquent admiration, and Max Planck's centennial essay quotes his Wagnerian account of the impression which the memoir made on him:

At first are developed majestically the Variations of the Velocities, then from one side enter the Equations of State, from the other the Equations of Motion in a Central Field; ever higher sweeps the chaos of Formulae; suddenly are heard the four words: "put $n = 5$." The evil spirit V (the relative velocity of two molecules) vanishes and the dominating figure in the bass is suddenly silent; that which had seemed insuperable being overcome as if by a magic stroke. There is no time to say why this or why that substitution was made; who cannot sense this should lay the book aside, for Maxwell is no writer of programme music, who is obliged to set the explanation over the score. Result after result is given by the pliant formulae till, as unexpected climax, comes the Heat Equilibrium of a heavy gas; the curtain then drops.

In fact another curtain was going up on the physics of particles, but we cannot stay to watch that play, which is still in progress. It will be enough to identify the character with which Maxwell had at the end invested the protagonists in the drama of classical physics, thus drawing to its close. Maxwell's influence throughout was nothing if not unexpected, and nowhere more than in the turn which he gave to the dialogue between atoms and the continuum. For he switched the roles. On the one hand, he took electromagnetism out of the domain of mechanistic formalism, and paired or even united it with the new optics in the physics of the continuum. On the other hand, he retrieved thermodynamics from the *Schwärmerei* of energetics, and turned the study of energy into a special case of the dynamics of particles of matter in motion.

In effect, therefore, Maxwell blocked the escape from mechanics into energetics. The astringency of his physical wit nowhere appears to more characteristic advantage than in his invention of a daemon to deflate the pretensions of the second law as a mystique of nature. In imagination he conjured up a box with two compartments, one containing a hot gas and the other a cold one. The partition contains a trap door, manned by a daemon whose senses and responses are of the same order of fineness and quickness as the darting molecules. Whenever a fast molecule approaches the door from the colder side, he might flip it open and let it through, and thus he would make heat pass from a cold body to a hot one. For it was clear to Maxwell that the second law had a status in science quite different from the first. The conservation of energy was his cornerstone, embedded in the nature of things. But the second law is statistical, expressing (as Maxwell thought) rather our incapacities to know or act than the nature of physical reality. Though he inaugurated statistical mechanics, he was still too close to the eighteenth century and the Enlightenment, still too far from the twentieth and positivism, to suppose that the order of things might itself be an order of chance, or that (like the second law) all science is about itself and its own operations. Maxwell could not foresee the predicament of the physicist as himself a daemon, intruding willy-nilly into nature, though without the compensating advantage of choice.

Well versed in the philosophical history of science, he knew the traditions which the two aspects of his physics carried on. "There are thus," he wrote in his paper on "Atoms" for the *Britannica,*

> two modes of thinking about the constitution of bodies, which have had their adherents both in ancient and in modern times. They correspond to the two methods of regarding quantity—the arithmetical and the geometrical. To the atomist the true method of estimating the quantity of matter in a body is to count the atoms in it. The void spaces between the atoms count for nothing. To those who identify matter with extension, the volume of space occupied by a body is the only measure of the quantity in it.

As for Maxwell, his atomism saved his own scientific soul alive from the Cartesian fallacy into which his mathematicization of Faraday's electromagnetic medium might otherwise so easily have led him. The opening sentences of the great memoir, "On the Dynamical Theory of Gases," clearly distinguish between the functions of geometry, concerned with extension, and the advantage of atomism, which allows science its grasp on the properties of bodies:

> Theories of the constitution of bodies suppose them either to be continuous and homogeneous, or to be composed of a finite number of distinct particles or molecules.

In certain applications of mathematics to physical questions, it is convenient to suppose bodies homogeneous in order to make the quantity of matter in each differential element a function of the co-ordinates, but I am not aware that any theory of this kind has been proposed to account for the different properties of bodies. Indeed the properties of a body supposed to be a uniform *plenum* may be affirmed dogmatically, but cannot be explained mathematically.

Nevertheless, Maxwell was not without hope of resolving the difference and arriving at a unification of aether and atoms, field physics and molecular physics. In the article on atoms, wherein he allowed his fancy freer reign, he explored a speculation which Sir William Thomson (Lord Kelvin to be) had spun out of certain mathematical investigations by Helmholtz into the theory of rotational motion in fluids. The discussion presents itself as a kind of aethereal hydrodynamics. He invests his fluid with convenient physical properties. It is to be homogeneous, incompressible, and devoid of viscosity. Its motion is to be continuous in space and time. It is, in fact, to be very like Maxwell's first mathematical fluid, except that it is a material substance. Nor need we worry whether such a fluid exists, so long as its properties are mutually consistent. Of these, the most significant is that such a fluid cannot be molecular. If it were, the volume would increase discontinuously with mass in any container being filled. And Helmholtz proved by mathematical analysis that when rotational motion is set up in such a fluid, the vortex lines are permanent and always contain the same points. Moreover, if a tube be formed bounded by a surface through which run vortex lines, such a tube must return into itself (unless it is infinite) and may be called a ring. These rings are invariant in volume, conservative in rotation, and discrete in identity within the fluid, though continuous in extension throughout it. They display, in other words, just those properties which are requisite in atoms: constant magnitude, capability of internal motion, and sufficient variety in characteristics of shape or size so that they might explain the properties of bodies. Such a medium would combine heterogeneity in its motion with homogeneity and continuity in its density and extensive properties. Might it not, fancied Maxwell, unify a physics still shifting between atoms and the void, now concerned with kinetics and again with fields?

> The disciple of Lucretius may cut and carve his solid atoms in the hope of getting them to combine into worlds; the followers of Boscovich may imagine new laws of force to meet the requirements of each new phenomenon; but he who dares to plant his feet in the path opened up by Helmholtz and Thomson has no such resources. His primitive fluid has no other properties than inertia, invariable density, and perfect mobility, and the method by which the motion of this fluid is to be traced is pure mathematical analysis. The difficulties of this method are enormous, but the glory of surmounting them would be unique.

Maxwell never quite said whether this fluid in which the vortex rings are atoms is the aether; but it is certainly very like the aether, as indeed it is in its mathematical properties very like the field-fluids of his papers on the lines of force. We do not, in any case, go beyond the evidence in seeing how the latter evolved in his scientific consciousness into the aether which filled all space. And historically at least, it is important to press the point a certain way. For it permits us to distinguish between Newton's aether and Maxwell's both in form and function. Newton's aether (to recall that argument) was hypothetical and ancillary to his physics, which would have suffered only in intelligibility but not in structure were it away. Not so Maxwell's, whose unification of light and electromagnetism in field theory worked the aether right into the bone and texture of classical physics. In Newton, space is the arena of physics, the housing of reality, and aether is only what he imagines to fill it. The distinction will impose itself when it is remembered that space in Newton is continuous Euclidean extension, but that aether is particulate.

Maxwell's aether has changed its texture. It is not particulate, not corpuscular, not atomic. The findings of kinetics alone would rule that out. "A molecular aether would be neither more nor less than a gas," he pointed out in a lecture on the "Molecular Constitution of Bodies." It would be subject to the gas laws. "Its presence, therefore, could not fail to be detected in our experiments on specific heat, and we may therefore assert that the constitution of the aether is not molecular." Its presence might, of course, be detected in another way, consistent with its function as the continuous medium filling all space and transmitting transverse vibrations of electromagnetic and luminous energy. And Maxwell evoked the image of the "great ocean of aether." Do dense bodies drag some portion of it with them in their passage? Or does aether simply stream through them, "as the water of the sea passes through the meshes of a net when it is towed along by a boat"? If it were possible to compare the velocities of light in opposite directions between two points on earth, that question might be answered. Unfortunately the theory predicted a difference of only about one hundred millionth of the time of transmission—too small to be detected. Maxwell had himself designed and performed a spectroscopic experiment by which he hoped to find a difference in the deviation of a ray from a distant star according as it traversed a prism in the direction of the earth's motion, and perpendicular thereto. The results were negative. But physics was only at the most preliminary state of its investigations of the theory of the aether, which would require unparalleled delicacy in conception as in execution. And near the time of his death Maxwell could say with confidence only this:

> Whatever difficulties we may have in forming a consistent idea of the constitution of the aether, there can be no doubt that the interplanetary and interstellar spaces are not empty, but are occupied by a material substance or body, which is certainly the largest, and probably the most uniform body of which we have any knowledge.

And in practice Maxwell's physics went further than he knew, all discernment though he was. For both in form and function—continuous in the one and energetic in the other—so full was Maxwell's space of aether, that no physical distinction might be saved between them. Any question asked about the one would implicate the other. And it would take a more critical, a more abstract Cartesianism than that of Thomson's vortex molecules to restore reason to this science, or vice versa. It is easy for us, wise after the event (and it will grow easier as elementary physics courses are revised and these ideas are taught to children) to see how the complications which all these different requirements were introducing into the aether were giving it the character of circles before Kepler, of light before Newton, of phlogiston before Lavoisier, and of adaptation before Darwin. It was time for a new look.

But that is to anticipate, and nothing is more unhistorical or unfair to those who went before. And it will be just, therefore, and in keeping with the duality of Maxwell's own creation, to leave him as the ultimate impresario of classical physics, who brought the chief characters, the atom and the aether, to the center of the stage, and there left them all exposed to the winds of criticism blowing up out of positivism.

CHAPTER XI

EPILOGUE

IN 1885 HEINRICH HERTZ devised an electrical oscillator in his laboratory at Karlsruhe and set up certain vibrations which were not visible, but which did exhibit the other properties of light as to frequency, refrangibility, interference, polarization, and so on, and which were later found to travel at the same velocity. Radio waves, therefore, passed Maxwell's prediction of the identity of light and electromagnetism triumphantly through the trial by which a theory stands or falls. It was as decisive an experimental victory as any in the history of science, and more strategic than most, for it forced physics to assume a new and different posture and to confront seriously the preoccupations with space and aether which Faraday had initiated and Maxwell had formulated. Hitherto, indeed, physicists had adopted an attitude to field theory compounded of bewilderment at the turns taken by Maxwell's imagination and disdain of the detailed naïveté exhibited in the models created and destroyed in his imagination. Disdain was the response of the French school. Pierre Duhem wrote an entire treatise on Maxwell's theories of light and electromagnetism wherein the reader may learn (as Duhem says in *The Aim and Structure of Physical Theory*), "to what degree the lack of concern for all logic and even for any mathematical exactitude went in Maxwell's mind. . . ." And Duhem quotes Poincaré's *Science and Hypothesis* on how it was that, "The first time a French reader opens Maxwell's book a feeling of discomfort, and often even of distrust, is at first mingled with his admiration. . . ." And it may have been the penalty attaching to the perpetual Cartesianism of French science in its critical provincialism that it thus rose above the battle of the laboratory, and looked down in diminishing vigor upon Anglo-Saxon handiness, of the earth earthy.

It is rather German taste which relishes bewilderment and builds upon it. The very obscurities of Maxwell's views drew Hertz and Boltzmann into their confirmation. Hertz was twenty-eight when he succeeded in detecting radio waves. Because of his early death, his was the last great career to transpire entirely within the confines of classical thought. Of the same generation, Boltzmann resolved the difficulties in Maxwell's kinetic approach to gases, and deduced the laws of thermodynamics by means of a statistical method. To carry the stories of fields and particles beyond Maxwell would require either a higher mathematical competence than this history has so far exacted of the reader (and the author), or else a wider departure from the texts than has been the policy by which the book has been composed. Nevertheless, a history of the ideas of classical science would be incomplete without an epilogue to indicate how the great themes had in Maxwell not only their climax but their turning; how when the last knots were pulled tight, the great web of Newtonian realism all unexpectedly unravelled; what it was (in short) that Albert Einstein had in mind when he recalled that during his student days Maxwell's theory had been the most fascinating subject in physics.

No doubt this second scientific revolution, which like other revolutions of our time begins to seem open-ended in its permanence, has been technical as well as intellectual and philosophical in genesis and content. Technical triumphs forced the pace: Willard Gibbs's reformation of physical chemistry in the guise of mechanistic thermodynamics, H. A. Lorentz's capture by prediction of the individual electron, Konrad Roentgen's recognition of X-rays, the Curies' laborious isolation of radium, Max Planck's deduction of a discontinuous element in the radiation of energy and his calculation of the constant of action, the creation by all hands of an anatomy (if not yet a physiology) of the atom itself. And it is with no wish to undervalue these fundamental factors that attention is here directed instead to the sector of general ideas and assumptions about the relation of science to the way the world is made, and specifically to the renaissance of positivism at the end of the nineteenth century. For certainly the criticism which its leading adepts among scientists—Pierre Duhem in France, Wilhelm Ostwald in Germany, and most notably, Ernst Mach in Vienna—directed, first against domination by mechanics, and then against fundamental Newtonian assumptions about the structure of reality, this above all was what nourished the transformation of classical science with ideas.

Positivism was an essential element, moreover, destructively at least. Constructively it did not quite suffice, in its phenomenalism and radical despair of approaching theory to reality. Indeed, in no episode in intellectual history does the errancy of philosophy mingle more intimately with its indispensability. For at the start of the century the positivists repudiated with a certain violence the physics of particles—precisely that physics in which their doctrine has since gone from strength to strength, what with quantum mechanics, probability, and uncertainty. And on the other hand, positivism inspired the train of thought which led Einstein to the special theory of relativity—that same epistemology which he for his part repudiated in later life as the besetting abdication of reason in his science. To reach the general theory, Einstein went beyond Mach into an ontology of his own creation, rising into higher regions where the real merged with the ideal in the bracing atmosphere of (non-Euclidean) geometry.

What is less clear is the philosophical situation in which positivism itself revived. Auguste Comte coined the word early in the nineteenth century. Comte retrieved the essential elements of Condillac's philosophy of science from the somewhat sterile verbalism of the *idéologues*, those last spokesmen of the Enlightenment who, having outlived it, continued to write of reason amid the vast unreason of revolutionary and Napoleonic Europe. Merging Condillac's educationism into historicism, Comte converted sociology from the science to the engineering of humanity. Since history then replaced nature as the norm, Comte had to do what no Condillacian had done and repudiate not only metaphysics but also ontology. Thus would he deprive science of any and every claim to deal with objective reality or with any truth deeper than consistency or efficacy. He would know in order to predict, and predict in order to control, and such was the program of positivism. In Comte it became a social gospel. For all his extravagances Comte was the founder of his school, and it is extraordinary that he is held in so little honor among his successors, however repellent the secular religiosity of his later years, however alarming the authoritarianism of his technocracy, and however tedious the prolixity of his literary style. But it must be left to the historians of philosophy to identify the missing links between Paris and Vienna, between Comte and Mach, across the middle decades of the century when Hegel, Spencer, and the great spinners of systems enjoyed the renaissance of metaphysics.

The problem invites inquiry, for it appears that the relation between the protagonists who revived positivism in the decades before 1900 was rather one of resonance than derivation. And the historian of science is likely to experience a fellow feeling for this generation which is denied him in the work of their successors, the logical positivists among his own contemporaries. For unlike these latter, who ignore the development of science in favor of its method and logic, Duhem, Ostwald, and Mach were historically minded. Not perhaps in quite the historian's own sense—they would study the history of scientific doctrine not for its own sake, as a record of culture and creation, but rather didactically, in order to forge of it an instrument of criticism of current science (or perhaps of current scientific error, since there lived on in them something of the attitude to history of the Enlightenment which was their ultimate philosophical inspiration). Mach's *Science of Mechanics in Its Historical Development* remains a classic to be reckoned with by every student of the history or philosophy of science. That book, said Einstein of his own youth, first shook him in the "dogmatic faith" of physicists in mechanics as the foundation of their science.

Duhem, for his part, belongs not only to the philosophy, but perhaps primarily to the history of science, and indeed to science itself for his work in thermodynamics. His multi-volume history of cosmology from Thalès to Kepler inaugurated the modern historiography of science, and his three volumes of essays on the predecessors of Leonardo da Vinci in mechanics established the vitality of medieval thought on that subject by tracing the tradition which Galileo consummated back to the fourteenth-century school of Paris. In contrast to Mach, Duhem's scholarship appears to better advantage in his truly imposing erudition, which was often proof against his philosophy, than in his criticism, where *parti pris* too often got the better of judgment. Ostwald, finally, interested himself more in the intellectual personalities and thought processes of the great moderns of science than in the remote reaches of culture. His set of biographical essays, *Grosse Männer*, is a kind of Plutarch's Lives of Modern Science. The collection of great texts which he edited under the rubric *Ostwalds Klassiker* still serves the scholar and the student very well. In short, the historian of science is bound to appreciate the service which the positivists rendered in this, their heroic generation, through their perception of current science as the surface of its past—even though he may feel equally bound to deliver Galileo and Newton from the hands of Mach, who would turn everyone into a positivist insofar as he was a scientist at all, or to retrieve the Platonic realism of Copernicus and Kepler from Duhem's passion for downgrading all of physics into a likely story.

What the most dispassionate historian will have no wish to do is to underestimate the degree of Ostwald's enthusiasm for the movement in physics started by Robert Mayer. Indeed, a positivist in philosophy at the end of the century was almost certain to be an energeticist in physics. Heat itself occupied those who, following first Clausius and Maxwell, and then Gibbs and Boltzmann, would reduce it to a problem in dynamics (hence the term thermodynamics), and their dissenting colleagues who, in the manner of Duhem and Ostwald, and after Mayer, would transcend mechanics in the study of energy—and hence the implication of energetics as the phase of science that would supersede the Newtonian. For these latter did mean to make a revolution in physics. And it was their misfortune that they stormed the wrong door—the atom instead of the laws themselves.

Seldom has history so rapidly and so drastically undone a scientific crusade led by men who were themselves responsible and eminent scientists. It is easy to sympathize with the discontent that mechanism aroused. Certain temperaments still glory in a kind of emancipation from some classical tyranny, though it is now as dead as slavery. The formulation of conservation of energy, of electrodynamics, of optics, ultimately of all science, in the language of Newton's laws could be seen as a triumph. On the other hand, it could be seen—and the positivists did so see it—as a forcing of language and a straining of definitions. Energy had to be arbitrarily divided between potential and kinetic to fit the equations. A universal body had to be admitted rigid enough to bear shear waves and rare enough to pass detectable bodies undetectably through its subtlety. Electricity had to be resolved into mass-points (for the equations of ordinary dynamics) or into a hypothetical but incompressible fluid (for the equations of hydrodynamics). In either case something imponderable had to be given substance in order to be set in motion. Everywhere, outside of the domain of mechanics proper, it was not the observations but the laws of motion which prescribed these propositions. Everywhere, it seemed to those who felt the structure of classical science rather as a prison-house than a house built upon the rocks, everywhere the extension of mechanics committed the fallacy which invests the formulations of theory (or the conventions of language) with the attributes of reality.

And everywhere false problems sprang up to obstruct the path. Since atomism, the subsistence of reality in ultimate particles whose motions the laws describe, had provided classical physics with its ontology ever since the seventeenth century, it is obvious why the atom appeared to the critics of mechanics as the villain of the piece. And it must be admitted that the atom which offended them was not the rich and complicated structure, full of interest, promise, and threat, which emerged to refute them in the twentieth century. The object of their scorn was rather that atom which was a minute ball-bearing in dynamics, a carrier for valence (which would be the same without it) in chemistry, an infinitesimal concretization of energy in electricity, a population of the unobservable in the statistical mechanics of gases, and everywhere the postulation as an image of reality of what was properly only an analytical technique. Such atoms were easy targets. And perhaps there lay the trouble. Perhaps they were too easy. Perhaps they were straw atoms.

For the rub was less in the content of these criticisms, which were obvious enough and which may be repeated against the reality of strange particles in contemporary physics, than in the mood in which the campaign was conducted and the purpose that it served. Reading Duhem and Ostwald on (or rather contra) atoms, one is struck by the sense of having been here before and found it a blind alley. In *The Aim and Structure of Physical Theory* Duhem feels compelled to qualify phenomenalism, in which no physicist can quite content his intuition that a real world exists outside his science. To be sure, mathematical laws are in themselves only economical statements of experience. Nevertheless, they chime so beautifully together, they give so persuasive a suggestion of transcendent order, that they must echo some harmony subsisting out there in reality. It is not given to mechanical models, or to pictorial images, or indeed to any form of hypothesis, truly to represent that harmony. Our access to it is indirect, to be worked toward by discerning natural classifications among the laws of nature. And by "natural" Duhem means such a one as will reflect the relations of things in the natural order, though not their actual structure. We are, indeed, assured that proper classification does hold such a promise by the very capacity of theory to guide experiment in the prediction of phenomena not yet observed and in the formulation of laws not yet expressed. This, of course, was simply the difficulty with which the fact of discovery and the aim of innovation always confronts a phenomenalistic logic. It was no new problem, and the resort to classification was no new answer.

But Ostwald's general treatise, *Energy*, remains the most characteristic statement of the position. The argument is congruent with the contemporary biological retreat from the theory of natural selection into Lamarckism. Ostwald was a chemist, and it makes a curious reprise that once again an indictment of the sterility of physics should proceed from one whose own outlook was formed in the science of palpable matter rather than matter in motion, and who would yearn to re-invest science with a sense of biological subjectivity and process. Thus, Ostwald objects against mechanism that it deprives bodies of the properties in which alone they do have reality for a science of perceived phenomena. Precisely because manifestations of energy are what we do perceive, and all we perceive, energy is the only concept which can express reality for a phenomenalistic science. Energy is that which does act, and whatever its transformations in an event are the content of the event. It is absurd, therefore, to consider energy changes as an extension of mechanics. The practice requires us (to choose the most radical example) to take inertial motion as a fundamental occurrence, although no one has ever observed it, and to dismiss friction as an accident, although nothing ever transpires without it. And instead of this, Ostwald would have us penetrate beyond Newton in the direction marked out by Mayer. Nature is to be numbered in the intensive magnitudes of energy, rather than the extensive magnitudes of geometry. Then mechanical problems might themselves become special cases of exchange of energy.

Not all *ad hominem* arguments are to be avoided, not at least when historical judgment is in play rather than logical analysis, and in the perspective of the history of science it will appear more evident than it could to the protagonists—all men of good will—that energetics confronts us with a phenomenon by now familiar: a dissatisfaction with science ever latent among scientists themselves who would serve some grateful purpose. Thus, Duhem plays the pundit as well as the philosopher in his preachments on abstracting in theory and ordering by natural classifications. His implication was ever that the future for France in science lay in strict, uncompromising adherence to the "deep and narrow" way of her own national style, not to be adulterated by the "weak and ample" thinking of the Anglo-Saxon. It was natural and worthy in Duhem to wish to rehabilitate the scientific merit of those medieval centuries which were the great ones for his church. But his reader may feel reservations about his enthusiasm for reducing physics to the status of a useful fiction, in the reflection that a religious metaphysics might then monopolize the realm of truth.

Extremes of fidelity and agnosticism could touch in positivism. The religion which fired Ostwald's fervor was Comte's godless sociology revived. Ostwald thought to serve humanity by refounding all of physics in energetics. Indeed, the intractability of the propositions of mechanics offended him more deeply than their implausibility. Mechanics impoverished science by refusing any handle to psychology. A new phenomenalism, on the other hand, built upon our intensive perceptions of energy, would return us a psychology continuous with physics, correspondingly certain and enlightened, and applicable to the reformation of society. Ostwald was as uncompromising as Condillac on theory as the syntax of experience, as definite as Comte on prediction as the verification of theory, and as utilitarian as Bacon on application as the justification for science. And the advantage of considering science first as language and then as act was that it might then become subjective in the proper, the operational or inter-subjective sense, rather than in the metaphysical.

No such overtones accompany the work of Ernst Mach, who could make, therefore, a cleaner, straighter thrust toward a science which would be the purest rationalization of experience. His was a conscious debt to the Enlightenment—only to its skepticism, however, and never to its sentimentality. "I see Mach's greatness," wrote Einstein, "in his incorruptible skepticism and independence . . ."— in Mach's daring, Einstein might well have added, and in his radicalism. The mood of lesser positivists partook, after all, of petulance. Theirs was less a criticism than a dismissal of mechanics. No more than they did Mach have patience with the reality of atoms, but instead of simply waving away the entities that figured in the equations of mechanics, he attacked its fundamental assumptions about absolute space, time, and motion. He went right to the heart of Newtonian doctrine, and called the principle of inertia itself to account before the bar of empiricism.

Mach inherited his discontent with existential space and time from an early phase of idealism in his personal development, first Kantian and then Berkeleyan. He was prepared, therefore, to consider space and time rather as categories of experience than attributes of reality. Mach's own development, indeed, recapitulated the evolutionary contribution of idealism as the subjective component in positivist phenomenalism, for which the ego and its sensations constitute reality. And in execution of his own epistemological program, that we make knowledge only of observed phenomena, he turned his criticism of Newtonian inertia on points which Berkeley had already handled metaphysically.

By Newton's second law of motion, force is measured as the product of mass times acceleration. Acceleration relative to what, however? Had Newton allowed himself to be deterred by this difficulty, he might never have founded classical physics. In effect, he simply passed around it. Not without qualms, one suspects—Newton's language is labored in the distinction which he introduced into the definitions of the *Principia* between the absolute space and time of physics and reality on the one hand, and on the other the relative space and time wherein the vulgar live and have their being and awareness. All he can say in justification is that "the thing is not altogether desperate," in virtue of a famous thought experiment. A bucket of water hangs upon a twisted rope. The physicist releases the bucket so that the untwisting rope may spin it. Gradually the water picks up rotation and climbs along the side. Now the pail is suddenly stopped—and for a time the motion of the water and the concavity of its surface persist. Since that concavity obtained both when the water was at rest relative to the pail and when it was moving within it, its form must be the consequence of a (circular but no matter) acceleration relative to absolute space, and an instance, therefore, of absolute motion.

It is far from obvious why this is unconvincing, and Berkeley's refutation seemed even farther fetched. Newton's experiment loses its point unless taken to mean that the water curved its surface independently, not just of the matter in the bucket, but of all other matter whatsoever. Only so could motion be abstracted into a purely relational state between a body and space. But this is meaningless, Berkeley had objected. Matter can displace only relative to matter, and not to extension. The water is affected by nothing less than all the matter in the universe, and its apparent inertia is only motion relative to the fixed stars which, speaking strictly, cause the curvature of the surface. In their absence it would remain flat. But this means in the absence of all other matter. Thus, if there were only one body in the universe, it would have no inertia.

In appreciating the appeal which this extremely metaphysical proposition held for Mach, himself the scourge of metaphysics, one must remember that it was not the laws of Newtonian physics which he questioned, but their existential value; not the principle of inertia, the citadel of classical physics, but its attribution to matter as an intrinsic property independent of measurement and specifications of reference. His criticism in mechanics constantly aimed to expose the delusions which had led a Lagrange, a Galileo, an Archimedes to claim for their reason what they had learned from experience. Thus, Archimedes had really exploited his sense of the symmetry of his own body in his appeal to considerations of equilibrium. The question is not one of the utility of the law of the lever, but of rightly understanding it. Similarly, à propos of interpreting Foucault's pendulum (in principle the same experiment as Newton's bucket), Mach agrees that we must admit either to absolute motion in the earth (and in general), or to a fallacy in the expression of the law of inertia. And he differed from all other serious physical thinkers—it is the mark of his radicalism—in so posing the problem that he must dispute what no physicist had found wanting: the validity of the principle of inertia as a local law. What can it mean for a body to persist in the same direction if we do not specify what direction? We cannot attribute inertia to this or that body, held Mach. We must restate the law so that it contains the full indifference of nature to the false problem of whether the earth or heavens revolve daily. It must make no difference to what our law predicts of the surface, whether we spin the bucket or the fixed stars.

Bodies are not indifferent to their motion, therefore. They are affected in their inertia by the distance and disposition of all other masses. Deep ideas, these—the relativity of motion; the re-definition of its parameters, time and space, as orders of functional dependence observed among events; the principle that statements about inertia and therefore gravity and acceleration are about nothing if not about the interaction of all the matter in the universe— these very difficult but highly logical consequences of complete empiricism moved physics as close to its revolution into relativity as philosophy alone could carry it. It could go no further unnourished by consideration of what Mach's principles forbade him to take seriously, the reality of the electromagnetic field and the constitution of the aether that seemed increasingly to fill the universe, or at least the universe of discourse. For the Michelson-Morley experiment did not concern Mach. He did not see that the undetectability of the aether (at which any positivist might have said, "But of course!") posed as a real problem the status of the field.

H. A. Lorentz was to Einstein in physics what Mach was in philosophy, the mentor who had half the story. "For me personally," wrote Einstein in the centennial tribute to this John-the-Baptist of relativity, "he meant more than all the others I have met on my life's journey." Einstein was then looking back from near the journey's end, and though the tribute is personal, it might equally well have been scientific, for he came after Lorentz in equations as in spirit. And Lorentz lacked only that ultimate quality of something like divinity in Einstein's mind, which would take the same evidence and quite transform the shape of the world that physics sees in nature.

Lorentz had the advantage forced upon many Dutchmen of being equally conversant with the languages and scientific traditions of England, France, and Germany. Sympathetic to and detached from all three styles, he worked in the finest tradition of European scholarship. His own exemplar was Fresnel. The doctoral dissertation which Lorentz defended at Leyden considered the wave theory of light in relation to Maxwell's fields, and showed that certain anomalies in Fresnel's formulations—the most serious was the absence of longitudinal vibrations accompanying the transverse—disappeared in the electromagnetic theory of light. A more troubling complexity remained in field theory itself. Maxwell had established the field to obviate action at a distance. Nevertheless, his equations did not liberate it from association with ordinary matter. A boundary rather than a continuum obtained between aether in free space and aether in (say) glass. In order to describe the whole field, therefore, one had to combine its strength with the dielectric displacement by means of a specific constant, the dielectric constant for glass.

Lorentz came to this unhappy situation by way of Fresnel's investigation of a coefficient of aether drag. Fresnel's theory too had predicted that a specific portion of the aether coincident with a body in space is captive, so to say. And though Lorentz took all physics for his province with a more serene catholicity than was usual even then, rendering a consistent and rational account of the aether always did attract him more winningly than any other problem. It appeared to him early in his career that field theory might be straitened in the unity of light and electromagnetism and further simplified by liberating aether from this partial but rather vague and arbitrary association with ponderable matter. He tried as a hypothesis, therefore, that the seat of the field precisely is the aether, imponderable, omnipresent, and stationary. He divorced it, indeed, as sharply from ordinary matter as Newton had done in his day in distinguishing between matter and extension. Thus, he eliminated the distinction in terms between field strength and dielectric displacement. Elementary electric charges on the atomic constituents of matter are what create the field. They remain distinct from the field, however, and it exerts forces on the charges in accord with Newton's laws. That is to say, the field interacts with matter as physics knows it, not by mechanical linkages, but only through the intimate association of atoms with electrons (if in using that word one may presume on the discovery to which Lorentz's particularization of charge did actually lead).

Worked out in detail, the theory gave a beautifully simplified account of all the phenomena of electrodynamics and exacted as the price of simplicity only a single, essential condition: that the aether remain stationary with respect to matter in motion, and that the two be kept distinct. Maxwell's laws, in other words, were about one thing: the field; and Newton's laws, about another: bodies in motion; and Lorentz would save both sets by arraying them against the aether. If this were so, the earth and all bodies move through aether. Maxwell had himself suggested the design of an experiment to detect that motion. Michelson first performed it in 1881, and together with Morley refined it in 1887. They sought to detect a difference in the apparent velocity of light, according to whether it be measured in the direction of the motion of the earth or at right angles. It was an interesting and famous question in its own right, of course, and not only as an arbiter of Lorentz's uniform and stationary aether. Nor, though he accepted the verdict of the measurements which gave a final and a negative result, did Lorentz feel bound to interpret this experiment as crucially against his theory. What he modified were his equations.

In physical or mathematical problems it may often be convenient to convert the quantities from one set of co-ordinates to another. A problem may be stated in rectilinear co-ordinates. Its solution may be simpler in polar co-ordinates, or in some other set of rectilinear co-ordinates moving relative to the first—after which the results may be reduced to the initial system. Lorentz originated his transformations simply as such a mathematical device for relating the aether which carries the field to the inertial systems in which matter interacts with charge. Since classical mechanics treated time as a dimension of motion, his expressions were more complicated than simple spatial compositions in a plane or in three dimensions. Nevertheless, the handling of time did not differ in principle from the conversions worked (say) between the two systems of sidereal and solar time. In optics these transformations gave Lorentz the results with a stationary aether for which Fresnel had introduced his dragging coefficient. Their application to electrodynamics permitted description of the motion of a charge through a stationary field by the same formulae as through a second field in uniform linear motion relative to the first. And always it remained the postulate of Lorentz's own world picture that the fundamental frame for inertial systems (including those of moving charges) is the aether at rest.

The Michelson-Morley experiment shook that cornerstone, for it seemed to show that the aether moves with the earth (and presumably with any body on which measurements might be made). There were two possibilities. Either Lorentz must abandon his stationary aether. Or he must extend his transformations. He chose the latter, and adapted his equations to the suggestion first advanced by an imaginative Irishman, G. F. Fitzgerald, in 1892. Lorentz himself had established that an electrostatic charge in motion is equivalent in its effects—that is, in setting up a magnetic field—to an ordinary current of electricity in a wire. For a physicist here on the earth, Fitzgerald argued, a charge of statical electricity will generate only an electrostatic but not a magnetic field. If he were on the sun, however, he would find the charge to be in motion, and he ought to detect a magnetic field. In rigid bodies the atoms bear just such charges; the form of an object depends upon the forces between its molecules; those forces will be affected by the state of electromagnetic fields; an object will or will not be accompanied by a magnetic field, according to whether it is in motion or at rest relative to electrical charges and vice versa; a body may be shorter, therefore, in the direction of its motion than it would be

at rest or turned cross-wise; and that may be the explanation of the Michelson-Morley failure to observe motion
through the aether—their instruments contracted in the
direction of the earth's motion through the aether by just
that tiny amount which would necessarily compensate for
the expected diminution which that motion should occasion in the apparent velocity of light.

It seemed a far-fetched notion. One must remember the
alternative which Lorentz faced: to admit that Maxwell's
laws of the field, the laws of light and electromagnetism,
do not hold in a world where matter in motion exhibits
Newtonian inertia. By the Lorentz transformations, containing now the Fitzgerald contraction, the invariance of
Maxwell's laws was saved, and so too were Newton's laws
in their domain—provided it be allowed (as Lorentz said)
that time and space be pulled askew a little in his equations. It must be remembered, too, that some phenomenal-
ists in physics still regard what Einstein called the special
theory of relativity as identical with the Lorentz transformations. In effect, it is identical. The difference is philosophical, not mathematical, and has to do with whether
it is thought pleasing or physically important to discard
the aether (now become undetectable in principle) as the
locus of the field, and to treat as the foundation of the
new physics the relativity of space and time.

It has become less clear than it once seemed in what
sense Einstein went beyond Lorentz. And it may be that
we should coin a new term, the Einsteinian synthesis, to
suggest as does the phrase, Newtonian synthesis, a revolutionary combination of physics with philosophy in a new
conception of natural reality. At the end of his life Einstein was asked to compose an intellectual autobiography
to serve as preface to a symposium on the standing of his
work. "Here I sit," he began, "in order to write, at the
age of sixty-seven, something like my own obituary." And

though no man's reminiscences may do substitute for history, it may be helpful in the undeveloped state of scholarly criticism to respect Einstein's recollections of the order in which his thoughts followed one upon another. He recalls his youthful sense of displeasure at a dogmatism there at the foundations of physics in mechanics: a displeasure not so much dispelled as repressed in the necessity to admire the achievements of mechanistic analysis in extraneous areas. Physics seemed only to spiral deeper into this awkwardness. Even Maxwell and Hertz adhered overtly to mechanistic thinking, though theirs was the thrust which had exposed the worm in the apple. For it cannot be too much emphasized that the identification of the laws of light and electromagnetism is what opened the way to relativity, by rendering optical experiments relevant to the description of the field. But Einstein had to find grounds for his skepticism in Mach's criticism of physics and not in the writings of physicists.

Physical theories may be criticized, Einstein thought, from two points of view. The first, which considers the fit with facts, is obvious in principle, though often ambiguous of application, since assumptions may and sometimes should be added to smooth the wrinkles. The second has to do rather with what Einstein called the "inner perfection" of the theory than with its "external confirmation." He was unhappy with the vagueness of this characterization of theories—"naturalness," he calls it elsewhere, admitting that exact formulation eludes him, or "the logical simplicity" of the premises. On the score of "inner perfection," moreover, the laws of thermodynamics made a deeper impression than any other on Einstein in his youth. They were universal in their realm, and from this point of view, "We are confining ourselves to such theories whose object is the totality of physical appearances." (Thus early does Einstein's bent appear.) And even at the end of his life he wrote of classical thermo-dynamics: "It is the only physical theory of universal content concerning which I am convinced that, within the framework of the applicability of its basic concepts, it will never be overthrown." And it is of crucial interest that thermodynamics should thus have stood before him as the exemplar of universal theory.

For it was not the main course of physical thinking at the turn of the century, nor even his own researches in that line, which issued into relativity. Quite independently of the embarrassments at the foundations of mechanics, Max Planck found in a function relating density of radiant energy to frequency and temperature the constant which argued the discontinuous transmission of "quanta" of energy. That occurred in 1900. The implications were grave indeed, as well for classical mechanics as for the school which would transcend its atomism in energetics. Einstein himself, if he were not the founder of relativity, would figure most honorably as a founder of quantum physics. He saw and investigated the consequences for the photo-electric effect, and in a paper of 1905 he identified the photon as the quantum of light.

Einstein must always have lived more to himself than most scientists do, whose ideas, less their own than his, are forged out of the dialogue of the laboratory amid the never-ending din of colloquia and shop-talk. Thus, he devised a statistical mechanics to express the laws of thermo-dynamics kinetically—without knowing that Boltzmann and Gibbs, who were anything but obscure, had already accomplished this. Thus, too, he found in his derivations the prediction that microscopic motes should dart about in suspensions accessible to the microscope—without knowing that the Brownian movement, of which he here gave the theory, had been a phenomenon well-known to physicists (though unexplained) for nearly a century.

Critical though Einstein was of mechanics, moreover, he did not content himself with some facile philosophical victory over atoms. On the contrary: "My major aim in this was to find facts which would guarantee as much as possible the existence of atoms of definite finite size." It is the signet of his genius. The appeal which energetics held for Einstein drew him deeper into science, not away from it, and it must be put down to the credit of Ostwald that Einstein's paper convinced him of the futility of his own energeticist crusade against the atom.

Indeed, Einstein's relation to particle physics begins in piquancy and ends in pathos. These papers appeared in 1905, as did that on special relativity. And thus at the very moment when he turned classical physics out of its Euclidean housing, he rehabilitated Newtonian views in subsidiary matters, restoring to light its element of corpuscularity and to mechanics its atom. For he was never insensible to the triumphs of particle physics in detail, but only in principle. That Bohr could build upon these shifting sands and relate chemical and spectral properties of matter to the number of electrons in the atomic shell seemed to him a miracle. "The highest form of musicality in the sphere of thought," he called it, and for Einstein it was a pity that by the end of his life, an increasingly instrumental physics played rather that music than his own.

Instead, his early inability to draw a general theory out of experimental physics led him to abandon hope "of the possibility of discovering the true laws by means of constructive efforts based on known facts. The longer and the more despairingly I tried, the more I came to the conviction that only the discovery of a universal formal principle could lead us to assured results. The example I saw before me was thermodynamics. The general principle was there given in the theorem: the laws of nature are such that it is impossible to construct a *perpetuum mobile* (of the first and second kind). How then could such a principle be found?" And Einstein turned for guidance from physics to positivism, from Lorentz to Mach and the philosophy which at last accounts would prefer quantum mechanics to the relativity it launched.

How could such a principle be found? For ten years Einstein thought on the matter. And like Descartes he found his insight in himself, in a paradox which had first come to mind when he was sixteen. Suppose he were able to travel with a beam of light at its own velocity. It would seem to be oscillating before his eyes, a standing electromagnetic field. Science knew no such field at rest, either in the laboratory or in Maxwell's equations. And yet things should transpire for an observer, even rushing along at this extreme velocity, just as they do were he upon earth. How otherwise might he determine that he was in motion, except the laws of nature be the same from one system to another? It was a paradox first deepened and then resolved by consideration of the Michelson-Morley experiment, which established that whether the observer travel downstream with the light or upstream against it, its velocity appears to him to be the same. With sound waves things happen otherwise. Their behavior confirms the predictions of classical mechanics, in that their velocity appears augmented or diminished by one's own according as one moves in the contrary or forward direction relative to their propagation. One may, perhaps, borrow a thought experiment from the popularization to which Einstein himself much later gave his name in collaboration with Leopold Infeld. They ask us to imagine a laboratory with transparent walls arranged on a truck. Observers stand ready inside and out—physicists with instruments. A device at the precise center of the laboratory emits flashes of light. Mounted on its truck, the laboratory rolls forward. The signal blinks. The physicists inside measure the velocity of light, find it to be 186,000 miles per second, and observe the front and rear walls illuminated simultaneously. Their colleagues outside on the ground measure the velocity of the same signal. Though its source is moving relative to them, they agree that its velocity is 186,000 miles per second, but they see it reach the rear wall an instant before it does the front.

What is simultaneous inside is not simultaneous outside. There is no reconciling these results with classical Newtonian physics, in which here is here, there is there, and now is now.

Einstein would, and often did, state in more sophisticated language the paradox that had occurred to him in germ as a boy. Two fundamental results of experience contradict each other in classical physics: the unchanging velocity of light and the invariance of the laws of nature in different inertial systems. Classical physics supposes that time and space are absolute, and transposes information from one system to another—ship to shore or earth to moon—by linear compositions of the quantity of length or duration. Galileo was quite capable of such compositions of motion. Just so might one move from one set of Cartesian co-ordinates to another by adding x′ to x and thereafter using the new abscissa X. But to save the Michelson-Morey experiment, a different method of getting from one inertial system to another must be employed. This is the Lorentz transformation. Now what remains constant is the velocity of light instead of the metrical meaning of the co-ordinates. Since the velocity of light does not change, then what measures it must. It was more than a mannerism, it was one legacy of his positivist phase, that Einstein preferred writing of clocks and measuring rods to time and space. Clocks in motion do slow down, rods in motion do contract—just enough so that the velocity of light does appear constant to all observers, regardless of their state of (uniform) motion relative to light.

Thus, the critique of simultaneity takes us to the heart of Einstein's position on special relativity, and permits specifying what the elements were which he put together there. And clearly the first of these was his determination to refound theoretical physics in a universal principle like those of thermodynamics. He paid a price for his admiration for that science. The special theory of relativity was rather a restriction upon science than an induction from positive phenomena. In his taste for "inner perfection" in theory, Einstein answered to an aesthetic which logicians of science have not yet reduced to empirical terms, or to inter-subjective agreement. And certain very eminent physicists long felt uncomfortable about the physical reasoning in the special theory of relativity. In 1941, for example, Professor Bridgman in his book on *Thermodynamics* objected, in passing, that the special theory, like the second law, rests a general statement about the way the world works upon the physicist's incapacity to perform certain operations, in the one case to construct a *perpetuum mobile*, in the other to detect the motion of the earth through aether. But what reason is there to think nature restricted by the disabilities of physicists? Statements about relativity and entropy, then, are really about science, not about nature, and since they say what science cannot do rather than what it can, they may scarcely take pride of place in a science which is nothing if not operational.

Einstein's position before this criticism could scarcely have been altogether easy, since in the second place he had combined the mode of reasoning drawn from thermodynamics with a precept about physics itself drawn from positivism. The restriction on which the critique of simultaneity rests is that no signal may travel faster than the velocity of light. What gives that velocity its standing as a universal limiting constant is, therefore, rather a principle of communication than of nature. No one may say that nothing travels faster than light—it is only that we cannot be informed of it more rapidly. Information about measurements may be transmitted only by signals, light signals in the fastest case, and there is no significant statement in physics except about measurements. There is no measurement without an instrument, or ultimately without a physicist. And on the face of it, this might seem to have restored a species of subjectivism to science. Nevertheless, this must not be taken for a reversion to idealism—Einstein lapsing back into some Greek posture of humanism. It is all very well to say that there is no physics without a physicist—or perhaps two physicists, one to make a measurement and his colleague to be told of it. But it would, after all, be more accurate to say "without an instrument," because for such purposes a physicist is an instrument. We are concerned, that is to say, not with a personal subjectivism, but with an instrumental subjectivism, the kind of which a computer is capable.

In the third place, in the principle that the laws of nature are the same in all co-ordinate systems, Einstein simply adhered to the assumption of the uniformity of nature, which is anterior to the very possibility of science. But what laws of nature? In posing that question, Einstein gave his measure as an innovator. Instead of seeking like Lorentz and the others to reconcile the laws of mechanics and of the field—Newton's laws and Maxwell's—he gave the precedence to Maxwell. It was a preference unheard of in itself, and full of consequence. The special theory did not involve Einstein in the non-Euclidean formalization of space, later worked for the general theory; and this, which is an epilogue to classical ideas about nature, will not attempt to follow his thoughts so far in verbal paraphrase. Suffice it to say that the general theory requires more complex transformations than Lorentz's in order to save the invariance of the laws, regardless of whether or not systems were related in uniform motion. Therein he moved beyond positivism to geometrization, in a vein more suited to embracing nature in a single rationale. Already, however, the special theory bore witness to the appeal which the physics of the continuum had exerted over Einstein ever since he was first roused to admiration by Maxwell's field theory. The original embarrassment of Newtonian physics, action-at-a-distance, disappeared in company with absolute simultaneity. Actions propagated with the speed of light might remain conceivable, but hardly plausible, for how might they be contained in statements about conservation of energy? Physical reality, therefore, would have to be described in continuous functions in space, and the material point would cease to figure as a fundamental entity in theory, along with the extended void across which Newton and classical mechanics had made it act.

Throughout, therefore, the main features of Einstein's physics—unification, the continuum, and ultimately geometrization—renewed the tradition of rationalism in science, ever abstracting beyond common sense towards a more general formulation of the real in the ideal. It was not relativity of motion which administered the shock, nor even of space—the first is in Newton and the second imaginable enough, perhaps from awareness of perspective (though this is not the same thing). Rather, what wrenched the common consciousness was the relativity of time, and in this the common consciousness judged aright, for the critique of simultaneity lies deeper in Einstein's physics than the redefinition of space. And thus, classical physics ended as it had begun in Galileo's law of falling bodies—with a redefinition of the physical meaning of time. Time seems the intimate aspect of the continuum, and the consequence was to move science one stage further into the impersonal generality of things. There is no privilege left for quality in Galileo's physics, no privilege left in circles by Kepler, no privilege left in life by Darwin, and now no privileged frame of reference or geometry left by Einstein.

Even if there were no other, this would be the decisive superiority of Einstein over Lorentz. Confronted with the inner demand for a consistent account, Lorentz identified the aether with Newtonian space as the absolute co-ordinate system which must still unify his science. Confronted with the same information, Einstein conjured the aether away, Since every effort to detect it ended in failure; since, indeed, it must possess just those properties which rendered its effects undetectable in principle; why then, the only necessity it served was intellectual, not physical, and Einstein would seek unity in the proper domain of the intellect: in the laws of nature rather than in an imaginary entity out in nature. And so disappeared the last of the imponderables, the last frontier of privilege in physics, and with it the space it had come to embody.

"Physics," wrote Einstein at the end of his autobiography "is an attempt conceptually to grasp reality as it is thought independently of its being observed." This, no doubt, is philosophically inconsistent with Einstein's earlier, indeed his continuing instrumentalism. Nevertheless, the loftiness of his thought, as over against the brutality of the times and of its applications, was such that even the public obscurely sensed in him the symbol of the cultural predicament of physics: in the sad, sweet face; in that simplicity more suited to some other civilization, some gentler world; in the strange, the often inappropriate moments chosen for speech; in the great, the profound, the somehow altogether impersonal benevolence; in what shames the spotted adult as the innocence of a wise child.

A passage at the beginning of his Autobiography tells what it was he sought. He is writing of the shock of disillusionment he experienced at the age of twelve, when he found that the stories of the Bible could not be true, and of how he then decided that youth "is intentionally being deceived by the state through lies. . . ."

> It is quite clear to me that the religious paradise of youth, which was thus lost, was a first attempt to free myself from the chains of the "merely-personal," from an existence which is dominated by wishes, hopes, and primitive feelings. Out yonder there was this huge world, which exists independently of us human beings and which stands before us like a great, eternal riddle, at least partially accessible to our inspection and thinking. The contemplation of this world beckoned like a liberation, and I soon noticed that many a man whom I had learned to esteem and to admire had found inner freedom and security in devoted occupation with it. The mental grasp of this extra-personal world within the frame of the given possibilities swam as highest aim half consciously and half unconsciously before my mind's eye. Similarly motivated men of the present and of the past, as well as the insights which they had achieved, were the friends which could not be lost. The road to this paradise was not as comfortable and alluring as the road to the religious paradise; but it has proved itself as trustworthy, and I have never regretted having chosen it.

And surely the poignancy in Einstein's destiny welled from disappointments deeper than the drawing off of physical interest from relativity into quanta? Surely the very generality of his liberation, rendering the perfectly benign perfectly irrelevant to the vast impersonality of nature, invested his inner freedom and security with the loneliness of a Greek tragedy, one inhering in the necessities of things rather than (like Galileo's) in the characters of men.

It has always seemed to me that the kind of preface in which an author pays off debts and disarms criticism ought to come at the end of a book, after he has done his work and run the risks for which he begs indulgence. This book is no attempt to recount in summary the whole history of science from Galileo to Maxwell and Mendel. Instead, its purpose is to set out in narrative form what I take to be the structure in the history of classical science. This I find in the route which the advancing edge of objectivity has in fact taken through the study of nature from one science to another. History is made by men, not by causes or forces, and I have tried to write with due attention to the intellectual personalities who have borne the battle, and not without sympathy for its casualties. And though I have written as closely to the texts as my competence permits, I want the tale to move unencumbered by the barnacles of scholarly apparatus. There are liberal quotations from the great literature of past science, but they are included to convey its style and spirit, and not to establish this or that point of fact. I hope that this book will help win for history of science a place in historiography comparable in interest and professionalism to that which the philosophy of science has for long held in philosophy. But history is critical narrative, and considering the range of this one, whatever effect it may have will depend on the interpretation and the vision of the subject.

The account does not rest on obscure or little known sources. With very few exceptions, and those quite insignificant, the memoirs and treatises I cite are well known to specialists, and so too will be many of the passages quoted. All are identified in the text, and the editions used are those mentioned below in the bibliography. In every case the editions are standard collections of the works, letters, and lives of the scientists concerned, or of the relevant societies, all to be found readily in large libraries. Where a suitable English translation exists, I have adopted it. Otherwise I have made my own. But except where critical editions are standard or contribute to the interpretation of a subject, I have not listed bibliographical detail on all the books discussed. It seems absurd, for example, to itemize the editions of the *Principia* or the *Origin of Species*. That information will be well known to specialists, and easily available to others in card catalogues.

I do not wish to be so cavalier about the scholars, my contemporaries and colleagues, upon whose studies I have freely drawn. The references which follow do not exhaust the literature on the subject of any chapter or section. I have tried to keep to a minimum, while including everything which I have used directly and consciously, together with the most important articles and monographs of recent years. In every case these will lead the reader into the older literature should he wish to pursue some subject. In the case of Chapter Six, for example, on the chemical revolution, the writings of M. Daumas and Professor Guerlac, which I do cite, will inform the reader about the monographs of A. N. Meldrum and Hélène Metzger, which I do not.

Beyond this, or rather before it, there are certain persons who have shaped my views fundamentally, or to whom I wish to express special thanks. First is my wife, whose sympathetic and critical eye for detail and for style never fails to improve my publications. Whatever my scholarship is, it would be far less without her help and devotion. My editor, John Boles, of the Princeton University Press, has been a tower of strength and a model of patience. I am immensely indebted to two fellow historians of science, Professor Marshall Clagett of the University of Wisconsin and Professor I. Bernard Cohen of Harvard University, who read the first nine chapters. Dr. Alistair C. Crombie of Oxford University gave helpful criticism of Chapter Eight. Chapters Seven and Eight have both benefited from discussion of the argument with Professor Colin S. Pittendrigh of the Department of Biology in Princeton University. My colleague in the Department of Physics, Professor George Reynolds, also took time from his laboratory to read Chapter Nine and to discuss Chapter Ten. Finally, Professor George Temple of Oxford University gave me his opinion of Chapter Eleven. All have saved me solecisms over which I blush (though not so hotly as if they had got into print), and no one of these guardian angels is responsible for any which remain.

That Professors Clagett and Cohen have dedicated their own major works to Alexandre Koyré of the Ecole pratique des hautes études, Sorbonne, and the Institute for Advanced Study, Princeton, will not prevent me from saying that I, too, owe more to him professionally than to anyone else. His writings have revealed to me wherein the intellectual content of the history of science consists. His is by far the greatest influence on this book, and if I make bold to differ here and there—in the remarks I venture in his own domain on Newton and the aether, or on Darwin and the nineteenth-century sense of process, or in attributing more in general to personality and less to philosophy—it is with a feeling of some daring. He is the master of us all.

I should also like to pay special tribute to Professor Giorgio de Santillana of the Massachusetts Institute of Technology. Not only have I adopted his Galileo for my first chapter, but his writings have done more than I realized until I came to write this book to shape my views on the relations of science, history, and culture in general. Two recent books on the philosophy of science are very encouraging in their sympathetic treatment of its history. Professor Gustav Bergmann of the State University of Iowa once had the kindness to comment on a paper I had given that it exemplified the structural history of ideas. I went home to his book, *Philosophy of Science* (Madison, 1957), to find out what I was doing, and learned that structural history of thought is, indeed, what I try to write. I mean a narrative which accepts the difference between the logical order and the historical order, and seeks to discern a structure in the latter inhering in the relation of philosophy, technicality, personality, and circumstance. Secondly, N. R. Hanson's *Patterns of Discovery* (Cambridge, 1958) exemplifies what it would be excellent for the philosophy of science to do: address itself critically to science as inquiry, not just logically or verbally to science as system, and treat current science as continuous with its history, thus illuminating and drawing on the history of science.

Finally, I have profited from the researches of certain of my undergraduate students in Princeton University to whom I owe a debt of gratitude going beyond the general pleasure I have drawn from them all: to Cornelius C. Bond, '56, for his paper on Maxwell; to Gordon Hammes, '56, for his paper on Joule; to LeRoy Riddick, Jr., '58, for his senior thesis on John Wilkins; to Arthur M. Jaffe, '59, for his paper comparing Gibbs and Duhem in their philosophies of science; and to Charles Thornton Murphy, '59, for his paper on the field concept in nineteenth-century physics.

I take license from the example of Lavoisier for attempting to influence the conception of a whole discipline by writing an elementary and educational treatise, weaving together the researches of others with a generalization of certain specialized studies of my own. The interpretations of the first four chapters are adapted for the most part from the work of others, with certain exceptions, of which the most important are my reserve about Pascal, the analysis of Harvey, my response to Bacon, the emphasis on Boyle as an atomic physicist rather than a chemist, and the discussions in the Newton chapter of the discreteness of light consisting in rays, not particles, and of the relation of the aether to his personal problem in communication and intelligibility. The remaining chapters owe much substance to the scholars cited in the bibliography. Nevertheless, the conception of them is my own, as far as I know. I set this out less to claim credits than to permit the reader to judge what authority attaches to the interpretations.

This book originated in a commission from the D. Van Nostrand Company. I lacked the skill to work within the compass we had contemplated, and I should like to acknowledge the kindness and courtesy with which they released me from my contract and allowed me to publish with our University Press. Other publishers and proprietors of journals have very generously allowed me to reprint passages which have previously appeared under their copyright in writings of mine mentioned in the bibliography. I am most grateful to Harvard University Press, The Johns Hopkins University Press, The University of Wisconsin Press, Dover Publications, *American Scientist, Archives internationales d'histoire des sciences, Behavioral Science, Isis, Proceedings of the National Academy of Sciences, Revue d'histoire des sciences,* and *Victorian Studies.* The American Council of Learned Societies, the Guggenheim Foundation, the National Science Foundation, and Princeton University have supported studies and leaves which made it possible for me to write this book, and indeed to become a historian of science in the first place.

CHAPTER I

Of the many writers on Greek science, S. Sambursky (*The Physical World of the Greeks*, London, 1956) concerns himself more immediately than any other with the parentage of modern science in Greece. For his influence on my own views, see my review article, "A Physicist Looks at Greek Science" (*American Scientist*, 46: 62-74, 1958). His perspective is that of a physicist as well as a classical scholar, and it might be well to set it off against Marshall Clagett's *Greek Science in Antiquity* (New York, 1956), and *A Source Book in Greek Science* (edited by Morris R. Cohen and I. E. Drabkin, 2nd ed., New York, 1959). A. C. Crombie's *Medieval and Modern Science* (2 vols., New York, 1959) has quickly established itself as the standard history of medieval science, which Dr. Crombie regards as more continuous with modern science than my own omission of that subject would suggest. In any case, a massive and magisterial monograph by Marshall Clagett (*The Science of Mechanics in the Middle Ages*, Madison, Wis., 1959) is indispensable reading for all students of the history of science, and particularly for the medieval background of Galilean kinematics. A classic article by J. H. Randall, "The Development of Scientific Method in the School of Padua (*Journal of the History of Ideas*, 1: 177-206, 1940) should be consulted for the positive Aristotelian element in the scientific revolution. The most comprehensive recent work on Copernicus is by Thomas S. Kuhn: *The Copernican Revolution* (Cambridge, Mass., 1957). Kuhn treats his subject as a case study in the formation of concepts. The chapter on Copernicus and his conservatism may be the best of many fine features in Herbert Butterfield's *The Origins of Modern Science* (London, 1949), which little book did more than any other to win the history of science a place in contemporary historiography. A new edition of Edward Rosen's *Three Copernican Treatises* (New York, 1939) is in preparation. There is, finally, considerable emphasis on the archaism of Copernicus in Derek J. de Solla Price's "Contra-Copernicus: A Critical Re-estimate of the Mathematical Planetary Theory of Ptolemy, Copernicus, and Kepler" (*Critical Problems in the History of Science*, ed. Marshall Clagett, Madison, Wis., 1959).

For the history of cosmology in general and of Kepler in particular, it is still useful to consult J. L. E. Dreyer's *A History of the Planetary Systems from Thales to Kepler* (Cambridge, 1905; re-issued, New York, 1953). Though my own interpretation of the history of science is diametrically the opposite of Arthur Koestler's, there is much to be learned from his account of Kepler and his astronomy, which seems to me the best part of *The Sleepwalkers* (New York, 1959). Two recent discussions treat Kepler as a study in the formation of theory: Gerald Holton's "Johannes Kepler's Universe, its Physics and Metaphysics" (*American Journal of Physics*, 24: 340-351, 1956); and Chapter IV of N. R. Hanson's *Patterns of Discovery* (Cambridge, 1958). Alexandre Koyré begins with Kepler's celestial mechanics in "La gravitation universelle de Kepler à Newton" (*Archives internationales d'histoire des sciences*, 4: 638-653, 1951). Brief excerpts from the correspondence are translated in Carola Baumgardt's *Johannes Kepler: Life and Letters* (London, 1952); and there is an excellent biography by Max Caspar, *Johannes Kepler* (Stuttgart, 1948), who is editing Kepler's *Gesammelte Werke* (Munich, 1937-   ).

The indispensable starting point for the study, not only of Galileo but of the entire scientific revolution of the seventeenth century, is Alexandre Koyré's *Etudes galiléennes* (3 parts, Paris, 1939). In *The Crime of Galileo* (Chicago, 1955), Giorgio de Santillana has written a study of

Galileo's difficulties with the Church and of the intellectual personality which entailed those troubles. Santillana also edited Thomas Salusbury's translation of the *Dialogue on the Two Chief Systems of the World* (Chicago, 1953), to which I am indebted for my quotations from that work. A complementary edition in modern translation is by Stillman Drake, *Dialogue concerning the Two Chief World Systems* (Berkeley and Los Angeles, 1953). Drake has translated a number of other writings under the title, *Discoveries and Opinions of Galileo* (New York, 1957); and I owe to him the anecdote about Galileo on wine, "Galileo Gleanings I" (*Isis, 48*: 393-397, 1957). The only translation of the Discorsi is *Dialogues Concerning Two New Sciences* (New York, 1914) by Henry Crew and Alfonso de Salvio, and it is less than satisfactory, though it did win the approval of the great Galileo scholar, Antonio Favaro, editor of the *Edizione nazionale* of the *Opere* (20 vols., Florence, 1890-1909).

CHAPTERS II AND III

Since these two chapters deal with science as a movement of thought between Galileo and Newton, it seems best to group relevant writings in a single section. The *Notebooks of Leonardo da Vinci* (New York, 1955) have been translated and collected by Edward MacCurdy, and Leonardo's technical drawings appear in the immense folio *Leonardo da Vinci* published in New York by Reynal, and copyrighted in Italy in 1956 by the Istituto Geografico de Agostini (Novara). The most important critical essays on Leonardo and science are in *Léonard de Vinci et l'expérience scientifique au XVI siècle* (Paris, Centre national de la recherche scientifique, 1952). My own interpretation of Vesalius has been shaped by an excellent discussion of *De fabrica* in A. R. Hall's *The Scientific Rev-*

olution (London, 1954). The ensuing succession of physiological discoveries appears in very clear relief in the selections translated in Henry Guerlac's *Selected Readings
in the History of Science* (multigraph, Ithaca, 1950, Vol. I,
fascicule 3). Sir Michael Foster's *Lectures on the History
of Physiology* (Cambridge, 1901) give an account of the
pre-Harveian doctrine on circulation. Kenneth V. Franklin's new translation of Harvey (*On The Motion of The
Heart*, Oxford, 1957), is as superb in its fidelity as in its
felicity, and Blackwell's has graced the work in a beautiful edition. Students of Harvey are fortunate in the translation of *De Motu Locali Animalium* (Cambridge, 1959)
by Gweneth Whitteridge, and in *Bibliography of the
Writings of Dr. William Harvey* (2nd ed., Cambridge,
1953), by Sir Geoffrey Keynes. Keynes's scholarship and
admiration for Harvey nowhere appear to better advantage
than in the charming vignette, *The Personality of William
Harvey* (Cambridge, 1949). The most recent biography,
finally, is by Louis Chauvois (*William Harvey*, trans. from
the French, New York, 1957), whose slight tendency to hero-
worship is a most sympathetic defect of the qualities of a
lifelong enthusiasm. Richard Foster Jones has made a useful selection from the vast corpus of Bacon in *Essays,
Advancement of Learning, New Atlantis, and other Pieces*
(1937). His introduction is an excellent essay in itself. It
might be well, moreover, to balance my rather critical
estimate of Bacon with Benjamin Farrington's *Francis
Bacon* (New York, 1949).

There is a very sensitive discussion of the *Discourse on
Method* in Alexandre Koyré's *Entretiens sur Descartes*
(New York and Paris, 1944). Koyré's *From the Closed
World to the Infinite Universe* (Baltimore, 1957) develops in full the implications of infinity for the problem
of man and nature, while the third fascicule of his *Etudes
galiléennes* studies Descartes and inertia. In addition there

is an excellent account of this, as of Cartesian mechanics in general, in René Dugas' *La Mécanique au XVIIᵉ Siècle* (Neuchâtel, 1954), and of the later physics of the Cartesian school in Paul Mouy's *Le développement de la physique cartésienne, 1646-1712* (Paris, 1934). Vasco Ronchi's *Storia della luce* (Bologna, 1939) is an authoritative history of optics. A French translation by Juliette Taton appeared in 1956. The critical edition of the *Oeuvres de Descartes* (13 vols., Paris, 1897-1913) was edited by Charles Adam and Paul Tannery, and is indispensable for any serious study.

The "Penguin Classics" contain a very workable translation of Lucretius by R. E. Latham (London, 1951) under the title *The Nature of the Universe*, and there is a fine discussion of ancient atomism in Sambursky's *The Physical World of the Greeks*. Robert Lenoble's *Mersenne, ou la naissance du mécanisme* (Paris, 1943) gives an account of the adoption of atomism in seventeenth-century natural philosophy. Papers contributed to a colloquium on Gassendi have appeared in *Pierre Gassendi, sa vie et son oeuvre* (Paris, Centre international de synthèse, 1955). *Blaise Pascal, l'homme et l'oeuvre* (Cahiers de Royaumont, Philosophie No. 1, Paris, 1956), edited by M. A. Bera, owes its origin to a similar occasion. I. H. B. and A. G. H. Spiers translated and edited *The Physical Treatises of Pascal* (New York, 1937), including a small selection of correspondence. James B. Conant opens *Harvard Case Histories in Experimental Science* (2 vols., Cambridge, 1957) with an account of Boyle's experiments on pneumatics. The place to begin a study of Boyle, however, is with the life prefixed by Thomas Birch to his edition of *The Works of the Honourable Robert Boyle* (5 vol., London, 1744). John Fulton's *A Bibliography of the Honourable Robert Boyle* (Oxford, 1933; Suppl. 1949) is a classic work of bibliophilia. Marie Boas treats Boyle as a mechanistic thinker in "The Establishment of the Mechanical

Philosophy" (*Osiris, 10*: 412-541, 1952), and she has recently published a more general study, *Robert Boyle and Seventeenth-Century Chemistry* (Cambridge, 1958).

Sprat's *History of the Royal Society* has been printed in a modern edition edited by J. I. Cope and H. W. Jones (St. Louis, 1958). A pioneer monograph in the institutional history of science still outshines many recent writings: Martha Ornstein's *The Role of Scientific Societies in the Seventeenth Century* (Chicago, 1928). Alfred Maury's *L'ancienne Académie des sciences* (Paris, 1864) is a semi-official and rather uncritical work, and one does better, perhaps, to turn to the historical introductions which Fontenelle prefixed to the *Mémoires de l'Académie royale des sciences depuis 1666 jusqu'à 1699*. The most recent history of the Royal Society is Dorothy Stimson's *Scientists and Amateurs* (New York, 1948). A standard (and superb) monograph on the sociology of the scientific movement in England is Robert K. Merton's *Science, Technology and Society in Seventeenth Century England* (Volume IV, Part 2 of *Osiris Studies*, Bruges, 1938). Merton may and should be supplemented by Basic Willey's *The Seventeenth-Century Background* (London, 1953); Richard Foster Jones' *Ancients and Moderns* (Saint Louis, 1936); and G. N. Clark's *Science and Social Welfare in the Age of Newton* (2nd ed., Oxford, 1949); and compared to R. H. Knapp and B. H. Goodrich's *Origins of American Scientists* (Chicago, 1952).

CHAPTER IV

The long and eagerly awaited publication by the Royal Society of *The Correspondence of Isaac Newton* (edited by H. W. Turnbull, Cambridge, 1959) has begun to appear. This chapter was in page proof before the first volume arrived, and readers may compare the interpretation to the record of Newton's letters as far as 1675. My account of the optics rests upon the documents introduced by Thomas S. Kuhn in I. Bernard Cohen's edition of *Isaac Newton's Letters and Papers on Natural Philosophy* (Cambridge, Mass., 1958), which gathers the writings Newton published on physics, other than the *Principia* and the *Optics*. A facsimile reproduction of the first edition of the former was published by William Dawson & Sons (London, 1957), and of the latter by Dover (New York, 1952) with an introduction by I. B. Cohen. Professor Cohen's *Franklin and Newton* (Philadelphia, The American Philosophical Society, 1956) is the most comprehensive work on Newton yet to appear, and it will surely serve as the point of departure for future scholarship, even as to bibliography. Its thoroughness in the latter respect makes superfluous the mention here of any but the most important essays in interpretation. Of these the foremost are (once again) by Alexandre Koyré: "The Significance of the Newtonian Synthesis" (*Archives internationales d'histoire des sciences, 11*: 291-311, 1950); "A Documentary History of the Problem of Fall from Kepler to Newton" (*Transactions of the American Philosophical Society, 45*: 329-395, 1955); and "L'hypothèse et l'expérience chez Newton" (*Bulletin de la Société française de Philosophie, 59-97*, avril-juin, 1956). Other important articles dealing with Newton and scientific explanation are A. C. Crombie's "Newton's Conception of Scientific Method" (*Bulletin of the Institute of Physics, 350-362*, Nov. 1957), and Stephen Toulmin's "Criticism in the History of Science: Newton on Absolute Space, Time and Motion" (*The Philosophical Review, 68*: 1-29 and 203-227, 1959). W. W. Rouse Ball's *An Essay on Newton's Principia* (London, 1893) gives an extremely useful synopsis of the great book. Finally, H. G. Alexander introduces *The Leibniz-Clarke Correspondence* (New York, 1956) with a discerning summary of the philosophical points at issue between Leibniz and Newton.

CHAPTER V

The argument of this chapter differs somewhat from other interpretations of the influence of science in the Enlightenment. I hope it will not be unseemly to mention some of the special studies which led me to it. They may serve to document the discussion, and I have adopted certain passages therefrom for the present work. I first became impressed with the current of eighteenth-century hostility to Newtonian science in researches for a social and intellectual history of science in the French Revolution, a work which is still in progress. An essay on "The Formation of Lamarck's Evolutionary Theory" (*Archives internationales d'histoire des sciences*, 323-338, Octobre-Décembre 1956) broached what now seems to me an example of romanticism in natural history. Another essay, "The *Encyclopédie* and the Jacobin Philosophy of Science," appeared in the symposium *Critical Problems in the History of Science* (edited by Marshall Clagett, Madison, Wis., 1959). It treats of the attack upon the *Académie des sciences* as a political instance of the attempt to substitute popular and organismic for abstract and mathematical science. The complementary story, the movement toward rationalizing science in institutions after Thermidor, is the subject of "Science and the French Revolution" (*Proceedings of the National Academy of Science, 45*: 677-689, 1959). My attention was led back to Diderot and technological Baconianism by research for an edition of *A Diderot Pictorial Encyclopedia of Trades and Industry: Manufacturing and the Technical Arts in Plates from l'Encyclopédie* (2 vols., New York, 1959). Two articles, finally, deal with the relations of science and industry in the Enlightenment: "The Discovery of the Leblanc Process" and "The Natural History of Industry" (*Isis, 48*: 152-170 and 398-407, 1957).

The most suggestive work for the positivistic tenor of scientific philosophy in the Enlightenment is Henri Gouhier's *La jeunesse de Comte et la naissance du positivisme* (3 vols., Paris, 1933-1941). Isaiah Berlin's *The Age of Enlightenment* (New York and Boston, 1956) serves to introduce Locke and the theory of ideas. It seems to me that the Voltaire of Peter Gay's stimulating *Voltaire's Politics* (Princeton, 1959) is congruent with the intellectual personality that I see in his Newtonianism. In any case, volumes 2 through 11 of *Voltaire's Correspondence* (edited by Theodore Besterman, Geneva, 1953-   ) contain expressions of his state of mind about physics during the 1730's. The *Oeuvres philosophiques de Condillac* (3 vols., Paris, 1947-51) have been edited with a brief critical and biographical introduction by Georges Le Roy for the series *Corpus général des philosophes français*. Condorcet's *Esquisse* has been given a fortunate translation by June Barraclough as *Sketch for a Historical Picture of the Progress of the Human Mind* (London, 1955). And Diderot, for his part, is the subject of an excellent biography by Arthur Wilson (*Diderot, the Testing Years*, New York, 1957), of which a second volume is to appear. Convenient editions of Diderot's writings appear as *Oeuvres philosophiques* (edited by Paul Vernière, Paris, 1956) and *Oeuvres romanesques* (edited by Henri Bénac, Paris, 1951). Of recent writings on the Enlightenment, the most suggestive seems to me to be Aram Vartanian's *Diderot and Descartes: A Study of Scientific Naturalism in the Enlightenment* (Princeton, 1953). I no longer agree with the thesis, which makes Descartes the fountainhead of eighteenth-century scientism. The argument requires overlooking the route from associationist psychology through Condillac's philosophy to positivism, by which Newtonian science was brought to bear on the human condition. Nevertheless, I do agree with much of Vartanian's criticism of the conventional historiography of ideas in the Enlightenment, and with his emphasis on a Cartesian strain running through the period.

As to Goethe, I feel a certain temerity, and I venture the interpretation only in the profound conviction that this reading of his science as anti-science makes sense. One may draw immense profit from books with which one profoundly disagrees, and no one (it seems to me) has so well understood the import of Goethe's science as a contemporary disciple, Ernst Lehrs, from whom I differ only in the ultimate judgment of the value of Goethe's science and not in its interpretation. I am much indebted to his *Man or Matter; Introduction to a Spiritual Understanding of Matter Based on Goethe's Method of Training, Observation, and Thought* (London, 1951). Among the immense Goethe literature, mention may also be made of René Berthelot's *Science et philosophie chez Goethe* (Paris, 1932), René Michéa's *Les travaux scientifiques de Goethe* (Paris, 1943), Martin Loesche's *Grundbegriffe in Goethes Naturwissenschaft* (Leipzig, 1944), and Marianne Trapp's *Goethes naturphilosophische Denkweise* (Stuttgart, 1949). The most convenient source for Goethe's writings on science is the Weimar edition, and the principal works with dates of composition are *Dem Menschen wie den Thieren ist ein Zwischenknochen der obern Kinnlade zuzuschreiben* (1784); *Die Metamorphose der Pflanzen* (1790); *Beitrage zur Optik* (1791-92); *Zur Farbenlehre* (1810-1823).

The work to which I allude on conservatism is R. J. White's *The Conservative Tradition* (London, 1950).

CHAPTER VI

The two foremost authorities on Lavoisier and the chemical revolution are Maurice Daumas and Henry Guerlac, to whose writings I am much indebted. Daumas summarized his researches to date in *Lavoisier, théoricien et expérimentateur* (Paris, 1955); and Guerlac is gathering into a book (which will be eagerly awaited) his series of articles, of which the most important are: "The Continental Reputation of Stephen Hales" (*Archives internationale d'histoire des sciences, 4*: 393-404, 1951); "Joseph Priestley's First Papers on Gases and their Reception in France" (*Journal of the History of Medicine and Allied Sciences, 12*: 1-12, 1957); "A Note on Lavoisier's Scientific Education" (*Isis, 47*: 211-216, 1956); "Joseph Black and Fixed Air" (*Isis, 48*: 124-151, 433-456, 1957); "Some French Antecedents of the Chemical Revolution" (*Chymia*, 73-112, 1959); "A Lost Memoir of Lavoisier" (*Isis, 50*: 125-129, 1959).

Mme. Lavoisier's account of Lavoisier's day comes from an article of my own, "Notice biographique de Lavoisier par Mme. Lavoisier" (*Revue d'histoire des sciences, 9*: 52-61, 1956). The interpretation of Lavoisier as a theorist is also mine, at least insofar as concerns the argument for the theoretical respectability of caloric and the attribution to Condillacian method of Lavoisier's over-extension of oxygen and his failure to reduce his concepts to atomism. The thesis was developed for presentation before the "Colloque international sur l'histoire de la chimie au XVIIIe siècle" held in Paris on 11, 12, and 13 September 1959, under the auspices of the Comité national français d'histoire et de philosophie des sciences. It is expected that the proceedings of that colloquium will be published.

An article by Uno Boklund advances the claims for Scheele's priority in the discovery of oxygen: "A Lost Letter from Scheele to Lavoisier" (*Lychnos*, 1-27, 1957). The history of Priestley's experimental discoveries and their relation to Lavoisier's conceptual thinking forms the subject of one of James B. Conant's case studies, "The Overthrow of the Phlogiston Theory" (*Harvard Case Studies, I*, 65-116). The same volume contains Leonard K. Nash's "The Atomic-Molecular Theory" (215-321), of which he modifies the argument while further emphasizing the physical quality of Dalton's thinking in "The Origin of Dalton's Chemical Atomic Theory" (*Isis*, *47*: 101-116, 1956).

Joseph Black's *Experiments upon Magnesia Alba* (1756) was reprinted by the Alembic Club (Edinburgh, 1898), and *The Scientific Papers of the Honourable Henry Cavendish* (2 volumes, Cambridge, 1921) were edited by James Clerk Maxwell and Sir Edward Thorpe. Joseph Priestley published his work in *Experiments and Observations upon Different Kinds of Air* (3 vols., London, 1774-1777), and *Experiments and Observations Relating to Various Branches of Natural Philosophy* (3 vols., London, 1779-1785). The French Ministry of Education published the *Oeuvres de Lavoisier*, vols. I-IV (1864-1868) edited by J. B. Dumas; and volumes V-VI (1892-1894) edited by Edouard Grimaux. Two volumes have appeared of the *Correspondance de Lavoisier* (edited by René Fric, Paris, 1955-). And finally, Dalton's *New System of Chemical Philosophy* (1808) is available in a facsimile reprint (2 vols., London, 1953).

CHAPTERS VII AND VIII

It will be best to handle the literature on nineteenth-century biology in a single section. The treatment of Lamarck and certain passages in the discussion of Darwin are drawn from two essays of my own: "The Formation of Lamarck's Evolutionary Theory" (*Archives internationales d'histoire des sciences*, *9*: 323-338, 1956); and "Lamarck and Darwin in the History of Science" (*Fore-*

runners of Darwin, edited by Bentley Glass, Baltimore, 1959); whereas the treatment of geology is condensed from my book, *Genesis and Geology* (Cambridge, Mass., 1951). By far the two best histories of evolutionary thought are Paul Ostoya's *Les théories de l'évolution* (Paris, 1951), and Loren C. Eiseley's *Darwin's Century* (New York, 1958). There is an excellent history of the taxonomic foundations of nineteenth-century biology which does not seem to have been much noticed by historians of science—Henri Daudin's *Cuvier et Lamarck: Les classes zoologiques et l'idée de série animale, 1790-1830* (2 vols., Paris, 1926). The remark about Cuvier's first utterance is quoted from (Mrs.) R. Lee's *Memoirs of Baron Cuvier* (New York, 1833). Eiseley has also published a very important paper, "Charles Darwin, Edward Blyth, and the Theory of Natural Selection" (*Proceedings of the American Philosophical Society, 103*: 94-158, 1959). He adduces evidence that Darwin adopted the theory without proper acknowledgment from the writings of the naturalist Blyth. The essay appeared too late to affect my own treatment, but I am not persuaded that Darwin suffers seriously from it— a discussion might turn on the difference between an idea (in Blyth) and a theory (in Darwin). Gertrude Himmelfarb's *Darwin and The Darwinian Revolution* (New York, 1959) denigrates Darwin and his work. The book is fuller biographically than any other. But the author's attitude to science and its history is like that of Arthur Koestler, and bespeaks the same offense which objective science and its founders seem to give the literary temperament, or at least the temperament which would merge psychologically into nature. Another, and a more widely acclaimed book, William Irvine's *Apes, Angels, and Victorians* (New York,

1955), treats the founders of evolutionary science as figures of fun.

The most useful book in the spate loosed by the centennial year is Nora Barlow's edition of her grandfather's *Autobiography* (London, 1959) which restores the passages omitted from the first publication in *The Life and Letters of Charles Darwin* (edited by Francis Darwin, 3 vols., London, 1888), and contains additional material on Samuel Butler and Darwin. Students will find convenient the selections and arrangement in *The Darwin Reader* (edited by Marston Bates and Philip S. Humphrey, New York, 1956), and a republication of the original Darwin and Wallace papers, *Evolution by Natural Selection* (with a foreword by Sir Gavin de Beer, Cambridge, 1958). David Lack's *Evolutionary Theory and Christian Belief* (London, 1957) is an admirably temperate statement of a view that it does not, however, seem necessary to me to hold— i.e. that there is a real conflict.

Major treatises relevant to the history of late nineteenth-century evolutionism have been translated: Hugo Iltis' *Life of Mendel* (trans. by Eden and Cedar Paul, London, 1932); Mendel's own original paper, *Experiments in Plant Hybridization* (edited by William Bateson, Cambridge, Mass., 1948); Ernst Haeckel's *The Riddle of the Universe* (New York, 1901); *The History of Creation* (London, 1876); *The Evolution of Man* (New York, 1892); August Weismann's *Essays upon Heredity* (2nd ed., 2 vols., Oxford, 1891-1892); and *The Germ-Plasm* (New York, 1893). Karl Wilhelm Nägeli's *Mechanisch-physiologische Theorie der Abstammungslehre* (Munich, 1884) does not seem to have been Englished, but there is an abstract of the argument, *A Mechanico-Physiolog-*

ical Theory of Organic Evolution (Chicago, 1898), trans. by V. A. Clark. Erik Nordenskjold's *The History of Biology* (New York, 1928) must still serve as the best general history of the science.

CHAPTER IX

The foremost essay on the history of thermodynamics is Thomas S. Kuhn's "Energy Conservation as an Example of Simultaneous Discovery" (*Critical Problems in the History of Science*). Professor Kuhn employs rather a taxonomic than an evolutionary approach to establishing the filiation of ideas, and his treatment is informed by a thorough mastery of the issues. A most suggestive book, one which is too little known, is particularly valuable in relating the work of Clausius to that of Carnot: Charles Brunold's *L'Entropie, son role dans le développement historique de la thermodynamique* (Paris, 1930). Throughout the whole discussion, I have found invaluable P. W. Bridgman's *The Nature of Thermodynamics* (Cambridge, Mass., 1941). Gerald Holton's *Introduction to Concepts and Theories in Physical Science* (Cambridge, 1952) is a textbook which makes liberal and intelligent use of historical materials. The older literature includes important positivist contributions: Ernst Mach's *Die Prinzipien der Wärmelehre, historisch-kritisch entwickelt* (Leipzig, 1923); Pierre Duhem's *L'évolution de la mécanique* (Paris, 1905), which is historical in purpose; and (more important) Duhem's *Traité d'énergetique* (2 volumes, Paris, 1911). Though of more recent date, Ernst Cassirer's *The Problem of Knowledge* (New Haven, 1950) seems immensely more remote because of its metaphysical approach. Nevertheless, Cassirer understood (perhaps because he approved of) the tenets of Mayer's energeticism better than other commentators. As for the original papers, Sadi Carnot's *Réflexions sur la puissance motrice du feu* (1824) was re-issued in Paris in 1878. Translations of Carnot, together with memoirs by Clausius and Thomson, are printed in *The Second Law of Thermodynamics* (New York, 1899) edited by W. F. Magie. A general treatise of Clausius has also been put into English: *The Mechanical Theory of Heat* (trans. by Walter Browne, London, 1879). Joule's experiments are gathered without benefit of serious editorial attention into two volumes, *The Scientific Papers of James Prescott Joule* (London, 1884-87). H. Helmholtz's *Über die Erhaltung der Kraft* became the first of Ostwalds Klassiker (Leipzig, 1889). Mayer's two basic papers followed under the somewhat inappropriate title *Die Mechanik der Wärme* (Leipzig, 1911). There is an interesting edition of Mayer's correspondence *Robert von Mayer über die Erhaltung der Energie*, edited by W. Preyer (Berlin, 1889). Finally, one wonders when the historiography of science will surpass that monument of nineteenth-century scholarship, John Theodore Merz's *A History of European Thought in the Nineteenth Century* (4 vols., Edinburgh, 1896-1914).

CHAPTERS X AND XI

Ronchi's *Storia della luce* remains a helpful guide in following the transition from the particle to the wave theory of light. The quotations of Young's correspondence are from Frank Oldham and Alexander Wood's *Thomas Young, Natural Philosopher, 1773-1829* (Cambridge, 1954). Young summarized his earlier papers in the optical sections of *A Course of Lectures on Natural Philosophy* (2 vols.; London, 1807), which work provides (incidentally) an excellent conspectus of the state of physics at the beginning of the nineteenth century. A lecture by Louis de Broglie interprets the work of Fresnel in "La physique moderne et l'oeuvre de Fresnel" (*Recueil d'exposés sur les ondes et les corpuscles*, Paris, 1930). Emile Verdet's introduction to the *Oeuvres d'Augustin Fresnel* (3 volumes, Paris, 1866-1869) is discursive but, by the same token, very full and informative. Passages from Fresnel are translated from this collection. Ernst Mach is always to be read with profit: *The Principles of Physical Optics* (trans. by John S. Anderson and A. F. A. Young, London, 1926). No book bears so directly on the subject of this whole chapter as E. T. Whittaker's *History of the Theories of Aether and Electricity* (2 volumes; London, 1951-1953). Nevertheless, Whittaker must be used with a certain caution, for he expresses the physics of the nineteenth century in the formalism of the twentieth, and he arranges his materials in accord with their relations in the modern order of physics rather than in the order of history and discovery.

Everyone should begin the study of Faraday with John Tyndall's *Faraday as a Discoverer* (London, 1868), a charming memoir and a standing rebuke to Strachey's strictures on Victorian biography. Two articles by Robert C. Stauffer deal with the beginnings of electromagnetic science: "Persistent Errors Regarding Oersted's Discovery" (*Isis*, *44*: 307-310, 1953) and "Oersted's Discovery of Electromagnetism" (*Isis*, *48*: 33-50, 1957). Thomas Martin, the editor of Faraday's *Diaries*, has a meticulous little study, *Faraday's Discovery of Electro-Magnetic Induction* (London, 1949). And there are a number of editions of Faraday's collected papers, *Experimental Researches in Electricity*, of which the most recent is in the Chicago "Great Books" collections, volume *45* (Chicago, 1952), together with Lavoisier's *Elements of Chemistry*.

The *Scientific Papers of James Clerk Maxwell* were edited in two volumes by W. D. Niven (Cambridge, 1890) and reprinted in a photographic edition by Hermann (Paris, 1927). All my quotations from Maxwell are drawn from the papers contained therein. Lewis Campbell and

William Garnett's *The Life of James Clerk Maxwell* (London, 1882) is a full-scale but uncritical biography. More helpful, really, is *James Clerk Maxwell, a Commemoration Volume, 1831-1931* (Cambridge, 1931), containing essays by J. J. Thomson, Max Planck, Albert Einstein, James Jeans, et al. The positivists are worth reading as always; but Pierre Duhem's *Les théories électriques de J. Clerk Maxwell* (Paris, 1902) seems to me in less happy vein than other of his writings.

The epilogue to the present book does not purport to be history of science proper. It is rather commentary upon the reading of Pierre Duhem's *The Aim and Structure of Physical Theory* (trans. by Philip P. Wiener, Princeton, 1954); and *L'évolution de la mécanique*; Wilhelm Ostwald's *Die Energie* (Leipzig, 1908); Ernst Mach's *The Science of Mechanics* (trans. by T. J. McCormack, 2nd ed., Chicago, 1902); C. B. Weinberg's *Mach's Empirio-Pragmatism in Physical Science* (New York, 1937); *H. A. Lorentz, Impressions of His Life and Work* (edited by G. L. de Haas-Lorentz, Amsterdam, 1957); Cornelius Lanczos' "Albert Einstein and the Role of Theory in Contemporary Physics" (*American Scientist, 47*: 41-59, 1959); Albert Einstein and Leopold Infeld's *The Evolution of Physics* (New York, 1938); Albert Einstein's *The Meaning of Relativity* (5th edition, Princeton, 1955); and the various contributions, especially Einstein's own, to *Albert Einstein, Philosopher-Scientist* (edited by Paul Arthur Schilpp, New York, 1951), from which the quotations of Einstein's words are made.

INDEX

Abano Terme, 57

Aberdeen, 478

absolute zero, 394, 396, 480

abstraction, 42, 52, 90, 109, 126, 195, 246, 353, 440, 441, 447, 460, 502; in mechanics, 384, 385

Accademia dei Lincei, 110

Accademia del Cimento, 110

Académie française, 112

Académie (royale) des sciences, see Academy of Sciences (French)

Academy of Dijon, 180

Academy of Sciences (French), 109, 112, 174, 176, 202, 211, 212, 216, 220, 225, 229, 232, 234, 238, 246, 268, 279, 387, 423

acids, 222-224, 225, 226-227, 234, 235, 241-245, 246, 447

acoustics, 383

acquired characters, 270, 322, 325

action at a distance, 144, 449, 453, 471, 474, 506, 518; medium for, 435, 460

Adam, Charles, 530

Adams, Henry, 405

adaptation, 266, 284, 302, 310, 317, 492

adiabatic changes, 365, 367, 368, 397

aether, 34, 129-131, 133, 145, 148, 149, 205, 237, 238, 240, 241, 355, 356-357, 367, 369, 380, 407, 408, 413, 421, 426-428, 429, 433, 434-435, 457, 465, 469, 471, 472, 475, 476, 477, 478, 489, 490-492, 493, 506, 507-510, 519

agnosticism, 347, 348, 349, 502

alchemy, 104, 106, 122, 205, 212

d'Alembert, 165, 188, 199, 361

Alexander, H. G., 532

algebra, 3, 87, 167, 245, 249, 375, 377, 462, 463

Ampère, 392, 430, 433, 442-443, 453, 459

Amsterdam, 335

analytical geometry, 37, 84, 86-91, 250, 355

anatomy, 55, 56-73, 111, 179, 278, 291, 325; comparative, 267, 279, 282-284, 285, 286, 287, 288, 307, 318

Anglicanism, 115, 205, 295, 410, 420

animism, 39, 130

anticlericalism, 278

Aquinas, St. Thomas, 159

Arago, 407, 410, 421, 423, 426, 430, 432, 439, 444

Archimedes, 3, 14, 16, 40, 42, 52, 68, 101, 142, 144, 505

Aristarchos of Samos, 17, 20, 21

Aristotelianism, 7, 11-16, 17, 27, 33, 34, 41, 49, 50, 74, 96, 99, 141, 183, 263, 267, 268

Aristotelian physics, 11-14, 65

Aristotle, 11-16, 42, 43, 50, 59, 63, 64, 72, 73, 74, 79, 80, 82, 85, 96, 144, 145, 170, 181, 249, 266, 285, 288

astronomy, 17-39, 49, 55, 111, 143, 346, 354

Atlantis, 81, 111

atmosphere, composition of, 213, 214, 217, 218, 219, 221, 224

atomism, 15, 88, 95-99, 108, 114, 133, 144, 146, 148, 181, 190, 201, 250, 251, 253-259, 311, 332, 340, 341, 345, 355, 392, 401, 409, 413, 421, 448, 450, 477, 480, 482, 484, 487, 488, 489, 490, 491, 495, 498, 507, 509, 512; of Greeks, 15, 96, 182-183; of Boyle, 104-107; of Locke, 163-164; of Lavoisier, 250; of Faraday, 454-457; of Maxwell, 466-492; and positivism, 499-500, 503; and Einstein, 513

attraction, 142, 143, 145, 146, 158, 163, 191, 480; between particles, 186, 237, 238, 461, 467. *See also* gravity

Auzout, Adrien, 125
Avogadro's Hypothesis, 254, 258; number, 480

Babbage, Charles, 439
Bacon, Francis, 69, 74-82, 83, 108, 109, 111, 113, 152, 166, 169, 181, 189, 264, 502, 529. *See also* Baconianism
Baconianism, 54, 75, 77, 78, 82, 107, 125, 136, 145, 153, 162, 166, 174, 197, 229, 246, 310, 312, 314, 318, 357
Bacon, Roger, 80
Baer, Karl Ernst von, 261
Bakerian lecture, 412, 414
Balfour, A. J., 349
Ball, W. W. Rouse, 532
Barlow, Nora, 539
barometry, 110
Barraclough, June, 534
Barrow, Isaac, 118
basic research, 445
Bates, Marston, 539
Bateson, William, 539
Batsch, 193
Baumgardt, Carola, 527
Bayen, 219
Beagle, 308, 309, 313, 319
Becquerel, Henri, 385
de Beer, Gavin, 539
Bellarmine, Robert Cardinal, 46, 346
Bénac, Henri, 534
Bentham, Jeremy, 151, 154, 246
Bera, M. A., 530
Bergson, Henri, 344
Berkeley, 162, 164, 503, 504
Berlin, 323, 382; Physical Society of, 390
Berlin, Isaiah, 534
Bernard, Claude, 261, 320-321, 322, 382
Bernoulli, Daniel, 480
Berthelot, René, 535
Berthollet, 177, 251, 252
Besterman, Theodore, 534
Bible, 264, 265, 281, 298, 300, 348, 520
Bichat, Xavier, 177, 261

biology, 13, 55, 57-59, 60, 73, 92, 179, 196, 203, 260-291, 303-351, 379, 380, 393, 402, 500-501; in Enlightenment, 156, 179, 184, 192, 197, 199; revolution in, 337, 340
biophores, 326-327
Biot, 425-426
Birch, Thomas, 530
Birmingham, 208, 252, 306
Black, Joseph, 206-207, 212, 232, 238, 293, 411, 536, 537
Blagden, Charles, 228
Blake, William, 178
Blois, 231
Blum, H. F., 339
Blyth, Edward, 538
Boas, Marie, 530
Boehme, Jakob, 128
Boerhaave, 411
Bohr, Niels, 513
Boklund, Uno, 536
Bologna, 20, 61
Boltzmann, Ludwig, 361, 483, 486, 493, 498, 512
Borelli, 110
Boscovich, 455, 490
botany, 59, 112, 170, 176, 180, 192-194, 267, 269, 312, 323
Boyle, Robert, 103-109, 113, 114, 130, 133, 136, 161, 185, 204, 250, 253, 337, 530-531; pneumatic experiments of, 103, 104, 105; corpuscular philosophy of, 104, 105, 106-107, 122; and method, 107-109
Boyle's Law, 104, 254, 364, 368, 480
Bridgman, P. W., 382, 516, 540
British Association for the Advancement of Science, 372, 478
Broglie, Louis de, 541
Brongniart, Alexandre, 268, 289, 295
Brougham, Henry, 420
Brouncker, Lord, 114
Browne, Thomas, 302
Brownian movement, 512
Brunel, 358
Brunn Society for the Study of Natural Science, 329
Bruno, Giordano, 27, 84
Brunold, Charles, 362, 540
Brussels, 60

Buckland, William, 297, 300
Buffon, 189, 268, 269, 279
Buridan, Jean, 41
Burke, Edmund, 178, 201, 216
Butler, Samuel, 345, 350, 539
Butterfield, Herbert, 526

Cabanis, 178
Caen, 278, 423
Calais, 445
calcination, 202, 212, 214, 218, 220, 222, 242
calculus, 87, 188, 368, 380, 422, 423, 475; discovered by Newton, 119, 120, 140, 149; differential, 368; of probability, 483
calendar, 18, 22
caloric, 202, 205, 227, 235, 236-241, 247, 248, 250, 255, 356, 357, 359, 360, 361, 362-370, 373, 380, 387, 391, 392, 395, 396, 406, 407, 408, 409, 462, 463
calorimetry, 238, 239, 370, 373, 383
Calvin, John, 66
Calvinism, 29, 114, 277
Cambridge, 61, 68, 118, 119, 139, 210, 297, 307, 310, 411, 460, 478
Campbell, Lewis, 542-543
Canterbury, 68
Carlyle, Thomas, 178
Carnegie, Andrew, 343
Carnot, Lazare, 358
Carnot, Sadi, 241, 352, 357-370, 373, 380, 387, 389, 391, 392, 393, 394, 395, 396, 397, 401, 406-407, 422, 462, 463, 540-541
Cartesianism, 84, 85, 86, 87, 91, 94, 108, 121, 142, 145, 146, 147, 158, 159, 161, 166, 186, 197, 198, 210, 307, 323, 335, 344, 354, 358, 360, 408, 458, 468, 488, 492, 493, 515
Carus, 320
Caspar, Max, 527
Cassirer, Ernst, 342, 540
Castlereagh, 308
catastrophism, 290, 295, 298, 300, 301
causality, 11, 93, 94, 145, 147, 162,

317, 318, 334, 335, 344, 350, 375, 381, 384, 385, 401, 403, 462
Cavendish, Henry, 207, 217, 228, 229
Cavendish Laboratory, 478
Cesi, Prince Federigo, 110
Champollion, 412
change, 6, 11, 14, 97, 106, 120, 241, 299-300, 338-339, 355, 376, 386
Charles I, 69
Charles II, 113
Charles V, 60
Chauvois, Louis, 529
Chelsea Hospital, 164
chemistry, 93, 97, 104, 106, 107, 111, 202-259, 260, 270, 274, 286, 338, 354, 370, 376, 378, 379, 385, 409, 436, 437, 438, 443, 446, 447-448, 451, 455, 479, 484, 494, 501; atomic, 97, 185, 204, 251, 252, 253-258, 354, 454-455, 499; Newton's work on, 122, 134; in Enlightenment, 157, 170, 184-187, 188, 189, 226; revolution in, 157, 202-203, 214, 230-231, 233, 258, 335, 337; and Lavoisier, 202-204, 209-218; and Priestley, 208-210; mathematicization of, 245-246, 249; and Dalton, 250, 252-258; and positivism, 501
Cheselden, Dr., 164
Christianity, 8-9, 13, 26, 29, 46, 61, 98, 101, 160, 181, 183, 188, 210, 263, 347-351; hostility toward, 114, 151
chromosome, 326, 328
circles, 14, 20, 24, 27, 28-29, 34, 37, 77, 84, 492, 519
circularity, 19, 24-25, 28, 51
circular motion, see motion, circular
circulation of the blood, 58, 64-73, 92, 262
Clagett, Marshall, 526, 533
Clapeyron, 365, 369, 373, 387, 396
Clark, G. N., 531
Clarke, Samuel, 148
classification, 12, 59, 79-80, 109, 153, 166, 169, 170, 173, 174, 175, 177, 193, 234, 246, 248, 250, 260, 262,

266, 277, 294, 295, 340, 500, 502; Linnaean, 170-171
Clausius, Rudolf, 369, 389, 395-401, 481-482, 498, 540, 541
Cohen, I. Bernard, 532
Cohen, Morris R., 526
Colbert, 111, 112
Coleridge, 178
Collège de France, 176, 177, 279
Collège Mazarin, 210
color, 106, 123, 124, 125, 126, 128-129, 131-133, 134, 179, 192, 193, 194-195, 205, 406, 409, 412, 413, 415, 416, 418, 421
Columbia University, 336
combination, chemical, 227, 230, 251, 252, 254, 255
combining volumes, law of, 484
combustion, 173, 202, 203, 204, 211-212, 213, 214, 217, 218, 219, 220, 222, 224-226, 229, 230, 235, 242, 248, 258, 274, 275, 361
communication, 48, 62, 109-113, 163, 166, 167, 168, 178, 189, 352, 404, 517
Comte, Auguste, 496, 502, 534
Conant, James B., 530, 537
Condillac, abbé de, 163, 164, 165-170, 171, 172, 175, 180, 203, 217, 236, 247, 250, 258, 448, 496, 502, 534
Condorcet, 165, 171, 172, 175, 534
Conduitt, John, 137
conservation, 51, 66, 96, 97, 218, 231, 235, 239, 241, 251, 283, 354, 356, 357, 361, 366, 367, 376, 383, 385, 386, 387, 388, 389, 391, 393, 394, 395, 397, 402; of energy, 283, 370, 374, 381, 385-386, 390-391, 396, 399, 400, 474, 475, 476, 486, 487, 498
conservatism, 200-201, 286-287, 420
Conservatoire des arts et métiers, 176, 216
continuum, 88, 90, 91, 95, 144, 183, 186, 187, 198-199, 240, 341, 357, 368, 434, 456, 461, 476, 487, 507, 518; biological, 193, 198, 271, 276
convertibility, 370, 371, 372, 394, 395, 396

Conybeare, William Daniel, 296
Cope, J. I., 531
Copernicanism, 3, 29, 30, 31, 46-47, 49, 50, 134
Copernicus, 18-27, 29, 33, 39, 41, 43, 47, 51, 57, 66, 72, 74, 77, 84, 88, 92, 268, 498, 526-527
Correns, Carl, 336
Cork, 372; Earl of, 103
corpuscular mechanics, 354, 356, 390, 394
corpuscular philosophy, 182, 258, 335, 380, 401, 454, 480, 491; of Boyle, 104, 105, 106-107, 122
corpuscular theory of light, 132, 354, 406, 413, 415, 418, 419, 420, 422, 428, 429, 431, 433, 434, 435, 459, 513
correlation of parts, 282-284
cosmogony, 291-292, 341
cosmology, 11, 14, 16-51, 84, 113, 143, 341, 497; Newtonian, 32, 143, 152; Cartesian, 91-93, 414; and Stoicism, 182-183; of Diderot, 191
cosmos, 9, 11, 14, 16, 27, 29, 32, 37, 39, 50, 51, 159, 182, 184, 188
Cotes, Roger, 148
Coulomb, 354, 459
Crew, Henry, 528
Crombie, A. C., 526, 532
Cromwell, Oliver, 112, 113
Curie, Pierre and Marie, 495
Cuvier, Georges, 177, 261, 267, 268, 269, 270, 277-291, 295, 298, 300, 341, 538

Dalton, John, 107, 250, 252-259, 335, 337, 339, 368, 371, 448, 537
Dantzig, 23
Darwin, Charles, 57, 59, 203, 217, 241, 260, 261, 266, 269, 270, 276, 283, 284, 297, 300, 301, 302, 303-351, 403, 451, 492, 519, 537-539; *On the Origin of Species*, 260, 269, 303, 306, 315-320, 330, 337, 347, 349. *See also* Darwinism
Darwin, Erasmus, 306, 307
Darwin, Francis, 539
Darwin, Dr. Robert, 307

Darwinism, 187, 191, 321, 326, 327, 340, 342, 343, 402, 404
Daudin, Henri, 538
Daumas, Maurice, 236, 535-536
Davy, Humphry, 370, 411, 437
Definite Proportions, Law of, 252, 448, 454
Delille, 288
Democritus, 96, 99, 107, 133, 361, 384
"dephlogisticated air" of Priestley, 220, 221, 224, 226. *See also* oxygen
Descartes, René, 73, 77, 82-95, 99, 107, 109, 111, 112, 118, 120, 122, 144, 150, 152, 153, 157, 158, 159, 169, 189, 198, 217, 367, 414, 456, 477, 514, 529-530, 534. *See also* Cartesianism
Devonshire, Duke of, 207
De Vries, Hugo, 335-336
Diderot, Denis, 164, 165, 173-174, 180-181, 182, 184, 187-188, 197, 200, 233, 261, 263, 269, 275, 307, 341, 345, 347, 438, 533, 534; *Encyclopedia*, 173, 174, 180, 184-187, 189, 233; natural philosophy of, 188-192, 262
Dijon, Parlement of, 233
Directory, 176
dissection, 61-63, 65, 262, 286, 340
diurnal motion, *see* motion, planetary
Dover, 445
Drabkin, I. E., 526
Drake, Stillman, 528
Dreyer, J. L. E., 527
Driesch, 344
du Châtelet, Marquise, 158
Dugas, René, 530
Duhem, Pierre, 493, 495, 497-498, 500, 501-502, 524, 540, 543
Dulong, 377
Dumas, J. B., 537
dynamics, 16-17, 32, 37, 41, 51, 54, 55, 99, 111, 120, 134, 135, 286, 379, 383, 384, 386-387, 390, 401, 420, 428, 443, 453, 474, 483, 484, 498, 499; of the body, 68; geological, 300; wave, 421; fluid, 432; and gases, 481-482

eccentric, 18, 24-25
Eckerman, 179
École des mines, 176
École des ponts et chaussées, 176
École normale, 176
École polytechnique, 176, 177, 358, 369, 421, 422
economy, 24, 93, 126, 171, 235, 285, 301
Eddington, Arthur, 396, 402
Edinburgh, 206, 208, 293, 307, 308, 310, 411
Edison, Thomas A., 80
educationism, 168, 169, 172, 175-177, 203, 246, 249, 258, 404, 496
Einstein, Albert, 9, 16, 34, 80, 85, 88, 95, 198, 304, 352, 353, 359, 422, 444, 456, 474, 477, 494, 495, 497, 503, 506, 510-520, 543
Eiseley, Loren C., 538
electricity, 148, 173, 204, 205, 354, 360, 361, 370, 371, 372, 374, 375, 379, 380, 381, 385, 391, 392, 400, 421, 435, 438, 439, 441-449, 451, 452, 454-455, 459, 464, 465, 467-472, 474, 475, 499, 507, 509; static, 392, 442, 458, 464; galvanic, 392, 442, 444; frictional, 442
electrochemistry, 437, 448
electrode, 447-448
electrodynamics, 376, 392, 442, 459, 472, 473, 478, 498, 508, 509
electrolysis, 446-448
electrolyte, 448, 449
electromagnetism, 140, 356, 357, 372, 426, 441-449, 453, 461-476, 487; and light, 471, 473, 490, 491, 494, 506, 507, 510, 511
electron, 155, 494, 507, 513
electronics, 389
electrostatics, 354, 465, 471, 472, 509
electrotonic state, 465-466
elements, chemical, 250, 258
Eliot, George, 349
Elizabeth I, 61, 103
empiricism, 54, 68, 77, 83, 156, 159, 160, 162, 166, 287, 318, 456, 503, 505

energetics, 352-405, 406, 477, 487, 498, 501, 502, 512, 513

energy, 242, 245, 361, 366, 372, 380, 383, 385, 389, 393, 400-401, 402, 403, 412, 470, 485, 487, 491, 495, 498, 499, 512; kinetic, 239, 377, 389, 390, 394, 470, 471, 481, 484, 499; conservation of, 283, 370, 374, 381, 385-386, 390-391, 396, 399, 400, 474, 475, 476, 486, 487, 498; potential, 376-377, 379, 389, 390, 471, 476, 499; chemical, 391, 392, 393; electrical, 392, 472; in extension, 455-456; mechanical, 476; of electromagnetic field, 476; and positivism, 498, 500-502

English Constitution, 174

Enlightenment, 77, 112, 114, 151-201, 206, 232, 233, 249, 263, 294, 342, 345, 354, 404, 420, 488, 496, 497, 502-503

Ent, George, 112

entropy, 339, 362, 369, 394, 396-401, 402, 404-405, 516

Epicureanism, 96, 97-98, 104, 164, 182, 183, 355

Epicurus, 96, 98, 99

epicycle, 18, 24-25, 33

d'Epinay, Mme, 165

epistemology, 162, 383, 495, 503

equant, 18, 25

equations: chemical, 245, 249; differential, 359, 384, 426; field, 472, 474-475, 485; kinetic, 483-484

l'Espinasse, Mlle. de, 165

ethics and science, 44, 154-155, 342, 348-350

Euclid, 3, 16, 40, 45, 88, 91, 118, 142, 151, 266, 414, 420, 453, 513; Euclidean space, 84, 87, 355, 491

Evelyn, John, 114

evolution, 55, 155, 177, 191, 198, 260, 231, 330, 335, 402, 404; time as coordinate of, 339, 351, 402. See also evolutionary theory

evolutionary theory, 261, 284, 300, 308, 312, 321-322, 323, 328, 346, 347, 348, 350; Lamarck, 261-262, 269-277; Darwin, 302, 313-320,

337-338, 342, 343; Nägeli, 323-325

experimentalism, 10, 54, 68, 71, 79, 83, 107-109, 188, 204, 217-218, 221, 223, 261, 286, 291, 384, 437, 439-441, 450, 483. See also Baconianism

extension, 86, 92-93, 95, 96, 146, 249, 356, 360, 367, 376, 379, 380, 381, 384, 400, 488, 490, 491, 504, 507; energy in, 455-456

Fabrici d'Acquapendente, 67-68

falling bodies law of, 3-7, 11, 15, 42, 46, 50, 51, 59, 87, 121, 134-135, 474, 518

Faraday, Michael, 352, 435-458, 459, 460, 462, 464, 465, 466, 468, 485, 493, 542; and electromagnetic induction, 439, 441-445, 449; and view of structure of matter, 441, 446, 448-449, 451-458; and atomism, 454-457; theory of space, 445, 454-458, 462

Farrington, Benjamin, 529

Favaro, Antonio, 528

Ferdinand II, Grand Duke (of Tuscany), 110

Fichte, 178

field: electromagnetic, 464, 466-467, 470, 473-476, 490, 506, 509, 511; theory, 357, 506, 507, 518; physics, 457-492, 493, 494; Maxwell's Laws of, 466, 474, 508, 510, 517; magnetic, 469, 471, 474, 475, 509; electrical, 474, 475; equations, 472, 474-475, 485, 507, 514, 524

fire, see pyrotic theory

Fitzgerald, G. F., 509-510

Fitzroy, 308

Fizeau, Hyppolyte, 427-428, 473

Flamsteed, John, 124

Florence, 48, 52, 110

fluid, 272, 276, 356, 430, 489; heat as, 235, 239, 391; and electricity, 448, 450, 499; and light, 407, 408, 414, 430, 432, 433; and electromagnetism, 461-465, 468, 471, 485, 490. See also aether; caloric

flux, 97, 198, 270, 277, 338, 339, 341, 368, 464, 467, 471, 473, 475
Fontenelle, 118, 124, 147, 335, 344, 531
Forbes, Edward, 313
force, 43, 343, 372, 373, 374, 376, 378-381, 383, 385, 386, 387, 388, 389, 390, 393, 400, 401, 442, 449, 477, 507, 509; law of, 32, 51, 469, 490; gravitational, 33, 51, 135, 136, 355, 360, 389, 457, 467; centrifugal, 93, 120, 136, 374, 468; and Newton, 120, 121, 142, 143, 144, 354, 360, 389, 390, 503; centripetal, 121; centers of, 142, 389, 449, 455, 458; electromagnetic field of, 240, 426, 459, 474; plastic, 271, 274; causal, 375, 376, 380; lines of, 436, 438, 449, 451, 452-453, 458, 459, 460, 461-469, 475, 490; tubes of, 457, 462, 463, 468. *See also* caloric
form, 12, 15, 16, 59, 96, 106, 108, 180, 246, 249, 270, 284-285, 291, 379, 381, 441, 461; organic, 269, 271, 302, 319, 320, 330
formulas, chemical, 242-245
Foster, Michael, 529
Foucault, 428, 505
Fourier, 358
Franklin, Benjamin, 114, 354
Franklin, Kenneth V., 529
Frauenburg (Poland), 20, 22, 23
Fréchines, 231
Freiberg-im-Sachsen, 292
French Republic, 55, 175, 176, 268
French Revolution, 174, 175, 178, 246, 251, 268, 278, 354
Fresnel, Augustin, 407-410, 421-435, 506, 507, 509, 541-542
Freud, Sigmund, 164, 307
Fric, René, 537
friction, 11, 371, 372, 373, 374, 387, 399, 446, 501
Froude, James Anthony, 349
Fulton, John, 530

Galen, 62, 63-66
Galileo, 3-8, 11, 13, 14, 15, 16, 19, 26, 27, 37, 39-53, 54, 57, 58, 66, 67, 68, 73, 74, 79, 81, 83, 85, 87, 90, 93, 99, 100, 101, 104, 109, 111, 112, 113, 118, 120, 133, 134, 135, 140, 142, 143, 181, 202, 217, 241, 305, 378, 497, 498, 505, 519, 520, 527-528; and motion, 3-7, 41-42, 44-45, 50-52, 89, 90-91, 106, 120, 133, 241, 249, 338, 339, 355, 515; and Platonism, 39-40, 49, 57
Galois, Evariste, 422
Galvani, 371
Garnett, William, 543
gas chemistry, 203, 205-209, 211-212, 213, 214, 219, 253, 254, 255
gases, kinetic theory of, 479-487, 494, 499
Gassendi, Pierre, 99, 530
Gauss's law of error, 483
Gay, Peter, 534
Gay-Lussac, 258, 368, 377, 426, 484
gene, 328, 334, 335, 337, 338
genetics, 190, 318, 320, 322, 326, 328, 334, 336-337, 338, 341, 346
Geneva, 66, 480
geography, 307, 319
geology, 55, 206, 211, 265, 266, 270, 289, 290, 291-302, 307, 308, 309, 312, 318, 319, 347, 403; pre-Darwinian, 268; religion and, 295, 297-299, 300. *See also* catastrophism; Neptunism; uniformitarianism; Vulcanism
geometry, 4-7, 16, 27, 32, 35-37, 38, 49, 55, 86-91, 111, 118, 380, 441, 456, 462, 488, 501, 518; Newton's use of, 140-141, 143, 160, 307, 343, 344; descriptive, 228; Cartesian, *see* analytical geometry
George III, 151
germ-plasm, 325-328, 335
Gibbs, Willard, 361, 494, 498, 512, 524
Gilbert, William, 32, 74, 77
Gladstone, 350
Glasgow, 206
Glass, Bentley, 538
Glisson, Francis, 112
Gobineau, 343
Goddard, Jonathan, 112
Goethe, 107, 125, 179, 180, 192-198,

200, 263, 277, 287, 288, 307, 320, 321, 341, 347, 357, 379, 535
Goodrich, B. H., 531
Göttingen, 411
Gouhier, Henri, 534
gravimetrics, 204, 207, 212, 258
gravitation, 33, 51, 135, 136, 354, 360, 388, 389, 457, 458, 467, 468, 480
gravity, 16, 32, 51, 58, 77, 93, 94, 119-120, 121-122, 135, 136, 137, 142, 143, 145-147, 154, 168, 186, 238, 294, 317, 318, 319, 334, 335, 337, 354, 355, 360, 372, 380, 385, 389, 391, 430, 441, 449, 451, 453, 457, 459, 461, 467, 474, 480, 505
Greek philosophy, 9, 16, 144, 182, 338, 341, 456. See also atomism; Stoicism
Greek science, 10-16, 51, 79, 81, 92, 94, 97, 155, 212
Grimaldi, 424
Grimaux, Edouard, 537
Guericke, Burgomaster Otto von, 100-101
Guerlac, Henry, 529, 535-536
Guyton de Morveau, 233-234, 236, 247

Haak, Theodore, 112
Haas-Lorentz, G. L. de, 543
Haeckel, Ernst, 321-322, 382, 539
Hales, Stephen, 205-206, 207, 212, 536
Hall, A. R., 528
Halley, Edmond, 114, 136, 137, 138
Hanson, N. R., 527
harmony, 15, 28, 38-39, 146, 152, 311, 342, 343, 381, 383, 500
Harvey, William, 41, 58, 66, 68-73, 74, 75, 79, 92, 108, 262, 529; and theory of circulation, 66, 68-73, 262
heat, 93, 106, 140, 148, 203, 204, 207, 222, 227, 235-241, 242, 253, 255, 276, 293, 356, 383, 385, 386, 387, 389, 391, 392, 394, 395, 396, 397, 402, 404, 406, 408, 409, 421, 441, 451, 455; as motion, 357-370, 376, 481; mechanical equivalent

of, 369, 374, 375, 378, 381, 391; as substance, 370-381; specific, 377, 378, 387, 394, 480, 484-485, 491; mechanical theory of, 387, 395, 400, 401; as energy, 391, 400, 470. See also convertibility; fluid
Hegel, 323, 328, 342, 383, 496
Heilbronn, 375
Heilsburg (Poland), 22
Helmholtz, 352, 374, 383-394, 396, 397, 401, 412, 449, 489, 490, 541
Henderson, L. J., 357
Henry, William, 253; Henry's Law, 254
Heraclitus, 182, 224, 341
Herapath, John, 480-481
Herder, 200
heredity, 322, 325, 328, 329, 331-337, 341
d'Héricy, Comte, 278
Herschel, William, 406, 439
Hertz, Heinrich, 493, 494, 511
Himmelfarb, Gertrude, 538
Hippocrates, 63, 186
Hitler, 343
Hobbes, 146
d'Holbach, 165
Holton, Gerald, 378, 527, 540
Home, Henry, 206
Hooke, Robert, 109, 114, 125, 126-127, 128, 129, 131, 132-133, 134, 135, 136, 137, 138-139, 145, 148
Hooker, J. D., 313, 315
Huguenots, 277
humanism, 10, 13, 20, 40, 46, 49, 54, 55, 61, 77, 197, 200, 517, 520
humanitarianism, 81, 82, 152, 153, 208, 249
human nature, 77, 82, 152, 159, 189; science of, 163-164, 168
Humboldt, Alexander von, 268
Hume, David, 162, 206
Humphrey, Philip S., 539
Hutton, James, 293-294, 295, 299, 402
Hutton, Joseph, 206
Huxley, Thomas Henry, 261, 302, 305, 313, 321, 349, 350, 382
Huygens, Christiaan, 120-121, 125,

136, 146, 160, 408, 414, 423, 429, 434

Hveen, 30

hybridization, 330, 331, 334

hydrodynamics, 142, 362, 463, 489, 499

hydrostatics, 102

hypothesis: of Newton, 126, 128-129, 147, 162; of Helmholtz, 384; of Young, 408, 412-413; of Fresnel, 422, 431; of Maxwell, 460-461, 479

Ibn al-Nafis, 67

idealism, 162, 260, 288, 321, 322, 326, 379, 383, 403, 503, 517

ideology: of Diderot, 174

idioplasm, 323-325, 326, 335, 344

Iltis, Hugo, 539

Imperial University (France), 280

Index (Congregation of the), 49

induction, electromagnetic, 357, 439, 441-445, 449, 452-465, 467, 471

industrial revolution, 173

inertia, 19, 33, 51, 55, 77, 83-84, 87, 89-90, 92, 120, 142, 283, 354, 503-505, 508, 509, 510, 514-515

inertial systems, 508, 509, 514-515

Infeld, Leopold, 514, 543

infinity, 27, 51, 84, 89-90, 96, 97, 143, 188, 402, 477, 480

Inquisition, 50, 52

Institut de France, 176, 423

interference, of light, 131-132, 410, 412, 414-421, 423, 424, 428, 430, 434, 493

Irvine, William, 539

isothermal change, 365, 367, 368, 397

Jacobin Terror, 175, 176, 278

James I, 69

Jansenism, 115

Jardin des Plantes, see *Muséum d'histoire naturelle*

Jeans, James, 543

Jefferson, Thomas, 151, 279

Joule, James Prescott, 370, 371-375, 377, 378, 384, 386, 387, 392, 394, 395, 397, 401, 481, 524, 541

Jones, H. W., 531

Jones, Richard Foster, 529, 531

Jung, 164

Jupiter, moons of, 37, 47

Kant, 375, 376, 379, 383, 384, 385, 503

Karlsruhe, 493

Kelvin, Lord, 394-395, 396, 403, 465, 467, 489, 490, 492, 541

Kepler, Johannes, 14, 25, 26, 27-39, 43-44, 47, 51, 73, 74, 77, 88, 118, 143, 217, 440, 441, 443, 492, 497, 498, 519, 527, 532; and planetary laws, 58, 119-120, 136, 143, 402

Keynes, Geoffrey, 529

Kielmeyer, 278

kinematics, 6, 14, 19, 27, 32, 41, 120, 202, 338, 355, 391

kinetics, 255, 311, 356, 357, 361, 381, 477, 480, 481, 490, 491; kinetic theory of gases, 479-487

King's College (London), 478

Knapp, R. H., 531

Koch, Robert, 261

Koestler, Arthur, 29, 43-44, 52, 527, 538

Koyré, Alexandre, 84, 147, 527, 529, 532

Kuhlrausch, 438, 472-473

Kuhn, Thomas S., 526, 532, 540

Lacépède, 268

Lack, David, 539

Lagrange, 117, 173, 177, 178, 211, 354, 359, 360, 466, 505

Lamarck, Jean Baptiste de, 177, 260, 261, 262, 263, 267-277, 285-286, 300, 301, 306, 307, 322, 344, 347, 378, 379, 533, 537-538; evolutionary theory, 261-262, 269-277, 283, 288, 338, 339

Lamarckism, 341, 346, 500

Lanczos, Cornelius, 543

language, 113, 169, 171, 172, 233, 258, 343, 384, 404, 499, 502; Condillac's theory of, 166-168, 172. *See also* nomenclature

Laplace, 165, 173, 177, 211, 228,

229, 238, 239, 245, 319, 354, 359, 420, 425-426, 431

Latham, R. E., 530

Latreille, 268

Lavoisier, 89, 165, 184, 202-204, 209-218, 219, 220-221, 222-250, 251, 252, 253, 258, 271, 274, 276, 305, 312, 337, 338, 339, 356, 359, 360, 369-370, 385, 411, 492, 535-536, 537, 542; and combustion, 202, 203, 220-221, 224-226, 230, 235, 242, 248, 258; philosophy of science, 202-204, 211, 214, 223, 224, 230-232, 235, 245-247, 249; and chemical revolution, 202-203, 214, 230-231; *Elements of Chemistry*, 203-204, 231, 235, 237, 238, 240, 247, 249, 253; personality, 209-210, 215-218; program of research, 213-214; and heat, 222, 235; and acids, 222-224, 226-227, 234, 235; rationalization of chemistry, 232-235

Lavoisier, Madame, 215, 231, 371, 536

Lee, R., 538

Lehrs, Ernst, 535

Leibniz, 128, 145, 146, 149, 158, 307, 355, 360, 372, 532

Lenoble, Robert, 530

Leonardo da Vinci, 7, 8, 54-56, 84, 441, 497, 528

Leopold, Prince, 110

LeRoy, Georges, 534

LeSage, 480

Leucippus, 96

lever, law of the, 16, 505

Leyden, 506

liberalism, 159, 174, 343, 403, 405

Liége, 125

life, 10, 12-13, 106, 260, 272, 275, 276, 291, 327, 339, 343, 360, 371

light, 32, 87, 88-89, 93, 111, 195, 205, 235, 237, 239, 241, 248, 276, 370, 375, 383, 391, 406-435, 436, 454, 456, 461, 492, 512; Newton's view of, 120, 123, 124, 125, 126, 127, 128-129, 130-134, 148, 385; wave theory of, 121, 132, 357, 407-435, 457, 461, 472, 506; Hooke's view of, 126-127, 131, 132-133; cor-

puscular theory of, 132, 354, 406, 413, 415, 418, 419, 420, 422, 426, 428, 429, 431, 433, 434, 435, 459, 513; velocity of, 413, 417, 427-428, 472-473, 474, 508, 510, 514-517; emission theory of, 406, 408, 428, 431; transverse waves, 428-435, 465, 491, 506; polarization of, 451-452, 468, 493; and electromagnetism, 471, 473, 474, 490, 493, 506, 507, 510, 511. *See also* aether; interference; reflection; refrangibility

Linnaean Society, 315

Linnaeus (Carl von Linné), 170-171, 177, 178, 180, 192, 193, 195, 198, 267, 269, 277, 341, 411

Linus, Franciscus, 125, 128

Linz, 37

Locke, John, 113, 124, 139, 159-164, 165, 166, 172, 174, 371-372, 534

Loesche, Martin, 535

London, 69, 109, 110, 113, 125, 137, 139, 148, 157, 208, 478

Lorentz, H. A., 494, 506-510, 514, 515, 517, 518, 519, 543

Louis XIV, 111, 112

Louis XVIII, 280

Louis Philippe, 280

Louvain, 60

Lucasian Chair of Mathematics, 118, 138

Lucretius, 96, 97, 98, 490, 530

Luther, Martin, 22

Lutheranism, 23

Lyell, Charles, 261, 299-302, 308, 309, 314, 315

Lysenko, 199, 347, 350

MacCurdy, Edward, 528

Mach, Ernst, 495, 496, 497, 498, 502-506, 511, 514, 540, 541, 543

Machiavelli, Niccolo, 8, 40, 98

Magdeburg, 100

Magendie, François, 261

magic, 80

Magie, W. F., 541

magnetism, 32, 74, 354, 360, 379, 385, 392, 400, 429, 435, 441, 442-446, 451, 452, 455, 456, 458, 459,

465, 467-472, 474, 475; of earth, 438, 443, 445, 448

Maistre, Joseph de, 178

Malebranche, 159

Malpighi, 72, 73, 109

Malthus, T. R., 304, 310, 311, 314, 339, 343, 348

Malus, 429

Manchester, 252, 371

Marat, 408

Marcet, Mrs., 436

Mariotte, 368

Mars, theory of, 25, 31-34, 35

Martin, Thomas, 542

Marxism, 75, 156, 342, 346, 347

mass, 360, 367, 384, 392, 400; defined by Newton, 141, 143; lack of, in Maxwell's fluid, 463

mass action, law of, 251

"masses" in nature, 272, 274, 275

materialism, 98, 146, 204, 231, 321, 345

mathematicization: of physics, 54, 113, 354, 356, 381, 390; of chemistry, 245-246, 249; of biology, 333, 342, 347; of thermodynamics, 367, 369; of wave theory of light, 423, 426; of electromagnetism, 464-465, 466, 475, 488

mathematics, 6-7, 10, 13, 14-16, 26-27, 39, 41, 45-46, 49-50, 51, 55, 85-88, 93, 109, 146, 187-188, 203, 347, 356, 383, 453, 456, 459, 460, 463-465, 483, 489, 508-509; Faraday's ignorance of, 439-440, 447, 450

mathematics as language of science, 43, 45, 93, 144, 167, 187-188, 276, 352-353, 384, 420, 438, 439-440, 450

matter, 46, 88, 91, 105-106, 130, 132, 141, 147, 186, 202-259, 260, 275, 276, 338, 341, 355, 356, 374, 376, 380, 384, 385, 401, 403, 406, 408, 441, 449, 477, 482, 501, 503, 507; in motion, 16, 58, 104, 144, 271, 385, 400, 426, 470, 474, 487, 501, 508, 510; Boyle's view of structure of, 104, 105-106; Newton's view of structure of, 105, 130, 144, 163,

355; Faraday's view of structure of, 441, 446, 448-449, 451-458. See also atomism

Maury, Alfred, 531

Maxwell, James Clerk, 14, 95, 352, 361, 426, 458-492, 493, 494, 498, 506-507, 511, 517, 518, 523, 537, 542-543

Mayer, J. R., 375-381, 383, 384, 385, 386, 387, 393, 394, 395, 401, 449, 498, 501, 540, 541

mechanical arts, 80, 85

mechanics, 16, 45, 52, 55, 58, 92, 101, 111, 113, 140, 185, 187, 203, 205, 239, 286, 338, 354, 355-357, 359, 360, 361, 366, 367, 374, 376, 377, 378, 379, 382, 384, 388, 391, 392, 394, 420, 428, 431, 450, 459, 479, 487, 495, 497, 498-499, 501, 502, 503, 505, 508, 511-513, 514, 517-518; fluid, 73; Newtonian, 92, 140, 153, 360; analytical, 174, 354, 407, 422, 432, 442, 456; wave, 410, 422; statistical, 479, 483, 488, 499, 512; quantum, 87, 381, 495, 514

mechanism, 73, 85, 91, 92, 93, 94, 95, 111, 120, 121, 126, 153, 156, 157, 192, 199, 201, 238, 253, 261, 262, 322, 323, 326, 338, 358, 359, 402, 434, 474, 481, 498-501

medicine, 58, 61-63, 106, 160, 262, 382, 411

Mendel, Gregor, 261, 328-337, 339, 340, 539

Mendeleev, 258; Periodic table, 247, 251

Mendelism, 341

Merret, Christopher, 112

Mersenne, abbé, 111, 113

Merton, Robert K., 531

Merton Rule, 6, 15, 41

Merz, John Theodore, 541

metabolism, 375, 376, 381, 383, 393

metaphysics, 14, 45, 91, 144, 158, 355, 371, 375, 376, 496, 502, 504; Cartesian, 94, 153; Stoic, 182-183

meteorology, 88, 253, 270, 275

method, 69, 74, 83, 107-109, 125, 156, 161, 282-283, 287, 338, 366,

445, 459, 483, 497; of Descartes, 86, 88, 89, 153; in Enlightenment, 169, 170, 171, 173, 177-178; of Lavoisier, 203-204, 217-218, 232, 233; of Fresnel, 424

Methodism, 347, 349

metrics, 141, 144

Meusnier, 228, 229

Mézières, 228

Michéa, René, 535

Michelet, 281

Michelson, 427, 505, 508-510, 514, 515

Michurin school, 347

Middle Ages, 16, 61, 74, 263, 502

Milesian philosophers, 96

Mill, John Stuart, 349, 451

Milton, John, 98, 308

mineralogy, 59, 291, 292

moment, 16, 359

momentum, 41, 121, 140, 360, 463, 476, 480, 486

Monge, Gaspard, 177, 211, 228, 229

Montaigne, 85

Montbéliard, 277

Montesquieu, 182

Montpellier, 176

Morgan, T. H., 336-337

Morley, 427, 505, 508-510, 514, 515

motion, 205, 341, 354, 357, 360, 377-378, 379, 386, 390, 403, 478; and Galileo, 3-7, 41-42, 44-45, 50-52, 89, 90-91, 106, 120, 133, 241, 249, 338, 339, 355, 515; uniform, 4-7, 25, 518; and Aristotle, 11-12, 33, 41, 141; circular, 14, 17, 19, 25, 51, 84, 89, 90-91, 92, 121, 249; matter in, 16, 58, 104, 144, 271, 385, 400, 426, 470, 474, 487, 501, 508, 510; planetary, 17-20, 23, 24-26, 31-39, 136, 137, 143; laws of, 32, 84, 143, 499; inertial, 51, 83-84, 90, 121, 355, 367, 461, 467, 501; local, 52, 99, 106; of the heart, 68, 71; and Descartes, 83-84, 86, 87, 90, 93, 120; absolute, 91, 141, 503-505; and Newton, 120, 130, 136, 137, 138, 141, 142, 143, 144, 145, 150, 355, 402, 433, 459, 498, 503-505, 507, 508, 510, 517;

and atomism, 96-97, 106; light and, 127, 406, 414, 415, 421, 424, 427, 430, 433, 461, 514-516; and aether, 130, 145, 357, 407, 427, 428, 509, 510, 516; heat and, 239, 358, 369, 380; perpetual, 366, 367, 368, 387, 394, 401, 514; and electricity, 392, 443, 444; and electromagnetism, 446, 459, 467, 474; of fluid, 462-464, 489; in gases, 482, 485; relative, 505, 518

Mouy, Paul, 530

Mozart, Wolfgang Amadeus, 27

Müller, Johannes, 261, 382

Munich, 323, 328

Murchison, Roderick, 297

Muséum d'histoire naturelle, 176, 177, 268, 269, 272, 286, 288

mutation, 306-307, 335-337

Nägeli, Karl von, 322-325, 326, 328, 334, 341, 344, 539

Napoleon, 55, 117, 179, 279, 280, 411, 422, 496

Nash, Leonard K., 537

natural history, 12, 58-59, 77, 78, 79-80, 110, 152, 166, 170, 171, 173, 174, 187, 246, 260, 262, 267, 268, 271, 273, 278, 286, 287, 288, 304, 318, 347

naturalism, 14, 54, 55, 56, 57, 63, 170, 171, 201, 203, 263, 311, 347; social, 191

natural philosophy, 16, 34, 40, 41, 46, 47, 48, 69, 78, 108, 161, 179, 184-186, 263, 269, 285, 304, 346, 347, 411; of Diderot, 188-192, 262; of Goethe, 192-198, 263; romantic, 199-200

natural selection, theory of, 260, 261, 283, 302, 303, 307, 314, 316-320, 323, 325, 326, 327, 329, 337-338, 339, 340, 342, 344, 346, 351, 402, 405, 501

natural theology, 66, 114, 263-266, 302

nature, laws of, 93, 154, 159, 262, 283, 287, 317, 319, 402, 500, 513, 514, 515, 517, 519

Naturphilosophie, 197, 278, 320, 379
navigation, 18, 22
neo-Lamarckism, 322, 324, 326, 338, 341
neo-Platonism, 26, 88, 195
Neptunism, 292-293, 294
neutrino, 381
Newman, Francis, 349
Newton, Isaac, 8, 16, 31, 32, 34, 45, 50, 51, 58, 77, 84, 85, 88, 89, 90, 91, 93, 96, 105, 108, 109, 112, 117-150, 151, 152, 153, 154, 157-158, 161, 162, 167, 168, 170, 179, 180, 182, 184, 192, 195, 203, 217, 218, 238, 239, 250, 254, 258, 259, 264, 291, 304, 305, 307, 313, 318, 319, 320, 334, 335, 337, 338, 340, 341, 343, 344, 354, 355, 360, 378, 380, 403, 409, 411, 440, 449, 454, 474, 475, 492, 494, 495, 498, 501, 507, 518, 528, 531-532; on space, 88, 90-91, 141-142, 144, 354, 355, 490-491, 503-505, 515, 519; view of structure of matter, 105, 130, 144, 163, 355; on gravity, 119-120, 121-122, 135-143, 144-145, 147; on optics, 122-126, 128-129, 130-132, 133-134, 147, 148, 385, 406, 407, 408, 413-414, 416, 417, 420, 424, 429, 431, 432; *Principia*, 122, 125, 136-143, 147, 148, 149, 150, 158, 160, 203, 254, 320; opposition to, 124, 125, 126, 127-129; as speculative thinker, 128-130, 148; aether, 129-131, 148-150, 237, 238, 241, 355, 369, 490-491; atomism, 132, 133, 144, 146; and theology, 134, 145-150, 265; and motion, 120, 130, 136, 137, 141, 143, 150, 355, 402, 433, 459, 498, 503-505, 507, 508, 510, 517; disagreements with Hooke 138-139; definition of physics, 140-142; on time, 141-142, 143, 503-504, 515
Newtonianism, 342, 412
Newtonian science, 144, 145, 146, 156, 157, 158, 159, 160, 187, 189, 192, 197, 198, 317, 360, 498, 503, 513
Newtonian synthesis, 88, 144, 335, 510

Nightingale, Florence, 451
Niven, W. D., 542
nomenclature, 169, 171, 177, 193, 203, 232-235, 236, 245, 247, 296
Nonconformity, 115, 252, 348, 410, 441, 450
Nordenskiold, Erik, 320, 540
Nostradamus, 15, 192
number, 15, 23, 42, 44, 86, 88, 104, 250, 251, 255, 341, 379, 461
Nuremberg, 23

objectivity, 10, 13, 40, 41-42, 44, 63, 73, 97, 106, 107, 133, 154-156, 161, 164, 202, 204, 231, 241, 246, 248, 260, 261, 328, 338, 341, 348, 385, 496
Oersted, Hans Christian, 370, 442, 542
Oken, 320
Oldenburg, Samuel, 112-113, 128
Oldham, Frank, 541
ontogeny, 323, 325, 327
ontology, 98, 144, 199, 249, 271, 311, 380, 385, 401, 495, 496, 499
Oppenheimer, Robert, 115
optics, 84, 88-89, 108, 118, 122-134, 140, 145, 147, 205, 383, 407-435, 459, 487, 498, 509. *See also* refraction, law of; Newton, on optics
organism, 85, 156, 184, 187, 192, 197, 199, 201, 261, 263, 284, 338, 379
organization of science, 55, 78, 81, 247-250, 262, 292, 361
Ornstein, Martha, 531
Osiander, Andreas, 23
Ostoya, Paul, 538
Ostwald, Wilhelm, 495, 497-498, 500-501, 502, 513, 543
Owen, Richard, 313
Oxford, 6, 103, 113, 160, 210, 297, 305, 350
Oxford Movement, 349
oxidation, 222, 229, 237, 241, 242, 249, 383
oxygen, 203, 208-209, 214, 217, 218-221, 223, 224, 225, 226, 227, 236, 248, 252, 254, 258, 260, 275, 375, 385, 447; oxygenic principle of Lavoisier, 202, 227, 242, 248

Padua, 20, 41, 47, 48, 57, 60, 67, 68-69, 72, 477

Palais Royal, 212, 220

paleontology, 267, 268, 269, 288, 292, 295, 300, 307, 312

Paley, William, 263, 266

Panckoucke, 233

Paracelsus, 184, 212

Pardies, Father Ignatius, 125

Paris, 55, 101, 125, 151, 157, 165, 176, 210, 216, 219, 228, 232, 233, 268, 274, 278, 279, 288, 442, 496, 497; University of, 41, 60; basin, geology of, 280, 289, 290, 295

particles, 96, 97, 237, 240, 249, 254, 255, 258, 259, 276, 369, 380, 434, 442, 449, 454, 455, 458, 461, 467, 469-470, 471, 476, 477, 479, 480, 485, 487, 488, 491, 499, 500; "contiguous," 449, 451. See also physics, particle

Pascal, Blaise, 82, 101, 102-103, 105, 111, 112, 140, 188

Pasteur, Louis, 261

Pemberton, Henry, 119

pendulum, isochronism of, 41

Pepys, Samuel, 114

perception, 41, 54, 93, 98, 162, 166, 183, 195, 383, 412

Perier, Francois, 101, 102

periodicity of light, 413, 414, 434

Petrarch, 57

Petty, William, 112

Phillips, William, 296

philosophes, 98, 152, 156, 165, 173, 180, 203, 232

Philosophical (Royal) Society of Edinburgh, 206

philosophy of science, 9, 54, 73, 74-82, 86, 94, 166-171, 177, 335, 354-357, 383, 401-405; Aristotelian, 12-14; assumptions about reality, 14, 15, 16, 30, 45-46, 54, 88, 91, 495, 499, 503, 510, 519; Platonic, 14-16; ontology, 98, 144, 199, 249, 271, 311, 380, 385, 401, 496, 499; educationism, 168, 169, 172, 175-177, 203, 246, 249, 258, 263, 404, 496; Diderot's, 189-190; Lavoisier's, 202-204, 211, 214, 223, 224, 230-232, 235, 245-247, 249; and laws

of nature, 262, 283, 287; Lamarck's, 269; Mayer's, 378-381; Helmholtz's, 383-385; Faraday's, 438, 449, 454-458; Maxwell's, 464, 488-489; Einstein's, 510-520. See also hypothesis; natural philosophy; natural theology; teleology; theory

phlogiston, 202, 204-205, 211, 216, 220, 224, 225, 228, 229, 230, 232, 236, 237, 242, 492

photon, 418, 512

phylogeny, 321, 323, 325, 327

physics, 7, 15-16, 27, 33, 46, 47, 58-59, 84-85, 158, 185, 211, 214, 239, 262, 276, 291, 294, 304, 320, 329, 338, 343; Aristotelian, 11-14, 65; classical, 50, 87, 88, 90, 99, 105, 117, 118, 137, 140, 141, 311, 353-355, 380, 381, 403, 420, 487, 490, 499, 503-504, 513, 515, 518; experimental, 96, 99, 103, 108, 170, 287, 363, 370, 385, 513; theoretical, 99, 287, 385, 516; atomic, 108, 250, 341, 476, 477, 478, 481, 482, 487, 489, 494, 495, 513; quantum, 143, 512; Newtonian, 203, 246, 266, 353, 504, 515, 518; abstraction in, 246, 353; nineteenth-century, 352-405; field, 405-492, 493, 495

physiology, 58, 63-73, 177, 178, 206, 214, 271, 274, 276, 313, 325, 375, 379, 382, 393, 412

Picasso, Pablo, 8

Planck, Max, 95, 476-477, 486, 495, 512, 543

planetary laws of Kepler, 28, 34-39, 58, 93, 94, 119-120, 136, 143

Plato, 11, 13-16, 64, 72, 81, 86, 104, 109, 144

Platonism, 13-16, 88, 96, 144, 198, 263, 498; of Galileo, 39-40, 49, 57

Pliny Society, 308

pneuma, 182-183, 323. See also Stoicism

Poincaré, Henri, 154-155, 157, 381, 493

Poisson, 420, 425-426

polarization of light, 428-434, 435-436, 451-452, 468

positivism, 10, 128, 156, 162, 171,

177, 180, 260, 358, 488, 492, 495-506, 514, 516, 518

potential, 392, 443

power, 357, 360, 362, 370, 388

Prague, 30, 31

Pre-socratics, 15

Preyer, W., 541

Price, Derek J. de Solla, 527

Priestley, Joseph, 184, 204, 208-210, 217, 218, 219-220, 221, 222, 223, 227, 237, 252, 253, 271, 274, 306, 337, 385, 536, 537

probability, 87, 249, 332, 341, 403, 479, 483, 495

process, 45, 97, 106, 198, 341, 342, 402; science not derived from, 341; Hegelian, 383; biological, 501

progress, 8, 74, 78, 152, 153, 162, 172, 174, 175, 176, 178, 246, 261, 311, 326, 402, 403, 404

Protestantism, 31, 115, 278, 280, 347

Proust, Marcel, 252

Providence, 13, 145, 146, 183, 264, 311

providentialism, 263, 265, 268, 283, 292, 298, 302, 341, 347

psychology, 113, 156, 159, 160-164, 166, 168, 271, 274, 276, 502

Ptolemy, Claudius, 17-18, 22, 24, 33, 268, 527; Ptolemaic system, 23, 25

purpose, 10, 13, 59, 64, 73, 77, 85, 183, 317, 344

Puritanism, 113, 114, 115, 208, 348

Puy-de-Dôme, 102

pyrotic theory, 275-276

Pythagoras, 14, 441

Pythagoreanism, 15, 17, 21, 22, 23, 26, 29, 39, 40, 43, 478

Quakerism, 410, 420

qualities, 16, 41, 44, 73, 93, 96, 98, 106, 108, 123, 185, 191, 205, 241, 519

quanta, 334, 357, 512

quantities, 4, 6, 15, 28, 43, 45, 87, 90, 144, 185, 204, 235, 245, 347, 380, 384, 390, 461, 488

quantum mechanics, 87, 381, 495, 514

quantum physics, 143, 512

Racine, 158

radiation, 406, 433, 434; of energy, 495, 512

radioactivity, 357, 385

radio waves, 493, 494

Randall, J. H., 526

Rankine, 389

rationalism, 10, 12, 13, 78, 83, 114, 173, 246, 260, 294, 296, 358, 383, 518; Cartesian, 93, 121, 153, 198, 354; in Enlightenment, 151, 152, 153, 156, 166, 168, 174, 175, 178, 180-181

reaction, chemical, 217, 222, 229, 237, 249, 251, 258, 358, 370, 381, 455

realism, 126, 322, 494, 498

reflection of light, 406, 414, 429

Reformation, 46

refraction, 426, 428, 434, 461; law of, 84, 87, 88-89, 123, 126, 384, 406, 414; double, 429-434, 452

refrangibility of light, 123, 406, 492

relativity, 15, 28, 87, 143, 240, 262, 357, 413, 475, 495, 505-506, 511-518; special theory of, 513-517; general theory of, 517-518; of motion, 505, 518; of space, 141, 142, 503-505, 510, 515, 518; of time, 141, 142, 143, 503-505, 515, 518

religion, 98, 99, 101, 146, 160, 174, 175, 219, 264, 265, 266, 502, 520; conflict with science, 46-50, 114-116, 158, 297-299, 342, 347-351

Renaissance, 7, 10, 17, 20, 21, 22, 43, 54, 56, 57, 60, 79, 94, 109

Restoration (England), 113, 114

Restoration (France), 280

reversibility, 365-370, 387, 398-400, 401

Revolution of 1688, 160

Reynal, 528

Rheticus, Georg Joachim, 22-23

Richelieu, Cardinal, 112

Riemann, 87

Ritter, J. W., 406

Robespierre, 175, 179

Roentgen, W. Konrad, 494

Rohan, 157

Roman Catholic Church, 3, 20, 31, 46-49, 66, 92, 99, 115, 159

Roman Curia, 46

romanticism, 151, 156, 178-181, 188, 338, 344, 379, 383; biological, 197, 198-201, 246, 260, 262, 268, 276, 287, 322-328, 345

Rome, 13, 48-49, 50, 110

Ronchi, Vasco, 530, 541

Rosen, Edward, 527

Rosetta Stone, 411-412

Rosicrucianism, 15

Rousseau, Jean-Jacques, 151, 165, 171, 176, 180, 191, 192, 199, 216, 346

Roux, 326

Royal College of Physicians, 69

Royal Institution, 411, 418, 437, 438, 443

Royal Society of London, 108, 109, 111, 112-114, 115, 122, 125, 134, 135, 138, 139, 148, 149, 160, 208, 412, 415, 435, 437, 441, 444, 458, 478

Rudolf II, 30, 37

Rumford, Count (Benjamin Thompson), 369, 371, 411

Russell, Bertrand, 95

Sahara, 281

Saint Bartholomew's Hospital, 69

Saint-Hilaire, Geoffroy, 268, 320

Salusbury, Thomas, 528

Salvio, Alfonso de, 528

Sambursky, S., 97, 526, 530

Santillana, Giorgio de, 527-528

Sarpi, Paolo, 3

Scheele, Carl Wilhelm, 219, 536

Schelling, 178

Schilpp, Paul Arthur, 543

Schleiden, Matthais Jakob, 261

Schoenberg, Cardinal, 47

scholasticism, 7, 19, 42-43, 46-47, 74, 76, 78, 80, 146

Schwann, Theodor, 261

science: Greek, 10-16, 51, 79, 81, 92, 94, 97, 155, 212; modern, 10, 171; and ethics, 44, 154-155, 342, 348-350; conflict with theology, 46-50, 114-116, 260-262, 263, 265, 295, 297-300, 347-351. See also philosophy of science

scientific explanation, 94, 102, 156, 162, 168, 169, 198, 307, 442

scientific revolution, 9-10, 16, 43, 54, 57, 73, 83, 92, 94, 300, 477; second, 353-354, 477, 494

Scopes Trial, 350

Sedgwick, Adam, 297, 298, 301, 308, 350

Seneca, 182

Serveto, Miguel, 66-67

Shakespeare, William, 192

Shaw, George Bernard, 344-346, 347, 350

Shelley, 178

Siberia, 281, 288, 290

Siemen, 470

simplicity, principle of, 254, 258

simultaneity, 515-516, 518

Smiles, Samuel, 280

Smith, Adam, 206, 311

Smith, William, 294-295, 296

Snell, 89, 109

social science, 153, 343, 358

Society of Friends, 252

Society of Jesus, 46, 125, 165, 455

sociology, 496, 502

Sorbonne, 159, 280

sound, 93, 383, 391, 415, 514

space, 27, 42, 87-91, 240, 267, 291, 294, 357, 356, 357, 381, 403, 409, 421, 435, 441, 443, 445, 480, 490-492, 493, 517; Euclidean, 84, 87, 355, 381, 491; Cartesian, 87-91; Newtonian, 88, 90-91, 141, 143, 144, 354, 355, 490-491, 503, 505, 515, 519; absolute vs. relative, 141, 142, 503-505, 515; Faraday's view of, 441, 443, 445, 452, 462; Maxwell's view of, 463, 466, 468, 470, 473-474, 475; relativity of, 510, 518

space-matter, 91, 92, 95, 367, 408

species, 12, 170, 191, 198, 234, 246, 266, 271, 272, 273, 276, 300, 301, 310, 314, 316, 321, 332, 336, 338, 340, 402

Spencer, Herbert, 403, 496

Spiers, I. H. B. and A. G. H., 530

Spinoza, 192

Sprat, Bishop, 108, 109, 114, 116, 531

Stahl, G. E., 204, 205, 212, 224, 232, 237
statics, 391
Stauffer, Robert C., 542
steam engine, 357, 358, 373, 401, 470
Stendhal, 422
Stephen, Leslie, 349
Stevin, Simon, 102, 103
Stewart, Dugald, 206
Stimson, Dorothy, 531
Stoicism, 88, 96, 130, 181-183, 195, 212, 224, 263, 323, 341
Stokes, 483
Strasbourg, 176
stratigraphy, 289, 294, 295
Stuttgart, 278
style in science, 108, 112, 407-410, 420, 506; in Germany, 197; of French, 205, 206, 250, 401, 502; of English, 205, 210, 250, 401; in Scotland, 206; Priestley, 210; Lavoisier, 217; Dalton, 253; Cuvier, 280-281; eighteenth vs. nineteenth century, 404
Swinburne, 405

Tannery, Paul, 530
Taton, Juliette, 530
taxonomy, 59, 170, 171, 173, 178, 180, 192, 217, 247, 249, 261, 267, 268, 269, 271, 275, 277, 285, 287, 313
Taylor, Bayard, 196
Taylor, Richard, 454
technocracy, 81, 358, 496
technology, 8-9, 10, 54, 78, 80, 15: 173-175, 181, 264, 353, 381
teleology, 59, 63-64, 77, 93, 183, 284-285
de Tencin, Mme, 165
Tessier, abbé, 278
Thalés, 497, 527
theology, 16-17, 29-30, 66, 82, 97-98, 102, 104, 114, 134, 155, 183, 190, 266, 343, 371; conflict with science, 46-50, 114-116, 260-262, 263, 265, 295, 297-300, 347-351; and Newton, 134, 145-150, 265. *See also* natural theology
Theophylactus, 20

theory, 22, 33, 41, 57-58, 68, 71, 77, 109, 125-126, 211, 223, 225, 237, 251, 302, 318, 334, 359, 369, 385, 438, 448, 457, 494-500, 502; Fresnel, 421, 424, 426, 431-432, 435, 507; Maxwell and electromagnetic field, 460, 462, 466-467, 468-476, 482-492; Lorentz, 507-510; Einstein, 511-512, 516-519
theory of machines, 388, 389
thermodynamics, 240-241, 339, 356, 357, 361, 363-403, 422, 441, 477, 481, 487, 493, 497, 498, 511-512, 513-514, 516; First Law of, 370, 381, 382, 386, 393, 402, 403, 487; Second Law of, 370, 382, 395, 396, 400, 402, 403, 487, 516
thermometry, 110, 253
Thompson, Benjamin (Count Rumford), 369, 371, 411
Thomson, J. J., 543
Thomson, William, *see* Kelvin, Lord
Thorpe, Edward, 537
tides, 51, 93, 143, 157, 438, 445
time, 4-5, 42, 87, 90, 267, 291, 294, 337, 338, 339, 402, 474, 475, 508-509; absolute vs. relative, defined by Newton, 141, 142, 143, 503-504, 515; by Mach, 503, 505; relativity of, 510, 518; Einstein's special theory of, 510
Titian, 57
Torn (Poland), 20
Torricelli, 100, 101, 102, 103, 105
Toulmin, Stephen, 532 363
Toulouse, University of, 66
Tour d'Auvergne, 212
Townshend, Joseph, 295
Tractarianism, 349
transcendentalism, 276
transmutation, 270, 298, 301
Trapp, Marianne, 535
Trinity College (Cambridge), 118
Tschermak, Erich, 336
Tübingen, 29, 336
Turnbull, H. W., 531
Tycho Brahe, 30-31, 33, 34
Tyndall, John, 382, 437-440, 441, 449-450, 457, 542

uniformitarianism, 275, 294, 295, 299, 300, 301, 402

Unitarianism, 145, 208, 209, 210, 252, 278, 348

unity of nature, 15, 88, 95, 197-198, 260, 283, 288, 321, 341, 344, 356, 381, 456, 519

Upsalla, 219

Uraniborg, 30

Urban VIII, Pope (Maffeo Barberini), 49

utilitarianism, 114, 151, 153, 154, 188, 246, 502

Utopia, 81

vacuum, 82, 99-105, 111

valence, 251, 258, 499

variation, 307, 311, 316, 318, 319, 325, 330, 333, 334, 338-339

Vartanian, Aram, 534

Venel, 184-187, 274

Venice, 57, 60

Verdet, Emile, 542

Vernière, Paul, 534

Vesalius, Andreas, 41, 56-58, 59-63, 64, 65, 66, 67, 528

vibration and light, 132-133, 406, 408, 413, 421, 430-431, 433-435, 461

Vienna, 336, 495, 496

Vincennes, 180

Virchow, Rudolf, 261

viscosity of a gas, 484-486

vis viva, 360, 361, 373, 374-375, 376, 377, 383, 385, 386, 387, 388, 389, 390, 392, 393, 401, 412

vitalism, 261, 262, 276, 287, 322, 357

Viviani, 110

void, 12, 92, 97, 99, 104, 105, 122, 133, 144, 145, 182, 198, 250 341, 354, 355, 401, 454-455, 488, 490, 518

Volta, 370

Voltaire, 157-159, 164, 165, 178, 180, 201, 534

vortex, 489, 492; Cartesian, 92, 93, 122, 142; of Maxwell, 468-470, 490-491

Vulcanism, 293, 294

Wadham College, 113

Wallace, Alfred Russel, 313-315, 339-340, 539

Wallis, John, 113

Ward, Seth, 113

water, composition of, 217, 227-229, 258

Watt, James, 206, 228, 293, 306, 358, 363

Watzelrode, Lucas, 20, 22

Weber, Max, 472-473

Wedgwood, Josiah, 306

Wedgwood, Josiah, II, 308

Weimar, 192

Weinberg, C. B., 543

Weismann, August, 322, 325-328, 334, 335, 539

Werner, Abraham Gottlob, 292, 295, 296

Wernerianism, see Neptunism

Wesel, 60

Westminster Abbey, 157

Whewell, William, 301

White, R. J., 535

Whitehead, Alfred North, 200, 265

Whittaker, E. T., 542

Whitteridge, Gweneth, 529

Wilberforce, Samuel, 305, 350

Wilkins, John, 113, 524

Willey, Basil, 531

Willis, Thomas, 112

Wilson, Arthur, 534

Wood, Alexander, 541

Woolthorpe, 119

Wordsworth, 200

work, 360, 361, 365, 370, 388, 389, 391, 394, 395, 396, 397, 400, 401

World War II, 253

Wren, Christopher, 136, 137

Wurtemberg, Duke of, 278, 279

X-rays, 494

Young, Thomas, 132, 389, 407-421, 423, 424, 428, 429, 430, 435, 541

zoology, 59, 171, 194, 267, 268, 269, 285, 288